# A First Course in Probability

Sheldon Ross
*University of California, Berkeley*

# A First Course in Probability

Third Edition

Macmillan Publishing Company
*New York*

Collier Macmillan Publishers
*London*

For Rebecca



Printed in the United States of America



Macmillan Publishing Company
866 Third Avenue, New York, New York 10022

Collier Macmillan Canada, Inc.

Library of Congress Cataloging in Publication Data

Ross, Sheldon M.
A first course in probability.

Bibliography: p.
Includes index.
I. Probabilities. I. Title.
QA273.R83 1988 519.2 87-18599
ISBN 0-02-403850-4 (Hardcover Edition)
ISBN 0-02-946600-8 (International Edition)

Printing: 8 Year: 2 3 4 5 6 7

# Preface

"We see that the theory of probability is at bottom only common sense reduced to calculation; it makes us appreciate with exactitude what reasonable minds feel by a sort of instinct, often without being able to account for it. . . . It is remarkable that this science, which originated in the consideration of games of chance, should have become the most important object of human knowledge. . . . The most important questions of life are, for the most part, really only problems of probability." So said the famous French mathematician and astronomer (the "Newton of France") Pierre Simon, Marquis de Laplace. Although many people might feel that the famous marquis, who was also one of the great contributors to the development of probability, might have exaggerated somewhat, it is nevertheless true that probability theory has become a tool of fundamental importance to nearly all scientists, engineers, medical practitioners, jurists, and industrialists. In fact, the enlightened individual has learned to ask not "Is it so?" but rather "What is the probability that it is so?"

This book is intended as an elementary introduction to the mathematical theory of probability for students in mathematics, engineering, and the sciences (including the social sciences and management science) who possess the prerequisite knowledge of elementary calculus. It attempts to present not only the mathematics of probability theory, but also, through numerous examples, the many diverse possible applications of this subject.

In Chapter 1 we present the basic principles of combinatorial analysis, which are most useful in computing probabilities.

In Chapter 2 we consider the axioms of probability theory and show how they can be applied to compute various probabilities of interest. This chapter includes a proof of the important (and, unfortunately, often neglected) continuity property of probabilities, which is then used in the study of a "logical paradox."

Chapter 3 deals with the extremely important subjects of conditional probability and independence of events. By a series of examples we illustrate how conditional probabilities come into play not only when some partial information is available, but also as a tool to enable us to compute probabilities more easily, even when no partial information is present. This extremely important technique of obtaining probabilities by "conditioning" reappears in Chapter 7, where we use it to obtain expectations.

In Chapters 4, 5, and 6 we introduce the concept of random variables. Discrete random variables are dealt with in Chapter 4, continuous random variables in Chapter 5, and jointly distributed random variables in Chapter 6.

Chapter 7 introduces the important concept of expectation. After defining the expected value of a random variable, we show how to compute the expected value of a function of a random variable by using the law of the unconscious statistician, for which an elementary proof is presented. Many examples illustrating the usefulness of the result that the expected value of a sum of random variables is equal to the sum of their expected values are also presented. This chapter also contains sections on conditional expectation, including its use in prediction, and moment generating functions.

In Chapter 8 we present the major theoretical results of probability theory. In particular, we prove the strong law of large numbers and the central limit theorem for independent and identically distributed random variables. Our proof of the strong law (via Kolmogorov's inequality and the continuity property of probabilities) is complete, whereas our proof of the central limit theorem assumes Levy's continuity theorem.

Chapter 9 presents some additional topics, such as Markov chains, the Poisson process, and an introduction to information and coding theory, and Chapter 10 considers simulation.

There are many examples worked out throughout the text, and there are also a large number of problems—divided into "theoretical exercises" and "problems"—to be worked by the student. Answers for most of the problems are provided at the end of the book. A solutions manual is available for instructors.

We should like to express our thanks to the following reviewers: Thomas R. Fischer, Texas A & M University; Jay Devore, California Polytechnic University, San Luis Obispo; Robb J. Muirhead, University of Michigan; David Heath, Cornell University; M. Samuels, Purdue University; I. R. Savage, Yale University; and R. Miller, Stanford University.

**S. R.**

# Contents

## 6
## Jointly Distributed Random Variables 194



## 7
## Expectation 246



## 8
## Limit Theorems 336

## 9
## Additional Topics in Probability 362



## 10
## Simulation 386

# 1

# Combinatorial Analysis

## 1 Introduction

Here is a typical problem of interest involving probability. A communication system is to consist of $n$ seemingly identical antennas that are to be lined up in a linear order. The resulting system will then be able to receive all incoming signals—and will be called functional—as long as no two consecutive antennas are defective. If it turns out that exactly $m$ of the $n$ antennas are defective, what is the probability that the resulting system will be functional? For instance, in the special case where $n = 4$ and $m = 2$ there are 6 possible system configurations—namely

$$
\begin{array}{cccc}
\varnothing & 0 & 0 & \varnothing \\
\varnothing & 0 & \varnothing & 0 \\
0 & \varnothing & 0 & \varnothing \\
\varnothing & \varnothing & 0 & 0 \\
0 & \varnothing & \varnothing & 0 \\
0 & 0 & \varnothing & \varnothing
\end{array}
$$

where 0 means that the antenna is working and $\varnothing$ that it is defective. As the resulting system will be functional in the first 3 arrangements and not functional in the remaining 3, it seems reasonable to take $\frac{3}{6} = \frac{1}{2}$ as the desired probability. In the case of general $n$ and $m$, we could compute the probability that the system is functional in a similar fashion. That is, we could count the number of configurations that result in the system being functional and then divide by the total number of all possible configurations.

From the above we see that it would be useful to have an effective method for counting the number of ways that things can occur. In fact, many problems in probability theory can be solved by simply counting the number of different ways that a certain event can occur. The mathematical theory of counting is formally known as *combinatorial analysis*.

## 2 Basic Principle of Counting

The following principle of counting will be basic to all our work. Loosely put, it states that if one experiment can result in any of $m$ possible outcomes and if another experiment can result in any of $n$ possible outcomes, then there are $mn$ possible outcomes of the two experiments.

---

*Basic Principle of Counting*

*Suppose that two experiments are to be performed. Then if experiment 1 can result in any one of $m$ possible outcomes and if, for each outcome of experiment 1, there are $n$ possible outcomes of experiment 2, then together there are $mn$ possible outcomes of the two experiments.*

---

**Proof of the Basic Principle:** The basic principle may be proved by enumerating all the possible outcomes of the two experiments as follows:

$$\begin{gathered}(1, 1), (1, 2), \ldots, (1, n)\\(2, 1), (2, 2), \ldots, (2, n)\\\vdots\\(m, 1), (m, 2), \ldots, (m, n)\end{gathered}$$

where we say that the outcome is $(i, j)$ if experiment 1 results in its $i$th possible outcome and experiment 2 then results in the $j$th of its possible outcomes. Hence the set of possible outcomes consists of $m$ rows, each row containing $n$ elements, which proves the result. ∎

**Example 2a.** A small community consists of 10 men, each of whom has 3 sons. If one man and one of his sons are to be chosen as father and son of the year, how many different choices are possible?

***Solution:*** By regarding the choice of the man as the outcome of the first experiment and the subsequent choice of one of his sons as the outcome of the second experiment, we see, from the basic principle, that there are $10 \times 3 = 30$ possible choices. ∎

When there are more than two experiments to be performed, the basic principle can be generalized as follows.

### *Generalized Basic Principle of Counting*

*If $r$ experiments that are to be performed are such that the first one may result in any of $n_1$ possible outcomes, and if for each of these $n_1$ possible outcomes there are $n_2$ possible outcomes of the second experiment, and if for each of the possible outcomes of the first two experiments there are $n_3$ possible outcomes of the third experiment, and if, . . . , then there are a total of $n_1 \cdot n_2 \cdots n_r$ possible outcomes of the $r$ experiments.*

**Example 2b.** A college planning committee consists of 3 freshmen, 4 sophomores, 5 juniors, and 2 seniors. A subcommittee of 4, consisting of 1 individual from each class, is to be chosen. How many different subcommittees are possible?

***Solution:*** We may regard the choice of a subcommittee as the combined outcome of the four separate experiments of choosing a single representative from each of the classes. Hence it follows from the generalized version of the basic principle that there are $3 \times 4 \times 5 \times 2 = 120$ possible subcommittees. ∎

**Example 2c.** How many different 7-place license plates are possible if the first 3 places are to be occupied by letters and the final 4 by numbers?

***Solution:*** By the generalized version of the basic principle the answer is $26 \cdot 26 \cdot 26 \cdot 10 \cdot 10 \cdot 10 \cdot 10 = 175{,}760{,}000$. ∎

**Example 2d.** How many functions defined on $n$ points are possible if each functional value is either 0 or 1?

***Solution:*** Let the points be $1, 2, \ldots, n$. Since $f(i)$ must be either 0 or 1 for each $i = 1, 2, \ldots, n$, it follows that there are $2^n$ possible functions. ∎

**Example 2e.** In Example 2c, how many license plates would be possible if repetition among letters or numbers were prohibited?

***Solution:*** In this case there would be $26 \cdot 25 \cdot 24 \cdot 10 \cdot 9 \cdot 8 \cdot 7 = 78{,}624{,}000$ possible license plates. ∎

## 3 Permutations

How many different ordered arrangements of the letters $a$, $b$, and $c$ are possible? By direct enumeration we see that there are 6; namely, *abc*, *acb*, *bac*, *bca*, *cab*, and *cba*. Each arrangement is known as a *permutation*. Thus there are 6 possible permutations of a set of 3 objects. This result could also have been obtained from the basic principle, since the first object in the permutation can be any of the 3, the second object in the permutation can then be chosen from any of the remaining 2, and the third object in the permutation is then chosen from the remaining 1. Thus there are $3 \cdot 2 \cdot 1 = 6$ possible permutations.

Suppose now that we have $n$ objects. Reasoning similar to that we have just used for the 3 letters shows that there are

$$n(n-1)(n-2)\cdots 3 \cdot 2 \cdot 1 = n!$$

different permutations of the $n$ objects.

**Example 3a.** How many different batting orders are possible for a baseball team consisting of 9 players?

***Solution:*** There are $9! = 362{,}880$ possible batting orders. ∎

**Example 3b.** A class in probability theory consists of 6 men and 4 women. An examination is given, and the students are ranked according to their performance. Assume that no two students obtain the same score.

1. How many different rankings are possible?
2. If the men are ranked just among themselves and the women among themselves, how many different rankings are possible?

***Solution:*** (a) As each ranking corresponds to a particular ordered arrangement of the 10 people, we see that the answer to this part is $10! = 3{,}628{,}800$.

(b) As there are 6! possible rankings of the men among themselves and 4! possible rankings of the women among themselves, it follows from the basic principle that there are $(6!)(4!) = (720)(24) = 17{,}280$ possible rankings in this case. ∎

**Example 3c.** Mr. Jones has 10 books that he is going to put on his bookshelf. Of these, 4 are mathematics books, 3 are chemistry books, 2 are history books, and 1 is a language book. Jones wants to arrange his books so that all the books dealing with the same subject are together on the shelf. How many different arrangements are possible?

***Solution:*** There are 4! 3! 2! 1! arrangements such that the mathematics books are first in line, then the chemistry books, then the history books, and then the language book. Similarly, for each possible ordering of the subjects, there are 4! 3! 2! 1! possible arrangements. Hence, as there are 4! possible orderings of the subjects, the desired answer is 4! 4! 3! 2! 1! = 6912. ∎

We shall now determine the number of permutations of a set of $n$ objects when certain of the objects are indistinguishable from each other. To set this straight in our minds, consider the following example.

**Example 3d.** How many different letter arrangements can be formed using the letters $PEPPER$?

***Solution:*** We first note that there are 6! permutations of the letters $P_1\,E_1\,P_2\,P_3\,E_2\,R$ when the 3 $P$'s and the 2 $E$'s are distinguished from each other. However, consider any one of these permutations—for instance, $P_1\,P_2\,E_1\,P_3\,E_2\,R$. If we now permute the $P$'s among themselves and the $E$'s among themselves, then the resultant arrangement would still be of the form $PPEPER$. That is, all 3! 2! permutations

$$
\begin{array}{ll}
P_1\,P_2\,E_1\,P_3\,E_2\,R & P_1\,P_2\,E_2\,P_3\,E_1\,R \\
P_1\,P_3\,E_1\,P_2\,E_2\,R & P_1\,P_3\,E_2\,P_2\,E_1\,R \\
P_2\,P_1\,E_1\,P_3\,E_2\,R & P_2\,P_1\,E_2\,P_3\,E_1\,R \\
P_2\,P_3\,E_1\,P_1\,E_2\,R & P_2\,P_3\,E_2\,P_1\,E_1\,R \\
P_3\,P_1\,E_1\,P_2\,E_2\,R & P_3\,P_1\,E_2\,P_2\,E_1\,R \\
P_3\,P_2\,E_1\,P_1\,E_2\,R & P_3\,P_2\,E_2\,P_1\,E_1\,R
\end{array}
$$

are of the form $PPEPER$. Hence there are 6!/3! 2! = 60 possible letter arrangements of the letters $PEPPER$. ∎

In general, the same reasoning as that used in Example 3d shows that there are

$$\frac{n!}{n_1!\,n_2!\cdots n_r!}$$

different permutations of $n$ objects, of which $n_1$ are alike, $n_2$ are alike, . . . , $n_r$ are alike.

**Example 3e.** A chess tournament has 10 competitors of which 4 are Russian, 3 are from the United States, 2 from Britain, and 1 from Brazil. If the tournament result just lists the nationalities of the players in the order in which they placed, how many outcomes are possible?

***Solution:*** There are

$$\frac{10!}{4!\,3!\,2!\,1!} = 12{,}600$$

possible outcomes. ∎

**Example 3f.** How many different signals, each consisting of 9 flags hung in a line, can be made from a set of 4 white flags, 3 red flags, and 2 blue flags if all flags of the same color are identical?

***Solution:*** There are

$$\frac{9!}{4!\,3!\,2!} = 1260$$

different signals. ∎

## 4 Combinations

We are often interested in determining the number of different groups of $r$ objects that could be formed from a total of $n$ objects. For instance, how many different groups of 3 could be selected from the 5 items $A$, $B$, $C$, $D$, and $E$? To answer this, reason as follows: Since there are 5 ways to select the initial item, 4 ways to then select the next item, and 3 ways to select the final item, there are thus $5 \cdot 4 \cdot 3$ ways of selecting the group of 3 when the order in which the items are selected is relevant. However, since every group of 3, say, the group consisting of items $A$, $B$, and $C$, will be counted 6 times (that is, all of the permutations $ABC$, $ACB$, $BAC$, $BCA$, $CAB$, and $CBA$ will be counted when the order of selection is relevant), it follows that the total number of groups that can be formed is

$$\frac{5 \cdot 4 \cdot 3}{3 \cdot 2 \cdot 1} = 10$$

In general, as $n(n-1)\cdots(n-r+1)$ represents the number of different ways that a group of $r$ items could be selected from $n$ items when the order of selection is relevant, and, as each group of $r$ items will be counted $r!$ times in this count, it follows that the number of different groups of $r$ items that could be formed from a set of $n$ items is

$$\frac{n(n-1)\cdots(n-r+1)}{r!} = \frac{n!}{(n-r)!\,r!}$$

---

### *Notation and Terminology*

*We define* $\binom{n}{r}$, *for* $r \leq n$, *by*

$$\binom{n}{r} = \frac{n!}{(n-r)!\,r!}$$

*and say that* $\binom{n}{r}$ *represents the number of possible combinations of n objects taken r at a time.*[1]

---

Thus $\binom{n}{r}$ represents the number of different groups of size $r$ that could be selected from a set of $n$ objects when the order of selection is not considered relevant.

**Example 4a.** A committee of 3 is to be formed from a group of 20 people. How many different committees are possible?

***Solution:*** There are $\binom{20}{3} = \frac{20 \cdot 19 \cdot 18}{3 \cdot 2 \cdot 1} = 1140$ possible committees. ∎

**Example 4b.** From a group of 5 men and 7 women, how many different committees consisting of 2 men and 3 women can be formed? What if 2 of the women are feuding and refuse to serve on the committee together?

***Solution:*** As there are $\binom{5}{2}$ possible groups of 2 men, and $\binom{7}{3}$ possible groups of 3 women, it follows from the basic principle that there are $\binom{5}{2}\binom{7}{3} = \left(\frac{5 \cdot 4}{2 \cdot 1}\right)\frac{7 \cdot 6 \cdot 5}{3 \cdot 2 \cdot 1} = 350$ possible committees consisting of 2 men and 3 women.

On the other hand, if 2 of the women refuse to serve on the committee together, then, as there are $\binom{2}{0}\binom{5}{3}$ possible groups of 3 women not containing either of the 2 feuding women and $\binom{2}{1}\binom{5}{2}$ groups of 3 women containing exactly 1 of the feuding women, it follows that there are $\binom{2}{0}\binom{5}{3} + \binom{2}{1}\binom{5}{2} = 30$ groups of 3 women not containing both of the feuding women. Since there are $\binom{5}{2}$ ways to choose the 2 men, it follows that, in this case, there are $30\binom{5}{2} = 300$ possible committees. ∎

[1] By convention 0! is defined to be 1. Thus $\binom{n}{0} = \binom{n}{n} = 1$.

**Example 4c.** Consider a set of $n$ antennas of which $m$ are defective and $n - m$ are functional and assume that all of the defectives and all of the functionals are considered indistinguishable. How many linear orderings are there in which no two defectives are consecutive?

***Solution:*** Imagine that the $n - m$ functional antennas are lined up among themselves. Now, if no two defectives are to be consecutive, then the spaces between the functional antennas must each contain at most one defective antenna. That is, in the $n - m + 1$ possible positions—represented in Figure 1.1 by carets—between the $n - m$ functional antennas we must select $m$ of these in which to put the defective antennas. Hence there are $\binom{n - m + 1}{m}$ possible orderings in which there is at least one functional antenna between any two defective ones. ▮

‸o‸o‸o . . ‸o‸o‸

o = functional
‸ = place for at most one defective

**Figure 1.1**

A useful combinatorial identity is

$$\binom{n}{r} = \binom{n-1}{r-1} + \binom{n-1}{r} \qquad 1 \le r \le n \tag{4.1}$$

Equation (4.1) may be proved analytically or by the following combinatorial argument. Consider a group of $n$ objects and fix attention on some particular one of these objects—call it object 1. Now, there are $\binom{n-1}{r-1}$ combinations of size $r$ that contain object 1 (since each such combination is formed by selecting $r - 1$ from the remaining $n - 1$ objects). Also, there are $\binom{n-1}{r}$ combinations of size $r$ that do not contain object 1. As there are a total of $\binom{n}{r}$ combinations of size $r$, Equation (4.1) follows.

The values $\binom{n}{r}$ are often referred to as binomial coefficients. This is so because of their prominence in the binomial theorem.

*Binomial Theorem*

$$(x+y)^n = \sum_{k=0}^{n} \binom{n}{k} x^k y^{n-k} \tag{4.2}$$

We shall present two proofs of the binomial theorem. The first is a proof by mathematical induction, and the second is a proof based on combinatorial considerations.

**Proof of Binomial Theorem by Induction:** When $n = 1$, Equation (4.2) reduces to

$$x + y = \binom{1}{0} x^0 y^1 + \binom{1}{1} x^1 y^0 = x + y$$

Assume Equation (4.2) for $n - 1$. Now,

$$\begin{aligned}
(x+y)^n &= (x+y)(x+y)^{n-1} \\
&= (x+y) \sum_{k=0}^{n-1} \binom{n-1}{k} x^k y^{n-1-k} \\
&= \sum_{k=0}^{n-1} \binom{n-1}{k} x^{k+1} y^{n-1-k} + \sum_{k=0}^{n-1} \binom{n-1}{k} x^k y^{n-k}
\end{aligned}$$

Letting $i = k + 1$ in the first sum and $i = k$ in the second sum, we find that

$$\begin{aligned}
(x+y)^n &= \sum_{i=1}^{n} \binom{n-1}{i-1} x^i y^{n-i} + \sum_{i=0}^{n-1} \binom{n-1}{i} x^i y^{n-i} \\
&= x^n + \sum_{i=1}^{n-1} \left[ \binom{n-1}{i-1} + \binom{n-1}{i} \right] x^i y^{n-i} + y^n \\
&= x^n + \sum_{i=1}^{n-1} \binom{n}{i} x^i y^{n-i} + y^n \\
&= \sum_{i=0}^{n} \binom{n}{i} x^i y^{n-i}
\end{aligned}$$

where the next to last equality follows by Equation (4.1). By induction the theorem is now proved. ∎

**Combinatorial Proof of the Binomial Theorem:** Consider the product

$$(x_1 + y_1)(x_2 + y_2) \cdots (x_n + y_n)$$

Its expansion consists of the sum of $2^n$ terms, each term being the product of $n$ factors. Further, each of the $2^n$ terms in the sum will contain as a factor either $x_i$ or $y_i$ for each $i = 1, 2, \ldots, n$. For example,

$$(x_1 + y_1)(x_2 + y_2) = x_1x_2 + x_1y_2 + y_1x_2 + y_1y_2$$

Now, how many of the $2^n$ terms in the sum will have as factors $k$ of the $x_i$'s and $(n - k)$ of the $y_i$'s? As each term consisting of $k$ of the $x_i$'s and $(n - k)$ of the $y_i$'s corresponds to a choice of a group of $k$ from the $n$ values $x_1, x_2, \ldots, x_n$, there are $\binom{n}{k}$ such terms. Thus, letting $x_i = x$, $y_i = y$, $i = 1, \ldots, n$, we see that

$$(x + y)^n = \sum_{k=0}^{n} \binom{n}{k} x^k y^{n-k}$$ ▮

**Example 4d.** Expand $(x + y)^3$.

***Solution***

$$(x + y)^3 = \binom{3}{0}x^0y^3 + \binom{3}{1}x^1y^2 + \binom{3}{2}x^2y + \binom{3}{3}x^3y^0$$
$$= y^3 + 3xy^2 + 3x^2y + x^3$$ ▮

**Example 4e.** How many subsets are there of a set consisting of $n$ elements?

***Solution:*** Since there are $\binom{n}{k}$ subsets of size $k$, the desired answer is

$$\sum_{k=0}^{n} \binom{n}{k} = (1 + 1)^n = 2^n$$

This result could also have been obtained by assigning to each element in the set either the number 0 or the number 1. To each assignment of numbers there corresponds, in a one-to-one fashion, a subset, namely, that subset consisting of all elements that were assigned the value 1. As there are $2^n$ possible assignments, the result follows.

Note that we have included as a subset the set consisting of 0 elements (that is, the null set). Hence the number of subsets that contain at least one element is $2^n - 1$. ▮

## 5 Multinomial Coefficients

In this section we consider the following problem: A set of $n$ distinct items is to be divided into $r$ distinct groups of respective sizes $n_1, n_2, \ldots, n_r$, where $\sum_{i=1}^{r} n_i = n$. How many different divisions are possible? To answer this, we note that there are $\binom{n}{n_1}$ possible choices for the first group; for each choice of the first group there are $\binom{n-n_1}{n_2}$ possible choices for the second group; for each choice of the first two groups there are $\binom{n-n_1-n_2}{n_3}$ possible choices for the third group; and so on. Hence it follows from the generalized version of the basic counting principle that there are

$$\binom{n}{n_1}\binom{n-n_1}{n_2}\cdots\binom{n-n_1-n_2-\cdots-n_{r-1}}{n_r}$$

$$= \frac{n!}{(n-n_1)!\,n_1!}\,\frac{(n-n_1)!}{(n-n_1-n_2)!\,n_2!}\cdots\frac{(n-n_1-n_2-\cdots-n_{r-1})!}{0!\,n_r!}$$

$$= \frac{n!}{n_1!\,n_2!\cdots n_r!}$$

possible divisions.

---

## Notation

*If $n_1 + n_2 + \cdots + n_r = n$, we define $\binom{n}{n_1, n_2, \ldots, n_r}$ by*

$$\binom{n}{n_1, n_2, \ldots, n_r} = \frac{n!}{n_1!\,n_2!\cdots n_r!}$$

*Thus $\binom{n}{n_1, n_2, \ldots, n_r}$ represents the number of possible divisions of $n$ distinct objects into $r$ distinct groups of respective sizes $n_1, n_2, \ldots, n_r$.*

---

**Example 5a.** A police department in a small city consists of 10 officers. If the department policy is to have 5 of the officers patrolling the streets, 2 of the officers working full time at the station, and 3 of the officers on reserve at the station, how many different divisions of the 10 officers into the 3 groups are possible?

***Solution:*** There are $\dfrac{10!}{5!\,2!\,3!} = 2520$ possible divisions. ∎

**Example 5b.** There are 10 boys who are to be divided into an $A$ team and a $B$ team of 5 boys each. The $A$ team will play in one league and the $B$ team in another. How many different divisions are possible?

***Solution:*** There are $\dfrac{10!}{5!\,5!} = 252$ possible divisions. ∎

**Example 5c.** In order to play a game of basketball, 10 boys at a playground divide themselves into two teams of 5 each. How many different divisions are possible?

***Solution:*** Note that this example is different from the previous one because now the order of the two teams is irrelevant. That is, there is no $A$ and $B$ team but just a division consisting of 2 groups of 5 boys each. Hence the desired answer is

$$\frac{10!/5!\,5!}{2!} = 126$$ ∎

The proof of the following theorem, which generalizes the binomial theorem, is left as an exercise.

---

*The Multinomial Theorem*

$$(x_1 + x_2 + \cdots + x_r)^n = \sum_{\substack{(n_1, \ldots, n_r): \\ n_1 + \cdots + n_r = n}} \binom{n}{n_1, n_2, \ldots, n_r} x_1^{n_1} x_2^{n_2} \cdots x_r^{n_r}$$

*That is, the sum is over all nonnegative integer-valued vectors* $(n_1, n_2, \ldots, n_r)$ *such that* $n_1 + n_2 + \cdots + n_r = n$.

---

The numbers $\dbinom{n}{n_1, n_2, \ldots, n_r}$ are known as multinomial coefficients.

**Example 5d**

$$(x_1 + x_2 + x_3)^2 = \binom{2}{2, 0, 0} x_1^2 x_2^0 x_3^0 + \binom{2}{0, 2, 0} x_1^0 x_2^2 x_3^0$$

$$+ \binom{2}{0, 0, 2} x_1^0 x_2^0 x_3^2 + \binom{2}{1, 1, 0} x_1^1 x_2^1 x_3^0$$

$$+ \binom{2}{1, 0, 1} x_1^1 x_2^0 x_3^1 + \binom{2}{0, 1, 1} x_1^0 x_2^1 x_3^1$$

$$= x_1^2 + x_2^2 + x_3^2 + 2x_1x_2 + 2x_1x_3 + 2x_2x_3$$ ∎

## 6 On the Distribution of Balls in Urns

There are $r^n$ possible outcomes when $n$ distinguishable balls are to be distributed into $r$ distinguishable urns. This follows because each ball may be distributed into any of $r$ possible urns. Let us now, however, suppose that the $n$ balls are indistinguishable from each other. In this case, how many different outcomes are possible? As the balls are indistinguishable, it follows that the outcome of the experiment of distributing the $n$ balls into $r$ urns can be described by a vector $(x_1, x_2, \ldots, x_r)$, where $x_i$ denotes the number of balls that are distributed into the $i$th urn. Hence the problem reduces to finding the number of distinct nonnegative integer-valued vectors $(x_1, x_2, \ldots, x_r)$ such that

$$x_1 + x_2 + \cdots + x_r = n$$

To compute this, let us start by considering the number of positive integer-valued solutions. Toward this end imagine that we have $n$ indistinguishable objects lined up and that we want to divide them into $r$ nonempty groups.

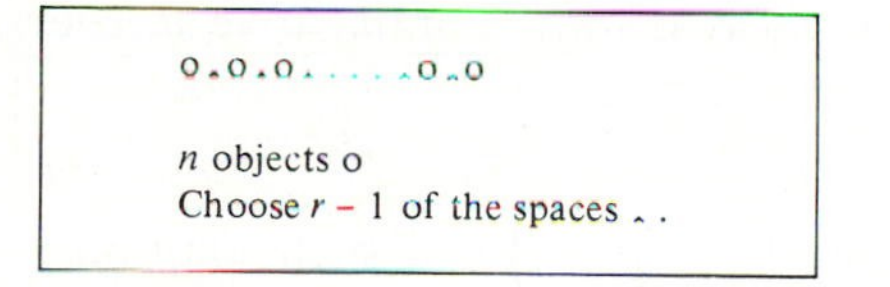


**Figure 1.2**

To do so, we can select $r - 1$ of the $n - 1$ spaces between adjacent objects as our dividing points. (See Figure 1.2.) For instance, if we have $n = 8$ and $r = 3$ and choose the 2 divisors as shown

$$\text{ooo}|\text{ooo}|\text{oo}$$

then the obtained vector is $x_1 = 3$, $x_2 = 3$, $x_3 = 2$. As there are $\binom{n-1}{r-1}$ possible selections, we obtain the following proposition.

---

*Proposition 6.1*

*There are* $\binom{n-1}{r-1}$ *distinct positive integer-valued vectors* $(x_1, x_2, \ldots, x_r)$ *satisfying*

$$x_1 + x_2 + \cdots + x_r = n, \qquad x_i > 0,\ i = 1, \ldots, r$$

---

To obtain the number of nonnegative (as opposed to positive) solutions, note that the number of nonnegative solutions of $x_1 + x_2 + \cdots + x_r = n$ is the same as the number of positive solutions of $y_1 + \cdots + y_r = n + r$ (seen by letting $y_i = x_i + 1$, $i = 1, \ldots, r$). Hence, from Proposition 6.1, we obtain the following proposition.

---

*Proposition 6.2*

*There are* $\binom{n+r-1}{n}$ *distinct nonnegative integer-valued vectors* $(x_1, x_2, \ldots, x_r)$ *satisfying*

$$x_1 + x_2 + \cdots + x_r = n \tag{6.1}$$

---

**Example 6a.** How many distinct nonnegative integer-valued solutions of $x_1 + x_2 = 3$ are possible?

***Solution:*** There are $\binom{3+2-1}{3} = 4$ such solutions: $(0, 3)$, $(1, 2)$, $(2, 1)$, $(3, 0)$. ∎

**Example 6b.** An investor has 20 thousand dollars to invest among 4 possible investments. Each investment must be in units of a thousand dollars. If the total 20 thousand is to be invested, how many different investment strategies are possible? What if not all the money need be invested?

***Solution:*** If we let $x_i$, $i = 1, 2, 3, 4$, denote the number of thousands invested in investment number $i$, then, when all is to be invested, $x_1$, $x_2$, $x_3$, $x_4$ are integers satisfying

$$x_1 + x_2 + x_3 + x_4 = 20, \qquad x_i \geq 0$$

Hence, by Proposition 6.2, there are $\binom{23}{3} = 1771$ possible investment strategies. If not all of the money need be invested, then, if we let $x_5$ denote the amount kept in reserve, a strategy is a nonnegative integer-valued vector $(x_1, x_2, x_3, x_4, x_5)$ satisfying

$$x_1 + x_2 + x_3 + x_4 + x_5 = 20$$

Hence, by Proposition 6.2, there are now $\binom{24}{4} = 10{,}626$ possible strategies. ∎

**Example 6c.** How many terms are there in the multinomial expansion of $(x_1 + x_2 + \cdots + x_r)^n$?

***Solution***

$$(x_1 + x_2 + \cdots + x_r)^n = \sum \binom{n}{n_1, \ldots, n_r} x_1^{n_1} \cdots x_r^{n_r}$$

where the sum is over all nonnegative integer-valued $(n_1, \ldots, n_r)$ such that $n_1 + \cdots + n_r = n$. Hence, by Proposition 6.2, there are $\binom{n + r - 1}{n}$ such terms. ∎

**Example 6d.** Let us reconsider Example 4c, in which we have a set of $n$ items, of which $m$ are (indistinguishable and) defective and the remaining $n - m$ are (also indistinguishable and) functional. Our objective is to determine the number of linear orderings in which no two defectives are next to each other. To determine this quantity, let us imagine that the defective items are lined up among themselves and the functional ones are now to be put in position. Let us denote $x_1$ as the number of functional items to the left of the first defective, $x_2$ as the number of functional items between the first two defectives, and so on. That is, schematically we have

$$x_1 \varnothing x_2 \varnothing \cdots x_m \varnothing x_{m+1}$$

Now there will be at least one functional item between any pair of defectives as long as $x_i > 0$, $i = 2, \ldots, m$. Hence the number of outcomes satisfying the condition is the number of vectors $x_1, \cdots, x_{m+1}$ that satisfy

$$x_1 + \cdots + x_{m+1} = n - m \qquad x_1 \geq 0, x_{m+1} \geq 0, x_i > 0, i = 2, \ldots, m$$

But on letting $y_1 = x_1 + 1$, $y_i = x_i$, $i = 2, \ldots, m$, $y_{m+1} = x_{m+1} + 1$, we see that this is equal to the number of positive vectors $(y_1, \ldots, y_{m+1})$ that satisfy

$$y_1 + y_2 + \cdots + y_{m+1} = n - m + 2$$

Hence, by Proposition 6.1, there are $\binom{n - m + 1}{m}$ such outcomes, which is in agreement with the results of Example 4c.

Suppose now that we are interested in the number of outcomes in which each pair of defective items is separated by at least 2 functional ones. By the same reasoning as that applied above, this would equal the number of vectors satisfying

$$x_1 + \cdots + x_{m+1} = n - m \qquad x_1 \geq 0, x_{m+1} \geq 0, x_i \geq 2, i = 2, \ldots, m$$

Upon letting $y_1 = x_1 + 1$, $y_i = x_i - 1$, $i = 2, \ldots, m$, $y_{m+1} = x_{m+1} + 1$, we see that this is the same as the number of positive solutions of

$$y_1 + \cdots + y_{m+1} = n - 2m + 3$$

Hence, from Proposition 6.1, there are $\binom{n - 2m + 2}{m}$ such outcomes. ∎

## *Theoretical Exercises*

**1.** Prove the generalized version of the basic counting principle.

**2.** Two experiments are to be performed. The first can result in any one of $m$ possible outcomes. If the first experiment results in outcome number $i$, then the second experiment can result in any of $n_i$ possible outcomes, $i = 1, 2, \ldots, m$. What is the number of possible outcomes of the two experiments?

**3.** In how many ways can $r$ objects be selected from a set of $n$ if the order of selection is considered relevant?

**4.** Give a combinatorial explanation of why $\binom{n}{r}$ equals $\binom{n}{n-r}$.

✓ **5.** There are $\binom{n}{r}$ permutations of $n$ balls of which $r$ are black and $n - r$ are white. Give a combinatorial explanation of this fact.

✓**6.** Give an analytic proof of Equation (4.1.).

✓**7.** Prove that

$$\binom{n+m}{r} = \binom{n}{0}\binom{m}{r} + \binom{n}{1}\binom{m}{r-1} + \cdots + \binom{n}{r}\binom{m}{0}$$

whenever $r \leq n, r \leq m$.

HINT: Consider a group of $n$ men and $m$ women. How many groups of size $r$ are possible?

**8.** Verify that for $n \geq 4$

$$\binom{n+1}{4} = \frac{\binom{\binom{n}{2}}{2}}{3}$$

Now present a combinatorial argument for the above.

HINT: Consider a group of $n + 1$ items of which one is considered special. Argue that both sides of the above identity represent the number of subsets of size 4.

**9.** Present a combinatorial explanation of why $\binom{n}{r} = \binom{n}{r, n-r}$.

**10.** Argue that

$$\binom{n}{n_1, n_2, \ldots, n_r} = \binom{n-1}{n_1 - 1, n_2, \ldots, n_r} + \binom{n-1}{n_1, n_2 - 1, \ldots, n_r} + \cdots + \binom{n-1}{n_1, n_2, \ldots, n_r - 1}$$

**11.** Prove the multinomial theorem.

**12.** Show that for $n > 0$

$$\sum_{i=0}^{n} (-1)^{n-i}\binom{n}{i} = 0$$

**13.** (a) Prove the following combinatorial identity by induction:

$$\sum_{k=1}^{n} k\binom{n}{k} = n2^{n-1}$$

(b) Present a combinatorial argument for the above by considering a set of $n$ people and determining, in two ways, the number of possible selections of a committee and a chairperson for the committee.

(c) Verify the following identity for $n = 1, 2, 3, 4, 5$:

$$\sum_{k=1}^{n} \binom{n}{k} k^2 = 2^{n-2} n(n+1)$$

For a combinatorial proof of the above consider a set of $n$ people, and argue that both sides of the above identity represent the number of different selections of a committee, its chairperson, and its secretary (possibly the same as the chairperson).

HINT: (i) How many different selections result in the committee containing exactly $k$ people?

(ii) How many different selections are there in which the chairperson and the secretary are the same? (Answer: $n2^{n-1}$.)

(iii) How many different selections result in the chairperson and the secretary being different?

(d) Now argue that

$$\sum_{k=1}^{n} \binom{n}{k} k^3 = 2^{n-3} n^2(n+3)$$

**14.** In how many ways can $n$ indistinguishable balls be distributed into $r$ urns in such a way that the $i$th urn contains at least $m_i$ balls? Assume that

$$n \geq \sum_{i=1}^{r} m_i$$

ANSWER: $\binom{n - \sum m_i + r - 1}{n - \sum m_i}$

**15.** Show that

$$\binom{n + r - 1}{n} = \sum_{i=0}^{n} \binom{n - i + r - 2}{n - i}$$

HINT: Make use of Proposition 6.2.

**16.** Prove that there are exactly $\binom{r}{k}\binom{n-1}{n-r+k}$ solutions of $x_1 + x_2 + \cdots + x_r = n$ for which exactly $k$ of the $x$'s equal 0.

**17.** Consider a function $f(x_1, x_2, \ldots, x_n)$ of $n$ variables. How many different partial derivatives of order $r$ are possible?

**18.** Use Theoretical Exercise 7 to prove that

$$\binom{2n}{n} = \sum_{k=0}^{n} \binom{n}{k}^2$$

**19.** (a) Use mathematical induction and the combinatorial identity

$$\binom{m}{k} = \binom{m-1}{k-1} + \binom{m-1}{k}$$

to prove the identity

$$\binom{n+r}{n} = \sum_{j=0}^{n} \binom{j+r-1}{j}$$

(b) Give a second proof by arguing that both sides of the above equal the number of distinct nonnegative integer-valued solutions of

$$x_1 + x_2 + \cdots + x_r \leq n$$

**20.** From a set of $n$ people a committee of size $j$ is to be chosen, and from this committee a subcommittee of size $i$, $i \leq j$, is also to be chosen.

(a) Derive a combinatorial identity by computing, in two ways, the number of possible choices of the committee and subcommittee—first by supposing that the committee is chosen first and then the subcommittee, and second by supposing that the subcommittee is chosen first and then the remaining members of the committee are chosen.

(b) Use (a) to prove the following combinatorial identity:

$$\sum_{j=i}^{n} \binom{n}{j}\binom{j}{i} = \binom{n}{i} 2^{n-i}, \qquad i \leq n$$

(c) Use (a) and Theoretical Exercise 12 to show that

$$\sum_{j=i}^{n} \binom{n}{j}\binom{j}{i}(-1)^{n-j} = 0, \qquad i \leq n$$

## *Problems*

**1.** (a) How many different 7-place license plates are possible if the first 2 places are for letters and the other 5 for numbers?

(b) Repeat (a) under the assumption that no letter or number can be repeated in a single license plate.

**2.** John, Jim, Jay, and Jack have formed a band consisting of 4 instruments. If each of the boys can play all 4 instruments, how many different arrangements are possible? What if John and Jim can play all 4 instruments, but Jay and Jack can each play only piano and drums?

**3.** If 4 Americans, 3 Frenchmen, and 3 Englishmen are to be seated in a row, how many seating arrangements are possible when people of the same nationality must sit next to each other?

**4.** (a) In how many ways can 3 boys and 3 girls sit in a row?
(b) In how many ways can 3 boys and 3 girls sit in a row if the boys and the girls are each to sit together?
(c) In how many ways if only the boys must sit together?
(d) In how many ways if no two people of the same sex are allowed to sit together?

**5.** How many different letter arrangements can be made from the letters (a) FLUKE, (b) PROPOSE, (c) MISSISSIPPI, and (d) ARRANGE?

**6.** A child has 12 blocks, of which 6 are black, 4 are red, 1 is white, and 1 is blue. If the child puts the blocks in a line, how many arrangements are possible?

**7.** In how many ways can 8 people be seated in a row if
(a) there are no restrictions on the seating arrangement;
(b) persons $A$ and $B$ must sit next to each other;
(c) there are 4 men and 4 women and no 2 men or 2 women can sit next to each other;
(d) there are 5 men and they must sit next to each other;
(e) there are 4 married couples and each couple must sit together?

**8.** In how many ways can 3 novels, 2 mathematics books, and 1 chemistry book be arranged on a bookshelf if
(a) the books can be arranged in any order;
(b) the mathematics books must be together and the novels must be together;
(c) the novels must be together but the other books can be arranged in any order?

**9.** A president, treasurer, and secretary, all different, are to be chosen from a club consisting of 10 people. How many different choices of officers are possible if
(a) there are no restrictions;
(b) $A$ and $B$ will not serve together;
(c) $C$ and $D$ will serve together or not at all;
(d) $E$ must be an officer;
(e) $F$ will only serve if he is president?

**10.** Five separate awards (best scholarship, best leadership qualities, and so on) are to be presented to selected students from a class of 30. How many different outcomes are possible if
(a) a student can receive any number of awards;
(b) each student can receive at most 1 award?

$30^5$

**11.** How many 5-card poker hands are there?

**12.** A committee of 7, consisting of 2 Republicans, 2 Democrats, and 3 Independents, is to be chosen from a group of 5 Republicans, 6 Democrats, and 4 Independents. How many committees are possible?

**13.** A student is to answer 7 out of 10 questions in an examination. How many choices has she? How many if she must answer at least 3 of the first 5 questions?

**14.** A woman has 8 friends, of whom she will invite 5 to a tea party. How many choices has she if 2 of the friends are feuding and will not attend together? How many choices has she if 2 of her friends will only attend together?

**15.** A psychology laboratory conducting dream research contains 3 rooms, with 2 beds in each room. If 3 sets of identical twins are to be assigned to these 6 beds so that each set of twins sleeps in different beds in the same room, how many assignments are possible?

**16.** Expand $(3x^2 + y)^5$.

**17.** The game of bridge is played by 4 players, each of whom is dealt 13 cards. How many bridge deals are possible?

**18.** Expand $(x_1 + 2x_2 + 3x_3)^4$.

**19.** If 12 people are to be divided into 3 committees of respective sizes 3, 4, and 5, how many divisions are possible?

**20.** In how many ways can a man divide 7 gifts among his 3 children if the eldest is to receive 3 gifts and the others 2 each?

**21.** If 8 identical blackboards are to be divided among 4 schools, how many divisions are possible? How many, if each school must receive at least 1 blackboard?

**22.** If 8 new teachers are to be divided among 4 schools, how many divisions are possible? What if each school must receive 2 teachers?

**23.** An elevator starts at the basement with 8 people (not including the elevator operator) and discharges them all by the time it reaches the top floor, number 6. In how many ways could the operator have perceived the people leaving the elevator if all people look alike to him? What if the 8 people consisted of 5 men and 3 women and the operator could tell a man from a woman?

**24.** An art collection on auction consisted of 4 Dalis, 5 Van Goghs, and 6 Picassos, and at the auction were 5 art collectors. The society page reporter only observed the number of Dalis, Van Goghs, and Picassos acquired by each collector. How many different results could have been recorded if all works were sold?

**25.** Ten weight lifters are competing in a team weight-lifting contest. Of the lifters, 3 are from the United States, 4 are from the USSR, 2 are from mainland China, and 1 is from Canada. If the scoring only takes account of the countries that the lifters represent and not their individual identities, how many different outcomes are possible from the point of view of scores? How many different outcomes correspond to results in which the United States has 1 competitor in the top three and 2 in the bottom three?

**26.** Delegates from 10 countries, including Russia, France, England, and the United States, are to be seated in a row. How many different seating arrangements are possible if the French and English delegates are to be seated next to each other, and the Russian and U.S. delegates are not to be next to each other?

**27.** We have 20 thousand dollars that must be invested among 4 possible opportunities. Each investment must be integral in units of 1 thousand dollars, and there are minimal investments that need to be made if one is to invest in these opportunities. The minimal investments are 2, 2, 3, and 4 thousand dollars. How many different investment strategies are available if

(a) an investment must be made in each opportunity;
(b) investments must be made in at least 3 of the 4 opportunities.

# 2

# Axioms of Probability

## 1 Introduction

In this chapter we first introduce the concept of the probability of an event and then show how these probabilities can be computed in certain situations. As a preliminary, however, we need the concept of the sample space and the events of an experiment.

## 2 Sample Space and Events

Consider an experiment whose outcome is not predictable with certainty in advance. However, although the outcome of the experiment will not be known in advance, let us suppose that the set of all possible outcomes is known. This set of all possible outcomes of an experiment is known as the *sample space* of the experiment and is denoted by $S$. Some examples follow.

1. If the outcome of an experiment consists in the determination of the sex of a newborn child, then

$$S = \{g, b\}$$

where the outcome $g$ means that the child is a girl and $b$ that it is a boy.

2. If the outcome of an experiment is the order of finish in a race among the 7 horses having post positions 1, 2, 3, 4, 5, 6, 7, then

$$S = \{\text{all } 7! \text{ permutations of } (1, 2, 3, 4, 5, 6, 7)\}$$

The outcome (2, 3, 1, 6, 5, 4, 7) means, for instance, that the number 2 horse comes in first, then the number 3 horse, then the number 1 horse, and so on.

3. If the experiment consists of flipping two coins, then the sample space consists of the following four points:

$$S = \{(H, H), (H, T), (T, H), (T, T)\}$$

The outcome will be $(H, H)$ if both coins are heads, $(H, T)$ if the first coin is heads and the second tails, $(T, H)$ if the first is tails and the second heads, and $(T, T)$ if both coins are tails.

4. If the experiment consists of tossing two dice, then the sample space consists of the 36 points

$$S = \{(i, j) \quad i, j = 1, 2, 3, 4, 5, 6\}$$

where the outcome $(i, j)$ is said to occur if $i$ appears on the leftmost die and $j$ on the other die.

5. If the experiment consists of measuring (in hours) the lifetime of a transistor, then the sample space consists of all nonnegative real numbers. That is

$$S = \{x: 0 \leq x < \infty\}$$

Any subset $E$ of the sample space is known as an *event*. That is, an event is a set consisting of possible outcomes of the experiment. If the outcome of the experiment is contained in $E$, then we say that $E$ has occurred. Some examples of events are the following.

In example 1 above, if $E = \{g\}$, then $E$ is the event that the child is a girl. Similarly, if $F = \{b\}$, then $F$ is the event that the child is a boy.

In example 2, if

$$E = \{\text{all outcomes in } S \text{ starting with a } 3\}$$

then $E$ is the event that horse 3 wins the race.

In example 3, if $E = \{(H, H), (H, T)\}$, then $E$ is the event that a head appears on the first coin.

In example 4, if $E = \{(1, 6), (2, 5), (3, 4), (4, 3), (5, 2), (6, 1)\}$, then $E$ is the event that the sum of the dice equals 7.

In example 5, if $E = \{x: 0 \leq x \leq 5\}$, then $E$ is the event that the transistor does not last longer than 5 hours.

For any two events $E$ and $F$ of a sample space $S$, we define the new event $E \cup F$ to consist of all points that are either in $E$ or in $F$ or in both $E$ and $F$. That is, the event $E \cup F$ will occur if *either* $E$ or $F$ occurs. For instance, in example 1 if event $E = \{g\}$ and $F = \{b\}$, then

$$E \cup F = \{g, b\}$$

That is, $E \cup F$ would be the whole sample space $S$. In example 3 if $E = \{(H, H), (H, T)\}$ and $F = \{(T, H)\}$, then

$$E \cup F = \{(H, H), (H, T), (T, H)\}$$

Thus $E \cup F$ would occur if a head appeared on either coin.

The event $E \cup F$ is called the *union* of the event $E$ and the event $F$.

Similarly, for any two events $E$ and $F$ we may also define the new event $EF$, called the *intersection* of $E$ and $F$, to consist of all outcomes that are both in $E$ and in $F$. That is, the event $EF$ will occur only if both $E$ and $F$

occur. For instance, in example 3 if $E = \{(H, H), (H, T), (T, H)\}$ is the event that at least 1 head occurs, and $F = \{(H, T), (T, H), (T, T)\}$ is the event that at least 1 tail occurs, then

$$EF = \{(H, T), (T, H)\}$$

is the event that exactly 1 head and 1 tail appear. In example 4 if $E = \{(1, 6), (2, 5), (3, 4), (4, 3), (5, 2), (6, 1)\}$ is the event that the sum of the dice is 7 and $F = \{(1, 5), (2, 4), (3, 3), (4, 2), (5, 1)\}$ is the event that the sum is 6, then the event $EF$ does not contain any outcomes and hence could not occur. To give such an event a name, we shall refer to it as the null event and denote it by $\varnothing$ (that is, $\varnothing$ refers to the event consisting of no points). If $EF = \varnothing$, then $E$ and $F$ are said to be *mutually exclusive.*

We also define unions and intersections of more than two events in a similar manner. If $E_1, E_2, \ldots$ are events, the union of these events, denoted by $\bigcup_{n=1}^{\infty} E_n$, is defined to be that event which consists of all points that are in $E_n$ for at least one value of $n = 1, 2, \ldots$. Similarly, the intersection of the events $E_n$, denoted by $\bigcap_{n=1}^{\infty} E_n$, is defined to be the event consisting of those points that are in all of the events $E_n$, $n = 1, 2, \ldots$.

Finally, for any event $E$ we define the new event $E^c$, referred to as the complement of $E$, to consist of all points in the sample space $S$ that are not in $E$. That is, $E^c$ will occur if and only if $E$ does not occur. In example 4, if event $E = \{(1, 6), (2, 5), (3, 4), (4, 3), (5, 2), (6, 1)\}$, then $E^c$ will occur when the sum of the dice does not equal 7. Also note that because the experiment must result in some outcome, it follows that $S^c = \varnothing$.

For any two events $E$ and $F$, if all of the points in $E$ are also in $F$, then we say that $E$ is contained in $F$ and write $E \subset F$ (or equivalently, $F \supset E$). Thus, if $E \subset F$, the occurrence of $E$ necessarily implies the occurrence of $F$. If $E \subset F$ and $F \subset E$, we say that $E$ and $F$ are equal and write $E = F$.

A graphical representation that is very useful for illustrating logical relations among events is the Venn diagram. The subspace $S$ is represented as consisting of all the points in a large rectangle, and the events $E$, $F$, $G, \ldots$ are represented as consisting of all the points in given circles within the rectangle. Events of interest can then be indicated by shading appropriate regions of the diagram. For instance, in the three Venn diagrams shown in Figure 2.1, the shaded areas represent, respectively, the events $E \cup F$, $EF$, and $E^c$. The Venn diagram in Figure 2.2 indicates that $E \subset F$.

The operation of forming unions, intersections, and complements of events obey certain rules not dissimilar to the rules of algebra. We list a few of these rules.

| | | |
|---|---|---|
| Commutative law | $E \cup F = F \cup E$ | $EF = FE$ |
| Associative law | $(E \cup F) \cup G = E \cup (F \cup G)$ | $(EF)G = E(FG)$ |
| Distributive law | $(E \cup F)G = EG \cup FG$ | $EF \cup G = (E \cup G)(F \cup G)$ |

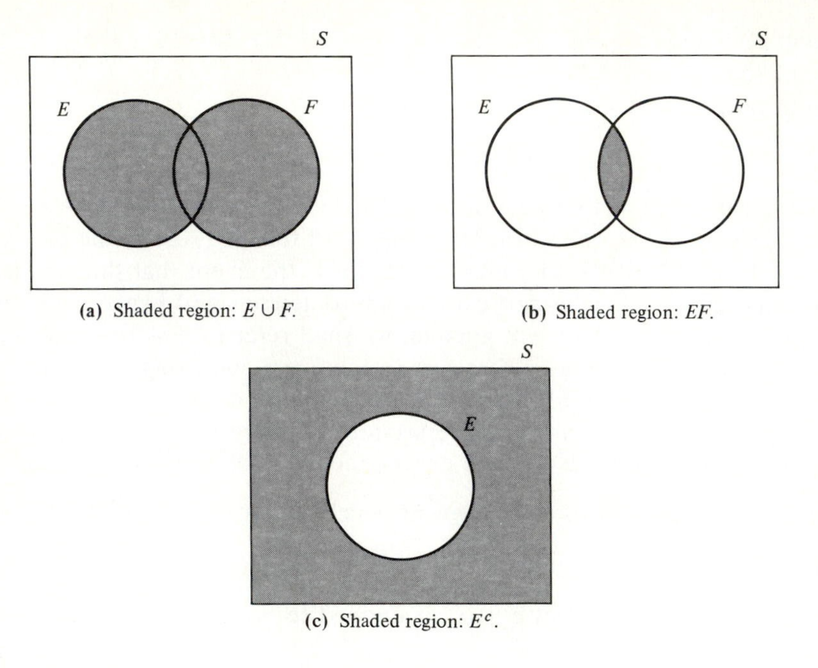


(a) Shaded region: $E \cup F$.

(b) Shaded region: $EF$.

(c) Shaded region: $E^c$.

**Figure 2.1**

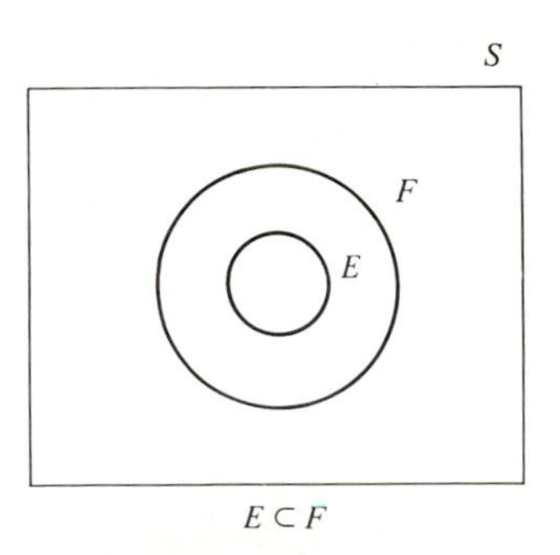


$E \subset F$

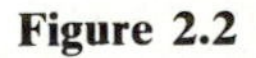

**Figure 2.2**

These relations are verified by showing that any outcome that is contained in the event on the left side of the equality sign is also contained in the event on the right side and vice versa. One way of showing this is by means of Venn diagrams. For instance, the distributive law may be verified by the sequence of diagrams in Figure 2.3.

The following useful relationship between the three basic operations of forming unions, intersections, and complements is known as *DeMorgan's laws*:

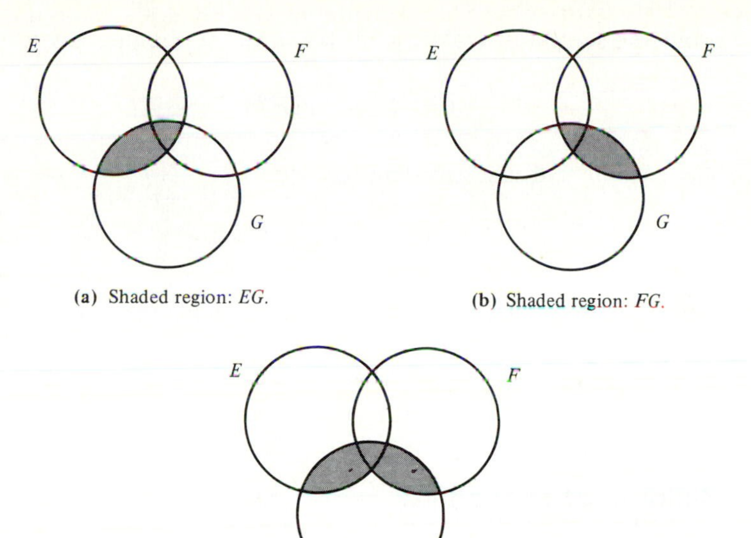


**(a)** Shaded region: $EG$.

**(b)** Shaded region: $FG$.

**(c)** Shaded region: $(E \cup F)G$.

$(E \cup F)G = EG \cup FG$

**Figure 2.3**

$$\left(\bigcup_{i=1}^{n} E_i\right)^c = \bigcap_{i=1}^{n} E_i^c$$

$$\left(\bigcap_{i=1}^{n} E_i\right)^c = \bigcup_{i=1}^{n} E_i^c$$

To prove DeMorgan's laws, suppose first that $x$ is a point of $\left(\bigcup_{i=1}^{n} E_i\right)^c$. Then $x$ is not contained in $\bigcup_{i=1}^{n} E_i$, which means that $x$ is not contained in any of the events $E_i$, $i = 1, 2, \ldots, n$, implying that $x$ is contained in $E_i^c$ for all $i = 1, 2, \ldots, n$ and thus is contained in $\bigcap_{i=1}^{n} E_i^c$. To go the other way, suppose that $x$ is a point of $\bigcap_{i=1}^{n} E_i^c$. Then $x$ is contained in $E_i^c$ for all $i = 1, 2, \ldots, n$, which means that $x$ is not contained in $E_i$ for any $i = 1, 2, \ldots, n$, implying that $x$ is not contained in $\bigcup_{i}^{n} E_i$, which yields that $x$ is contained in $\left(\bigcup_{1}^{n} E_i\right)^c$. This proves the first of DeMorgan's laws.

To prove the second of DeMorgan's laws, we use the first law to obtain

$$\left(\bigcup_{i=1}^{n} E_i^c\right)^c = \bigcap_{i=1}^{n} (E_i^c)^c$$

which, since $(E^c)^c = E$, is equivalent to

$$\left(\bigcup_{1}^{n} E_i^c\right)^c = \bigcap_{1}^{n} E_i$$

Taking complements of both sides of the above yields the result, namely,

$$\bigcup_{1}^{n} E_i^c = \left(\bigcap_{1}^{n} E_i\right)^c$$

## 3 Axioms of Probability

One possible way of defining the probability of an event is in terms of its relative frequency. Such a definition usually goes as follows: We suppose that an experiment, whose sample space is $S$, is repeatedly performed under exactly the same conditions. For each event $E$ of the sample space $S$, we define $n(E)$ to be the number of times in the first $n$ repetitions of the experiment that the event $E$ occurs. Then $P(E)$, the probability of the event $E$, is defined by

$$P(E) = \lim_{n\to\infty} \frac{n(E)}{n}$$

That is, $P(E)$ is defined as the (limiting) percentage of time that $E$ occurs. It is thus the limiting frequency of $E$.

Although the preceding definition is certainly intuitively pleasing and should always be kept in mind by the reader, it possesses a serious drawback: How do we know that $n(E)/n$ will converge to some constant limiting value that will be the same for each possible sequence of repetitions of the experiment? For example, suppose that the experiment to be repeatedly performed consists of flipping a coin. How do we know that the proportion of heads obtained in the first $n$ flips will converge to some value as $n$ gets large? Also, even if it does converge to some value, how do we know that, if the experiment is repeatedly performed a second time, we shall again obtain the same limiting proportion of heads?

Proponents of the relative frequency definition of probability usually answer this objection by stating that the convergence of $n(E)/n$ to a constant limiting value is an assumption, or an *axiom*, of the system. However, to assume that $n(E)/n$ will necessarily converge to some constant value seems to be a very complex assumption. For, although we might indeed hope that such a constant limiting frequency exists, it does not at all seem to be a priori evident that this need be the case. In fact, would it not be more reasonable to assume a set of simpler and more self-evident axioms about probability and then attempt to prove that such a constant limiting frequency does in some sense exist? This latter approach is the modern axiomatic approach to probability theory that we shall adopt in this text. In particular, we shall assume that for each event $E$ in the sample space $S$ there exists a value $P(E)$, referred to as the probability of $E$. We shall then assume that these probabilities satisfy a certain set of axioms, which, hopefully the reader will agree, is in accordance with our intuitive notion of probability.

Consider an experiment whose sample space is $S$. For each event $E$ of the sample space $S$ we assume that a number $P(E)$ is defined and satisfies the following three axioms.

---

*Axiom 1*

$$0 \leq P(E) \leq 1$$

---

*Axiom 2*

$$P(S) = 1$$

---

*Axiom 3*

*For any sequence of mutually exclusive events $E_1, E_2, \ldots$ (that is, events for which $E_iE_j = \varnothing$ when $i \neq j$),*

$$P\left(\bigcup_{i=1}^{\infty} E_i\right) = \sum_{i=1}^{\infty} P(E_i)$$

*We refer to $P(E)$ as the probability of the event $E$.*

---

Thus Axiom 1 states that the probability that the outcome of the experiment is a point in $E$ is some number between 0 and 1. Axiom 2 states that, with probability 1, the outcome will be a point in the sample space $S$. Axiom

3 states that for any sequence of mutually exclusive events the probability of at least one of these events occurring is just the sum of their respective probabilities.

If we consider a sequence of events $E_1, E_2, \ldots$, where $E_1 = S$, $E_i = \varnothing$ for $i > 1$, then, as the events are mutually exclusive and as $S = \bigcup_{i=1}^{\infty} E_i$, we have from Axiom 3 that

$$P(S) = \sum_{i=1}^{\infty} P(E_i) = P(S) + \sum_{i=2}^{\infty} P(\varnothing)$$

implying that

$$P(\varnothing) = 0$$

That is, the null event has probability 0 of occurring.

It should also be noted that it follows that for any finite sequence of mutually exclusive events $E_1, E_2, \ldots, E_n$,

$$P\left(\bigcup_{1}^{n} E_i\right) = \sum_{i=1}^{n} P(E_i) \tag{3.1}$$

This follows from Axiom 3 by defining $E_i$ to be the null event for all values of $i$ greater than $n$. Axiom 3 is equivalent to Equation (3.1) when the state space is finite (why?). However, the added generality of Axiom 3 is necessary when the state space consists of an infinite number of points.

**Example 3a.** If our experiment consists of tossing a coin and if we assume that a head is equally likely to appear as a tail, then we would have

$$P(\{H\}) = P(\{T\}) = \tfrac{1}{2}$$

On the other hand, if the coin were biased and we felt that a head were twice as likely to appear as a tail, then we would have

$$P(\{H\}) = \tfrac{2}{3} \qquad P(\{T\}) = \tfrac{1}{3}$$ ∎

**Example 3b.** If a die is rolled and we suppose that all six sides are equally likely to appear, then we would have $P(\{1\}) = P(\{2\}) = P(\{3\}) = P(\{4\}) = P(\{5\}) = P(\{6\}) = \frac{1}{6}$. From Axiom 3 it would thus follow that the probability of rolling an even number would equal

$$P(\{2, 4, 6\}) = P(\{2\}) + P(\{4\}) + P(\{6\}) = \tfrac{1}{2}$$ ∎

The assumption of the existence of a set function $P$, defined on the events of a sample space $S$, and satisfying Axioms 1, 2, and 3, constitutes the modern mathematical approach to probability theory. Hopefully, the reader will agree that the axioms are natural and in accordance with our intuitive concept of probability as related to chance and randomness. Furthermore, using these axioms we shall be able to prove that if an experiment is repeated over and over again then, with probability 1, the proportion of time during which any specific event $E$ occurs will equal $P(E)$. This result, known as the strong law of large numbers, will be presented in Chapter 8. In addition, we will present another possible interpretation of probability—as being a measure of belief—in Section 7 of this chapter.

TECHNICAL REMARK. We have supposed that $P(E)$ is defined for all the events $E$ of the sample space. Actually, when the state space is an uncountably infinite set $P(E)$ is only defined for the so-called measurable events. However, this restriction need not concern us as all events of any practical interest are measurable.

## 4 Some Simple Propositions

In this section we shall prove some simple propositions regarding probabilities. We first note that as $E$ and $E^c$ are always mutually exclusive and since $E \cup E^c = S$, we have by Axioms 2 and 3 that

$$1 = P(S) = P(E \cup E^c) = P(E) + P(E^c)$$

Or equivalently, we have the statement given in Proposition 4.1.

---

*Proposition 4.1*

$$P(E^c) = 1 - P(E)$$

---

In words, Proposition 4.1 states that the probability that an event does not occur is 1 minus the probability that it does occur. For instance, if the probability of obtaining a head on the toss of a coin is $\frac{3}{8}$, the probability of obtaining a tail must be $\frac{5}{8}$.

Our second proposition states that if the event $E$ is contained in the event $F$, then the probability of $E$ is no greater than the probability of $F$.

---

*Proposition 4.2*

*If $E \subset F$, then $P(E) \leq P(F)$.*

---

**Proof:** Since $E \subset F$, it follows that we can express $F$ as

$$F = E \cup E^c F$$

Hence, as $E$ and $E^c F$ are mutually exclusive, we obtain from Axiom 3 that

$$P(F) = P(E) + P(E^c F)$$

which proves the result, since $P(E^c F) \geq 0$. ∎

Proposition 4.2 tells us, for instance, that the probability of rolling a 1 with a die is less than or equal to the probability of rolling an odd value with the die.

The next proposition gives the relationship between the probability of the union of two events in terms of the individual probabilities and the probability of the intersection.

---

*Proposition 4.3*

$$P(E \cup F) = P(E) + P(F) - P(EF)$$

---

**Proof:** To derive a formula for $P(E \cup F)$, we first note that $E \cup F$ can be written as the union of the two disjoint events $E$ and $E^c F$. Thus from Axiom 3 we obtain that

$$\begin{aligned} P(E \cup F) &= P(E \cup E^c F) \\ &= P(E) + P(E^c F) \end{aligned}$$

Furthermore, since $F = EF \cup E^c F$, we again obtain from Axiom 3 that

$$P(F) = P(EF) + P(E^c F)$$

or equivalently,

$$P(E^c F) = P(F) - P(EF)$$

thus completing the proof. ∎

Proposition 4.3 could also have been proved by making use of the Venn diagram in Figure 2.4.

Let us divide the diagram into three mutually exclusive sections, as shown in Figure 2.5. In words, Section I represents all the points in $E$ that are not in $F$ (that is, $EF^c$); Section II represents all points both in $E$ and in $F$ (that is, $EF$); and Section III represents all points in $F$ that are not in $E$ (that is, $E^c F$).

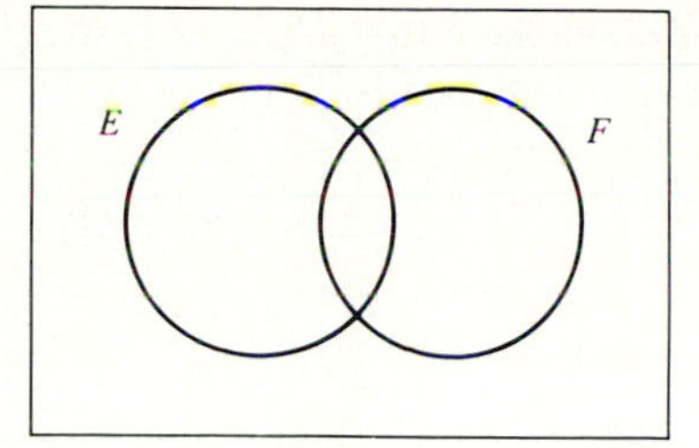


**Figure 2.4** A Venn diagram

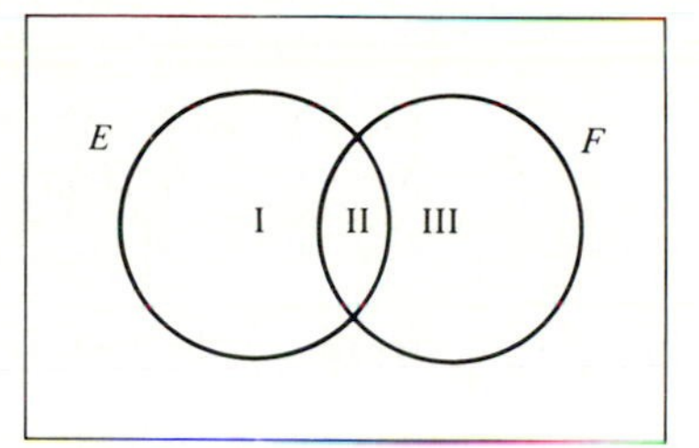


**Figure 2.5** The Venn diagram in sections

From Figure 2.5 we see that

$$\begin{aligned} E \cup F &= \mathrm{I} \cup \mathrm{II} \cup \mathrm{III} \\ E &= \mathrm{I} \cup \mathrm{II} \\ F &= \mathrm{II} \cup \mathrm{III} \end{aligned}$$

As I, II, and III are mutually exclusive, it follows from Axiom 3 that

$$\begin{aligned} P(E \cup F) &= P(\mathrm{I}) + P(\mathrm{II}) + P(\mathrm{III}) \\ P(E) &= P(\mathrm{I}) + P(\mathrm{II}) \\ P(F) &= P(\mathrm{II}) + P(\mathrm{III}) \end{aligned}$$

which shows that

$$P(E \cup F) = P(E) + P(F) - P(\mathrm{II})$$

and Proposition 4.3 is proved, since $\mathrm{II} = EF$.

**Example 4a.** Suppose that we toss two coins and suppose that each of the four points in the sample space $S = \{(H, H), (H, T), (T, H), (T, T)\}$ is equally likely and hence has probability $\frac{1}{4}$. Let

$$E = \{(H, H), (H, T)\} \qquad \text{and} \qquad F = \{(H, H), (T, H)\}$$

That is, $E$ is the event that the first coin falls heads, and $F$ is the event that the second coin falls heads.

By Proposition 4.3 we have that $P(E \cup F)$, the probability that either the first or second coin falls heads, is given by

$$\begin{aligned} P(E \cup F) &= P(E) + P(F) - P(EF) \\ &= \tfrac{1}{2} + \tfrac{1}{2} - P(\{(H, H)\}) \\ &= 1 - \tfrac{1}{4} \\ &= \tfrac{3}{4} \end{aligned}$$

This probability could, of course, have been computed directly because

$$P(E \cup F) = P(\{(H, H), (H, T), (T, H)\}) = \tfrac{3}{4}$$ ∎

We may also calculate the probability that any one of the three events $E$ or $F$ or $G$ occurs:

$$P(E \cup F \cup G) = P[(E \cup F) \cup G]$$

which by Proposition 4.3 equals

$$P(E \cup F) + P(G) - P[(E \cup F)G]$$

Now, it follows from the distributive law that the events $(E \cup F)G$ and $EG \cup FG$ are equivalent, and hence we obtain from the preceding equations that

$$\begin{aligned} &P(E \cup F \cup G) \\ &\quad = P(E) + P(F) - P(EF) + P(G) - P(EG \cup FG) \\ &\quad = P(E) + P(F) - P(EF) + P(G) - P(EG) - P(FG) + P(EGFG) \\ &\quad = P(E) + P(F) + P(G) - P(EF) - P(EG) - P(FG) + P(EFG) \end{aligned}$$

In fact, the following proposition can be proved by induction.

---

*Proposition 4.4*

$$\begin{aligned} &P(E_1 \cup E_2 \cup \cdots \cup E_n) \\ &\quad = \sum_{i=1}^{n} P(E_i) - \sum_{i_1 < i_2} P(E_{i_1} E_{i_2}) + \cdots \\ &\quad + (-1)^{r+1} \sum_{i_1 < i_2 < \cdots < i_r} P(E_{i_1} E_{i_2} \cdots E_{i_r}) \\ &\quad + \cdots + (-1)^{n+1} P(E_1 E_2 \cdots E_n) \end{aligned}$$

*The summation* $\sum_{i_1 < i_2 < \cdots < i_r} P(E_{i_1} E_{i_2} \cdots E_{i_r})$ *is taken over all of the* $\binom{n}{r}$ *possible subsets of size r of the set* $\{1, 2, \ldots, n\}$.

---

In words, Proposition 4.4 states that the probability of the union of $n$ events equals the sum of the probabilities of these events taken one at a time, minus the sum of the probabilities of these events taken two at a time, plus the sum of the probabilities of these events taken three at a time, and so on.

## 5 Sample Spaces Having Equally Likely Outcomes

For many experiments it is natural to assume that all outcomes in the sample space are equally likely to occur. That is, consider an experiment whose sample space $S$ is a finite set, say $S = \{1, 2, \ldots, N\}$. Then it is often natural to assume that

$$P(\{1\}) = P(\{2\}) = \cdots = P(\{N\})$$

which implies from Axioms 2 and 3 (why?) that

$$P(\{i\}) = \frac{1}{N} \qquad i = 1, 2, \ldots, N$$

From this it follows from Axiom 3 that for any event $E$

$$P(E) = \frac{\text{number of points in } E}{\text{number of points in } S}$$

In words, if we assume that all outcomes of an experiment are equally likely to occur, then the probability of any event $E$ equals the proportion of points in the sample space that are contained in $E$.

**Example 5a.** If two dice are rolled, what is the probability that the sum of the upturned faces will equal 7?

***Solution:*** We shall solve this problem under the assumption that all of the 36 possible outcomes are equally likely. Since there are 6 possible outcomes, namely, (1, 6), (2, 5), (3, 4), (4, 3), (5, 2), (6, 1) that result in the sum of the dice equaling 7, the desired probability is $\frac{6}{36} = \frac{1}{6}$. ∎

**Example 5b.** If 2 balls are "randomly drawn" from a bowl containing 6 white and 5 black balls, what is the probability that one of the drawn balls is white and the other black?

***Solution:*** If we regard the order in which the balls are selected as being significant, then the sample space consists of $11 \cdot 10 = 110$ points. Furthermore, there are $6 \cdot 5 = 30$ ways in which the first ball selected is white and the second black. Similarly, there are $5 \cdot 6 = 30$ ways in which the first ball is black and the second white. Hence, assuming that "randomly drawn" means that each of the 110 points in the sample space is equally likely to occur, we see that the desired probability is

$$\frac{30 + 30}{110} = \frac{6}{11}$$

This problem could also have been solved by regarding the outcome of the experiment as the (unordered) set of drawn balls. From this point of view, there would be $\binom{11}{2} = 55$ points in the sample space. It is easy to see that the assumption that all possible outcomes are equally likely when the order is considered relevant implies that all possible outcomes are equally likely when the order of selection is not considered relevant. (Reason it out.) Hence, using this second representation of the experiment, we see that the desired probability is

$$\frac{\binom{6}{1}\binom{5}{1}}{\binom{11}{2}} = \frac{6}{11}$$

which, of course, agrees with the earlier answer. ∎

**Example 5c.** A committee of 5 is to be selected from a group of 6 men and 9 women. If the selection is made randomly, what is the probability that the committee consists of 3 men and 2 women?

***Solution:*** Let us assume that "randomly selected" means that each of the $\binom{15}{5}$ possible combinations is equally likely to be selected. Hence the desired probability equals

$$\frac{\binom{6}{3}\binom{9}{2}}{\binom{15}{5}} = \frac{240}{1001}$$ ∎

**Example 5d.** A poker hand consists of 5 cards. If the cards have distinct consecutive values and are not all of the same suit, we say that the hand is a straight. For instance, a hand consisting of the five of spades, six of spades, seven of spades, eight of spades, and nine of hearts is a straight. What is the probability of being dealt a straight?

***Solution:*** We start by assuming that all $\binom{52}{5}$ possible poker hands are equally likely. To determine the number of outcomes that are straights, let us first determine the number of possible outcomes for which the poker hand consists of an ace, two, three, four, and five (the suits being irrelevant). Since the ace can be any 1 of the 4 possible aces, and similarly for the two, three, four, and five, it follows that there are $4^5$ outcomes leading to exactly one ace, two, three, four, and five. Hence, since in 4 of these outcomes all the cards will be of the same suit (such a hand is called a straight flush), it follows that there are $4^5 - 4$ hands that make up a straight of the form ace, two, three, four, and five. Similarly, there are $4^5 - 4$ hands that make up a straight of the form ten, jack, queen, king, and ace. Hence there are $10(4^5 - 4)$ hands that are straights. Thus the desired probability is

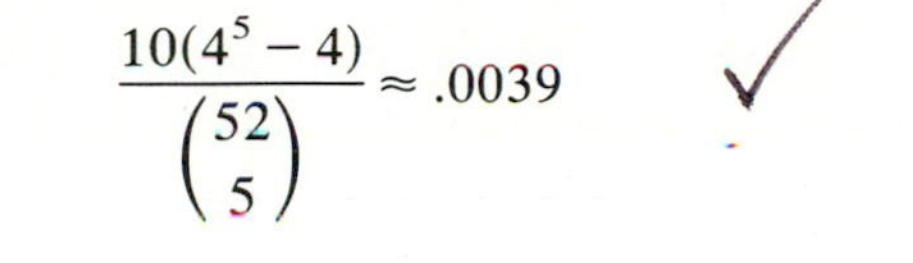

$$\frac{10(4^5 - 4)}{\binom{52}{5}} \approx .0039$$ ∎

**Example 5e.** A 5-card poker hand is said to be a full house if it consists of 3 cards of the same denomination and 2 cards of the same denomination. (That is, a full house is three of a kind plus a pair.) What is the probability of being dealt a full house?

***Solution:*** Again we assume that all $\binom{52}{5}$ possible hands are equally likely. To determine the number of possible full houses, we first note that there are $\binom{4}{2}\binom{4}{3}$ different combinations of, say, 2 tens and 3 jacks. Because there are 13 different choices for the kind of pair and, after a pair has been chosen, there are 12 other choices for the denomination of the remaining 3 cards, it follows that the probability of a full house is

$$\frac{13 \cdot 12 \cdot \binom{4}{2}\binom{4}{3}}{\binom{52}{5}} \approx .0014$$ ∎

**Example 5f.** In the game of bridge the entire deck of 52 cards is dealt out to 4 players. What is the probability that 1 of the players receives 13 spades?

***Solution:*** There are $\binom{52}{13, 13, 13, 13}$ possible divisions of the cards among the 4 distinct players. As there are $\binom{39}{13, 13, 13}$ possible divisions of the cards leading to a fixed player having all 13 spades, it follows that the desired probability is given by

$$\frac{4\binom{39}{13, 13, 13}}{\binom{52}{13, 13, 13, 13}} \approx 6.3 \times 10^{-12}$$ ∎

The next example illustrates the fact that results in probability are sometimes quite surprising when they are first encountered.

**Example 5g.** If $n$ people are present in a room, what is the probability that no two of them celebrate their birthday on the same day of the year? How large need $n$ be so that this probability is less than $\frac{1}{2}$?

***Solution:*** As each person can celebrate his or her birthday on any one of 365 days, there are a total of $(365)^n$ possible outcomes. (We are ignoring the possibility of someone's having been born on February 29.) Assuming that each outcome is equally likely, we see that the desired probability is $(365)(364)(363) \cdots (365 - n + 1)/(365)^n$. It is a rather surprising fact that when $n = 23$, this probability is less than $\frac{1}{2}$. That is, if there are 23 people in a room, the probability that at least two of them have the same birthday exceeds $\frac{1}{2}$. Many people find this surprising. Perhaps even more surprising, however, is that this probability increases to .970 when there are 50 people in the room. And with 100 persons in the room, the odds are better than 3,000,000 to 1 [that is, the probability is greater than $(3 \times 10^6)/(3 \times 10^6 + 1)$] that at least two people have the same birthday. ∎

**Example 5h.** A football team consists of 20 black and 20 white players. The players are to be paired in groups of 2 for the purpose of determining roommates. If the pairing is done at random, what is the probability that there are no white-black roommate pairs? What is the probability that there are $2i$ white-black roommate pairs, $i = 1, 2, \ldots, 10$?

***Solution:*** There are

$$\binom{40}{2, 2, \ldots, 2} = \frac{(40)!}{(2!)^{20}}$$

ways of dividing the 40 players into 20 *ordered* pairs of two each. [That is, there are $(40)!/2^{20}$ ways of dividing the players into a *first* pair, a *second* pair, and so on.] Hence there are $(40)!/2^{20}(20)!$ ways of dividing the players into (unordered) pairs of 2 each. Furthermore, since a division will result in no white-black pairs if the blacks (and whites) are paired among themselves, it follows that there are $[(20)!/2^{10}(10)!]^2$ such divisions. Hence the probability of no white-black roommate pairs, call it $P_0$, is given by

$$P_0 = \frac{\left(\dfrac{(20)!}{2^{10}(10)!}\right)^2}{\dfrac{(40)!}{2^{20}(20)!}} = \frac{[(20)!]^3}{[(10)!]^2(40)!}$$

To determine $P_{2i}$, the probability that there are $2i$ white-black pairs, we first note that there are $\binom{20}{2i}^2$ ways of selecting the $2i$ whites and the $2i$ blacks who are to be in the white-black pairs. These $4i$ players can then be paired up into $(2i)!$ possible white-black pairs. (This is so because the first black can be paired with any of the $2i$ whites, the second black with any of the remaining $2i - 1$ whites, and so on.) As the remaining $20 - 2i$ whites (and blacks) must be paired among themselves, it follows that there are

$$\binom{20}{2i}^2 (2i)! \left[\frac{(20-2i)!}{2^{10-i}(10-i)!}\right]^2$$

divisions, which lead to $2i$ white-black pairs. Hence

$$P_{2i} = \frac{\binom{20}{2i}^2 (2i)! \left[\dfrac{(20-2i)!}{2^{10-i}(10-i)!}\right]^2}{\dfrac{(40)!}{2^{20}(20)!}} \qquad i = 0, 1, \ldots, 10$$

The $P_{2i}$, $i = 0, 1, \ldots, 10$, can now be computed or they can be approximated by making use of a result of Stirling which shows that $n!$ can be approximated by $n^{n+1/2}e^{-n}\sqrt{2\pi}$. For instance, we obtain that

$$P_0 = 1.3403 \times 10^{-6}$$
$$P_{10} = .345861$$
$$P_{20} = 7.6068 \times 10^{-6}$$

∎

The next example in this section not only possesses the virtue of giving rise to a somewhat surprising answer, but is also of theoretical interest.

**Example 5i.** The Matching Problem. Suppose that each of $N$ men at a party throws his hat into the center of the room. The hats are first mixed up, and then each man randomly selects a hat.

1. What is the probability that none of the men selects his own hat?
2. What is the probability that exactly $k$ of the men select their own hats?

***Solution:*** We shall answer part (1) by first calculating the complementary probability of at least one man's selecting his own hat. Let us denote by $E_i$, $i = 1, 2, \ldots, N$ the event that the $i$th man selects his own hat. Now, by Proposition 4.4 $P\left(\bigcup_{i=1}^{N} E_i\right)$, the probability that at least one of the men selects his own hat, is given by

$$P\left(\bigcup_{i=1}^{N} E_i\right) = \sum_{i=1}^{N} P(E_i) - \sum_{i_1<i_2} P(E_{i_1}E_{i_2}) + \cdots$$
$$+ (-1)^{n+1} \sum_{i_1<i_2\cdots<i_n} P(E_{i_1}E_{i_2}\cdots E_{i_n})$$
$$+ \cdots + (-1)^{N+1} P(E_1E_2\cdots E_N)$$

If we regard the outcome of this experiment as a vector of $N$ numbers, where the $i$th element is the number of the hat drawn by the $i$th man, then there are $N!$ possible outcomes. [The outcome $(1, 2, 3, \ldots, N)$ means, for example, that each man selects his own hat.] Furthermore, $E_{i_1}E_{i_2}\cdots E_{i_n}$, the event that each of the $n$ men $i_1, i_2, \ldots, i_n$ selects his own hat, can occur in any of $(N-n)[N-(n+1)]\cdots 3\cdot 2\cdot 1 = (N-n)!$ possible ways; for, of the remaining $N-n$ men, the first can select any of $N-n$ hats, the second can then select any of $N-(n+1)$ hats, and so on. Hence, assuming that all $N!$ possible outcomes are equally likely, we see that

$$P(E_{i_1}E_{i_2}\cdots E_{i_n}) = \frac{(N-n)!}{N!}$$

Also, as there are $\binom{N}{n}$ terms in $\sum_{i_1<i_2\cdots<i_n} P(E_{i_1}E_{i_2}\cdots E_{i_n})$, we see that

$$\sum_{i_1<i_2\cdots<i_n} P(E_{i_1}E_{i_2}\cdots E_{i_n}) = \frac{N!(N-n)!}{(N-n)!n!N!} = \frac{1}{n!}$$

and thus

$$P\left(\bigcup_{i=1}^{N} E_i\right) = 1 - \frac{1}{2!} + \frac{1}{3!} - \cdots + (-1)^{N+1}\frac{1}{N!}$$

Hence the probability that none of the men selects his own hat is

$$1 - 1 + \frac{1}{2!} - \frac{1}{3!} + \cdots + \frac{(-1)^N}{N!}$$

which for $N$ large is approximately equal to $e^{-1} \approx .36788$. In other words, for $N$ large, the probability that none of the men selects his own hat is approximately .37. (How many readers would have incorrectly thought that this probability would go to 1 as $N \to \infty$?)

To obtain the probability that exactly $k$ of the $N$ men select their own hats, we first fix attention on a particular set of $k$ men. The number of ways in which these and only these $k$ men can select their own hats is equal to the number of ways in which the other $N - k$ men can select among their hats in such a way that none of them selects his own hat. But, as

$$1 - 1 + \frac{1}{2!} - \frac{1}{3!} + \cdots + \frac{(-1)^{N-k}}{(N-k)!}$$

is the probability that not one of $N - k$ men, selecting among their hats, selects his own, it follows that the number of ways in which the set of men selecting their own hats corresponds to the set of $k$ men under consideration is

$$(N-k)!\left[1 - 1 + \frac{1}{2!} - \frac{1}{3!} + \cdots + \frac{(-1)^{N-k}}{(N-k)!}\right]$$

Hence, as there are $\binom{N}{k}$ possible selections of a group of $k$ men, it follows that there are

$$\binom{N}{k}(N-k)!\left[1 - 1 + \frac{1}{2!} - \frac{1}{3!} + \cdots + \frac{(-1)^{N-k}}{(N-k)!}\right]$$

ways in which exactly $k$ of the men select their own hats. The desired probability is thus

$$\frac{\binom{N}{k}(N-k)!\left[1 - 1 + \frac{1}{2!} - \frac{1}{3!} + \cdots + \frac{(-1)^{N-k}}{(N-k)!}\right]}{N!}$$

$$= \frac{1 - 1 + \frac{1}{2!} - \frac{1}{3!} + \cdots + \frac{(-1)^{N-k}}{(N-k)!}}{k!}$$

which for $N$ large is approximately $e^{-1}/k!$. The values $e^{-1}/k!$, $k = 0, 1, \ldots,$ are of some theoretical importance as they represent the values associated with the Poisson distribution. This will be elaborated upon in Chapter 4.[1] ∎

For another illustration of the usefulness of Proposition 4.4, consider the following example.

**Example 5j.** If 10 married couples are seated at random at a round table, compute the probability that no wife sits next to her husband.

[1] See Example 5c of Chapter 3 for another approach to this problem.

***Solution:*** If we let $E_i$, $i = 1, 2, \ldots, 10$ denote the event that the $i$th couple sit next to each other, it follows that the desired probability is $1 - P\left(\bigcup_{i=1}^{10} E_i\right)$. Now, from Proposition 4.4,

$$P\left(\bigcup_{1}^{10} E_i\right) = \sum_{1}^{10} P(E_i) - \cdots + (-1)^{n+1} \sum_{i_1 < i_2 < \cdots < i_n} P(E_{i_1} E_{i_2} \cdots E_{i_n}) + \cdots - P(E_1 E_2 \cdots E_{10})$$

To compute $P(E_{i_1} E_{i_2} \cdots E_{i_n})$, we first note that there are 19! ways of arranging 20 people around a round table (why?). The number of arrangements that result in a specified set of $n$ men sitting next to their wives can most easily be obtained by first thinking of each of the $n$ married couples as being single entities. If this were the case, then we would need to arrange $20 - 2n + n = 20 - n$ entities around a round table, and there are clearly $(20 - n - 1)!$ such arrangements. Finally, since each of the $n$ married couples can be arranged next to each other in one of two possible ways, it follows that there are $2^n(20 - n - 1)!$ arrangements that result in a specified set of $n$ men each sitting next to their wives. Therefore,

$$P(E_{i_1} E_{i_2} \cdots E_{i_n}) = \frac{2^n(19 - n)!}{(19)!}$$

Thus, from Proposition 4.4, we obtain that the probability that at least one married couple sits together equals

$$\binom{10}{1} 2^1 \frac{(18)!}{(19)!} - \binom{10}{2} 2^2 \frac{(17)!}{(19)!} + \binom{10}{3} 2^3 \frac{(16)!}{(19)!} - \cdots - \binom{10}{10} 2^{10} \frac{9!}{(19)!} \approx .6605$$

and the desired probability is .3395. ∎

**Example 5k.** Runs. Consider an athletic team that had just finished its season with a final record of $n$ wins and $m$ losses. By examining the sequence of wins and losses, we are hoping to determine whether the team had stretches of games in which it was more likely to win than at other times. One way to gain some insight into this question is to count the number of runs of wins and then see how likely that result would be when all $(n + m)!/(n!\, m!)$ orderings of the $n$ wins and $m$ losses are assumed equally likely. By a run of wins we mean a consecutive sequence of wins. For instance, if $n = 10$, $m = 6$ and the sequence of outcomes was *WWLLWWWLWLLLWWWW*, then there would be 4 runs of wins—the first run being of size 2, the second of size 3, the third of size 1, and the fourth of size 4.

Suppose now that a team has $n$ wins and $m$ losses. Assuming that all $(n + m)!/(n!\, m!) = \binom{n + m}{n}$ orderings are equally likely, let us determine the probability that there will be exactly $r$ runs of wins. To do so, consider first any vector of positive integers $x_1, x_2, \ldots, x_r$ with $x_1 + \cdots + x_r = n$, and let us see how many outcomes result in $r$ runs of wins in which the $i$th run is of size $x_i$, $i = 1, \ldots, r$. For any such outcome, if we let $y_1$ denote the number of losses before the first run of wins, $y_2$ the number of losses between the first 2 runs of wins, $\ldots$, $y_{r+1}$ the number of losses after the last run of wins, then the $y_i$ satisfy

$$y_1 + y_2 + \cdots + y_{r+1} = m \qquad y_1 \geq 0,\ y_{r+1} \geq 0,\ y_i > 0,\ i = 2, \ldots, r$$

and the outcome can be represented schematically as

$$\underbrace{LL\ldots L}_{y_1}\ \underbrace{WW\ldots W}_{x_1}\ \underbrace{L\ldots L}_{y_2}\ \underbrace{WW\ldots W}_{x_2}\ \ldots\ \underbrace{WW}_{x_r}\ \underbrace{L\ldots L}_{y_{r+1}}$$

Hence the number of outcomes that result in $r$ runs of wins—the $i$th of size $x_i$, $i = 1, \ldots r$—is equal to the number of integers $y_1, \ldots, y_{r+1}$ that satisfy the above, or equivalently, to the number of positive integers

$$\bar{y}_1 = y_1 + 1, \qquad \bar{y}_i = y_i,\ i = 2, \ldots, r,\ \bar{y}_{r+1} = y_{r+1} + 1$$

that satisfy

$$\bar{y}_1 + \bar{y}_2 + \cdots + \bar{y}_{r+1} = m + 2$$

By Proposition 6.1 in Chapter 1 there are $\binom{m+1}{r}$ such outcomes. Hence the total number of outcomes that result in $r$ runs of wins is $\binom{m+1}{r}$ multiplied by the number of positive integral solutions of $x + \cdots + x_r = n$. Hence, again from Proposition 6.1, there are thus $\binom{m+1}{r}\binom{n-1}{r-1}$ outcomes resulting in $r$ runs of wins. As there are $\binom{n+m}{n}$ equally likely outcomes, we thus see that

$$P(\{r \text{ runs of wins}\}) = \frac{\binom{m+1}{r}\binom{n-1}{r-1}}{\binom{m+n}{n}}, \qquad r \geq 1$$

For example, if $n = 8$, $m = 6$, then the probability of 7 runs is $\binom{7}{7}\binom{7}{6}\Big/\binom{14}{8} = 1/429$ if all $\binom{14}{8}$ outcomes are equally likely. Hence, if the outcome were *WLWLWLWLWWLWLW*, then we might suspect that the team's win probability was changing over time. (In particular, the probability that the team wins seems to be quite high when it lost its last game and quite low when it won its last game.) On the other extreme, if the outcome were *WWWWWWWLLLLLLL*, then there would have been only 1 run, and as $P(\{1 \text{ run}\}) = \binom{7}{1}\binom{7}{0}\Big/\binom{14}{8} = 1/429$, it would thus again seem unlikely that the team's win probability remained unchanged over its 14 games.

## 6 Probability Is a Continuous Set Function

A sequence of events $\{E_n, n \geq 1\}$ is said to be an increasing sequence if

$$E_1 \subset E_2 \subset \cdots \subset E_n \subset E_{n+1} \subset \cdots$$

whereas it is said to be a decreasing sequence if

$$E_1 \supset E_2 \supset \cdots \supset E_n \supset E_{n+1} \supset \cdots$$

If $\{E_n, n \geq 1\}$ is an increasing sequence of events, then we define a new event, denoted by $\lim_{n\to\infty} E_n$, by

$$\lim_{n\to\infty} E_n = \bigcup_{i=1}^{\infty} E_i$$

when $E_n \subset E_{n+1}$ for all $n$. Similarly, if $\{E_n, n \geq 1\}$ is a decreasing sequence of events, we define $\lim_{n} E_n$ by

$$\lim_{n\to\infty} E_n = \bigcap_{i=1}^{\infty} E_i$$

when $E_n \supset E_{n+1}$ for all $n$.

We now prove Proposition 6.1.

---

*Proposition 6.1*

*If $\{E_n, n \geq 1\}$ is either an increasing or decreasing sequence of events, then*

$$\lim_{n\to\infty} P(E_n) = P(\lim_{n\to\infty} E_n)$$

---

**Proof:** Suppose, first, that $\{E_n, n \geq 1\}$ is an increasing sequence and define the events $F_n, n \geq 1$ by

$$F_1 = E_1$$

$$F_n = E_n\left(\bigcup_{1}^{n-1} E_i\right)^c = E_n E_{n-1}^c \qquad n > 1$$

where we have used the fact that $\bigcup_1^{n-1} E_i = E_{n-1}$, since the events are increasing. In words, $F_n$ consists of those points in $E_n$ that are not in any of the earlier $E_i$, $i < n$. It is easy to verify that the $F_n$ are mutually exclusive events such that

$$\bigcup_{i=1}^{\infty} F_i = \bigcup_{i=1}^{\infty} E_i \qquad \text{and} \qquad \bigcup_{i=1}^{n} F_i = \bigcup_{i=1}^{n} E_i \qquad \text{for all } n \geq 1$$

Thus

$$\begin{aligned}
P\left(\bigcup_1^{\infty} E_i\right) &= P\left(\bigcup_1^{\infty} F_i\right) \\
&= \sum_1^{\infty} P(F_i) \qquad \text{(by Axiom 3)} \\
&= \lim_{n\to\infty} \sum_1^{n} P(F_i) \\
&= \lim_{n\to\infty} P\left(\bigcup_1^{n} F_i\right) \\
&= \lim_{n\to\infty} P\left(\bigcup_1^{n} E_i\right) \\
&= \lim_{n\to\infty} P(E_n)
\end{aligned}$$

which proves the result when $\{E_n, n \geq 1\}$ is increasing.

If $\{E_n, n \geq 1\}$ is a decreasing sequence, then $\{E_n^c, n \geq 1\}$ is an increasing sequence; hence, from the preceding equations,

$$P\left(\bigcup_1^{\infty} E_i^c\right) = \lim_{n\to\infty} P(E_n^c)$$

But, as $\bigcup_1^{\infty} E_i^c = \left(\bigcap_1^{\infty} E_i\right)^c$, we see that

$$P\left(\left(\bigcap_1^{\infty} E_i\right)^c\right) = \lim_{n\to\infty} P(E_n^c)$$

Or equivalently,

$$1 - P\left(\bigcap_1^{\infty} E_i\right) = \lim_{n\to\infty} [1 - P(E_n)] = 1 - \lim_{n\to\infty} P(E_n)$$

or

$$P\left(\bigcap_1^{\infty} E_i\right) = \lim_{n\to\infty} P(E_n)$$

which proves the result. ∎

**Example 6a.** Probability and a Paradox. Suppose that we possess an infinitely large urn and an infinite collection of balls labeled ball number 1, number 2, number 3, and so on. Consider an experiment performed as follows. At 1 minute to 12 P.M., balls numbered 1 through 10 are placed in the urn, and ball number 10 is withdrawn. (Assume the withdrawal takes no time.) At $\frac{1}{2}$ minute to 12 P.M., balls numbered 11 through 20 are placed in the urn, and ball number 20 is withdrawn. At $\frac{1}{4}$ minute to 12 P.M., balls numbered 21 through 30 are placed in the urn, and ball number 30 is withdrawn. At $\frac{1}{8}$ minute to 12 P.M., and so on. The question of interest is, how many balls are in the urn at 12 P.M.?

The answer to this question is clearly that there is an infinite number of balls in the urn at 12 P.M., since any ball whose number is not of the form $10n$, $n \geq 1$, will have been placed in the urn and will not have been withdrawn before 12 P.M. Hence the problem is solved when the experiment is performed as described.

However, let us now change the experiment and suppose that at 1 minute to 12 P.M. balls numbered 1 through 10 are placed in the urn, and ball number 1 is withdrawn; at $\frac{1}{2}$ minute to 12 P.M., balls numbered 11 through 20 are placed in the urn, and ball number 2 is withdrawn; at $\frac{1}{4}$ minute to 12 P.M., balls numbered 21 through 30 are placed in the urn, and ball number 3 is withdrawn; at $\frac{1}{8}$ minute to 12 P.M., balls numbered 31 through 40 are placed in the urn, and ball number 4 is withdrawn, and so on. For this new experiment how many balls are in the urn at 12 P.M.?

Surprisingly enough, the answer now is that the urn is *empty* at 12 P.M. For, consider any ball—say, ball number $n$. At some time prior to 12 P.M. [in particular, at $(\frac{1}{2})^{n-1}$ minutes to 12 P.M.], this ball would have been withdrawn from the urn. Hence for each $n$ ball number $n$ is not in the urn at 12 P.M.; therefore, the urn must be empty at this time.

Thus we see from the preceding discussion that the manner in which the withdrawn balls are selected makes a difference. For, in the first case only balls numbered $10n$, $n \geq 1$, are ever withdrawn; whereas in the second case all of the balls are eventually withdrawn. Let us now suppose that whenever a ball is to be withdrawn that ball is randomly selected from among those present. That is, suppose that at 1 minute to 12 P.M. balls numbered 1 through 10 are placed in the urn, and a ball is randomly selected and withdrawn, and so on. In this case how many balls are in the urn at 12 P.M.?

***Solution:*** We shall show that, with probability 1, the urn is empty at 12 P.M. Let us first consider ball number 1. Define $E_n$ to be the event that ball number 1 is still in the urn after the first $n$ withdrawals have been made. Clearly,

$$P(E_n) = \frac{9 \cdot 18 \cdot 27 \cdots (9n)}{10 \cdot 19 \cdot 28 \cdots (9n+1)}$$

[To understand this equation, just note that if ball number 1 is still to be in the urn after the first $n$ withdrawals, the first ball withdrawn can be any one of 9, the second any one of 18 (there are 19 balls in the urn at the time of the second withdrawal, one of which must be ball number 1), and so on. The denominator is similarly obtained.]

Now, the event that ball number 1 is in the urn at 12 P.M. is just the event $\bigcap_{n=1}^{\infty} E_n$. As the events $E_n$, $n \geq 1$, are decreasing events, it follows from Proposition 6.1 that

$$\begin{aligned} &P\{\text{ball number 1 is in the urn at 12 P.M.}\} \\ &\quad = P\left(\bigcap_{n=1}^{\infty} E_n\right) \\ &\quad = \lim_{n\to\infty} P(E_n) \\ &\quad = \prod_{n=1}^{\infty} \left(\frac{9n}{9n+1}\right) \end{aligned}$$

We now show that

$$\prod_{n=1}^{\infty} \frac{9n}{9n+1} = 0$$

Since

$$\prod_{n=1}^{\infty} \left(\frac{9n}{9n+1}\right) = \left[\prod_{n=1}^{\infty} \left(\frac{9n+1}{9n}\right)\right]^{-1}$$

this is equivalent to showing that

$$\prod_{n=1}^{\infty} \left(1 + \frac{1}{9n}\right) = \infty$$

Now, for all $m \geq 1$,

$$\begin{aligned} \prod_{n=1}^{\infty} \left(1 + \frac{1}{9n}\right) &\geq \prod_{n=1}^{m} \left(1 + \frac{1}{9n}\right) \\ &= \left(1 + \frac{1}{9}\right)\left(1 + \frac{1}{18}\right)\left(1 + \frac{1}{27}\right) \cdots \left(1 + \frac{1}{9m}\right) \\ &> \frac{1}{9} + \frac{1}{18} + \frac{1}{27} + \cdots + \frac{1}{9m} \\ &= \frac{1}{9} \sum_{i=1}^{m} \frac{1}{i} \end{aligned}$$

Hence letting $m \to \infty$ and using the fact that $\sum_{i=1}^{\infty} 1/i = \infty$ yields

$$\prod_{n=1}^{\infty} \left(1 + \frac{1}{9n}\right) = \infty$$

Hence, letting $F_i$ denote the event that ball number $i$ is in the urn at 12 P.M., we have shown that $P(F_1) = 0$. Similarly, we can show that $P(F_i) = 0$ for all $i$. (For instance, the same reasoning shows that $P(F_i) = \prod_{n=2}^{\infty} [9n/(9n + 1)]$ for $i = 11, 12, \ldots, 20$.) Therefore, the probability that the urn is not empty at 12 P.M., $P\left(\bigcup_1^{\infty} F_i\right)$, satisfies

$$P\left(\bigcup_1^{\infty} F_i\right) \leq \sum_1^{\infty} P(F_i) = 0$$

by Boole's inequality [see Theoretical Exercises (for Sections 3–6) 3 and 15].

Thus, with probability 1, the urn will be empty at 12 P.M. ∎

## 7 Probability as a Measure of Belief

Thus far we have interpreted the probability of an event of a given experiment as being a measure of how frequently the event will occur when the experiment is continually repeated. However, there are also other uses of the term *probability*. For instance, we have all heard such statements as, "it is 90 percent probable that Shakespeare actually wrote *Hamlet*," or "the probability that Oswald acted alone in assassinating Kennedy is .8." How are we to interpret these statements?

The most simple and natural interpretation is that the probabilities referred to are measures of the individual's belief in the statements that he or she is making. In other words, the individual making the above statements is quite certain that Oswald acted alone and is even more certain that Shakespeare wrote Hamlet. This interpretation of probability as being a measure of one's belief is often referred to as the *personal* or *subjective* view of probability.

It seems logical to suppose that a "measure of belief" should satisfy all of the axioms of probability. For example, if we are 70 percent certain that Shakespeare wrote *Julius Caesar* and 10 percent certain that it was actually Marlowe, then it is logical to suppose that we are 80 percent certain that it was either Shakespeare or Marlowe. Hence, whether we interpret probability as a measure of belief or as a long-run frequency of occurrence, its mathematical properties remain unchanged.

**Example 7a.** Suppose that in a 7-horse race you feel that each of the first 2 horses has a 20 percent chance of winning, horses 3 and 4 each has a 15 percent chance, and the remaining 3 horses, a 10 percent chance each. Would it be better for you to wager at even money, that the winner will be one of the first three horses, or to wager, again at even money, that the winner will be one of the horses 1, 5, 6, 7?

***Solution:*** Based on your personal probabilities concerning the outcome of the race, your probability of winning the first bet is $.2 + .2 + .15 = .55$, whereas, it is $.2 + .1 + .1 + .1 = .5$ for the second. Hence the first wager is more attractive. ∎

It should be noted that in supposing that an individual's subjective probabilities are always consistent with the axioms of probability we are dealing with an idealized rather than an actual person. For instance, if we were to ask someone what he or she thought the chances were of

(a) rain today,
(b) rain tomorrow,
(c) rain both today and tomorrow,
(d) rain either today or tomorrow,

it is quite possible that, after some deliberation, this person might give 30 percent, 40 percent, 20 percent, and 60 percent as answers. Unfortunately, however, such answers (or such subjective probabilities) are not consistent with the axioms of probability (why not?). We would of course hope that after this was pointed out to the respondent he or she would change the answers. (One possibility we could accept is 30 percent, 40 percent, 10 percent, and 60 percent.)

## *Theoretical Exercises*

### Sections 1–2

Prove the following relations.

**1.** $EF \subset E \subset E \cup F$.

**2.** If $E \subset F$, then $F^c \subset E^c$.

**3.** $F = FE \cup FE^c$, and $E \cup F = E \cup E^cF$.

**4.** $\left(\bigcup_1^\infty E_i\right)F = \bigcup_1^\infty E_iF$, and $\left(\bigcap_1^\infty E_i\right) \cup F = \bigcap_1^\infty (E_i \cup F)$.

**5.** For any sequence of events $E_1, E_2, \ldots$, define a new sequence $F_1, F_2, \ldots$ of disjoint events (that is, events such that $F_iF_j = \varnothing$ whenever $i \neq j$) such that for all $n \geq 1$

$$\bigcup_1^n F_i = \bigcup_1^n E_i$$

**6.** Let $E$, $F$, and $G$ be three events. Find expressions for the events so that of $E$, $F$, and $G$:
(a) only $E$ occurs;
(b) both $E$ and $G$ but not $F$ occurs;
(c) at least one of the events occurs;
(d) at least two of the events occur;
(e) all three occur;
(f) none of the events occurs;
(g) at most one of them occurs;
(h) at most two of them occur;
(i) exactly two of them occur;
(j) at most three of them occur.

**7.** Find the simple expression for the following events:
(a) $(E \cup F)(E \cup F^c)$;
(b) $(E \cup F)(E^c \cup F)(E \cup F^c)$;
(c) $(E \cup F)(F \cup G)$.

**8.** Let $S$ be a given set. If, for some $k > 0$, $S_1, S_2, \ldots, S_k$ are mutually exclusive nonempty subsets of $S$ such that $\bigcup_{i=1}^{k} S_i = S$, then we call the set $\{S_1, S_2, \ldots, S_k\}$ a *partition* of $S$. Let $T_n$ denote the number of different partitions of $\{1, 2, \ldots, n\}$, and so $T_1 = 1$ (the only partition being $S_1 = \{1\}$), and $T_2 = 2$ (the 2 partitions being $\{\{1, 2\}\}$, $\{\{1\}, \{2\}\}$).
(a) Show, by computing all partitions, that $T_3 = 5$, $T_4 = 15$.
(b) Show that

$$T_{n+1} = 1 + \sum_{k=1}^{n} \binom{n}{k} T_k$$

and use this to compute $T_{10}$.

**Sections 3–6**

**1.** Suppose that an experiment is performed $n$ times. For any event $E$ of the sample space let $n(E)$ denote the number of times that event $E$ occurs, and define $f(E) = n(E)/n$. Show that $f(\cdot)$ satisfies Axioms 1, 2, and 3.

**2.** Prove $P(E \cup F \cup G) = P(E) + P(F) + P(G) - P(E^cFG) - P(EF^cG) - P(EFG^c) - 2P(EFG)$.

**3.** Prove Boole's inequality

$$P\left(\bigcup_1^n E_i\right) \le \sum_1^n P(E_i)$$

**4.** If $P(E) = .9$ and $P(F) = .8$, show that $P(EF) \ge .7$. In general, prove Bonferroni's inequality, namely,

$$P(EF) \ge P(E) + P(F) - 1$$

**5.** Show that the probability that exactly one of the events $E$ or $F$ occurs equals $P(E) + P(F) - 2P(EF)$.

**6.** Prove $P(EF^c) = P(E) - P(EF)$.

**7.** Prove $P(E^cF^c) = 1 - P(E) - P(F) + P(EF)$.

**8.** Prove Proposition 4.4.

**9.** An urn contains $M$ white and $N$ black balls. If a random sample of size $r$ is chosen, what is the probability that it will contain exactly $k$ white balls? What if $M = k = 1$?

**10.** Use induction to generalize Bonferroni's inequality to $n$ events, namely, show that

$$P(E_1E_2 \cdots E_n) \ge P(E_1) + \cdots + P(E_n) - (n - 1)$$

**11.** Consider the matching problem, Example 5i, and define $A_N$ to be the number of ways in which the $N$ men can select their hats so that no man selects his own. Argue that

$$A_N = (N - 1)(A_{N-1} + A_{N-2})$$

This formula, along with the boundary conditions $A_1 = 0$, $A_2 = 1$, can then be solved for $A_N$, and the desired probability of no matches would be $A_N/N!$.

HINT: After the first man selects a hat that is not his own, there remain $N - 1$ men to select among a set of $N - 1$ hats that does not contain the hat of one of these men. Thus there is one extra man and one extra hat. Argue that we can get no matches either with the extra man selecting the extra hat or with the extra man not selecting the extra hat.

**12.** Let $f_n$ denote the number of ways of tossing a coin $n$ times such that successive heads never appear. Argue that

$$f_n = f_{n-1} + f_{n-2} \qquad n \ge 2, \text{ where } f_0 = 1, f_1 = 2$$

If $P_n$ denoted the probability that successive heads never appear when a coin is tossed $n$ times, find $P_n$ (in terms of $f_n$) when all possible outcomes of the $n$ tosses are assumed equally likely. Compute $P_{10}$.

ANSWER: $P_{10} = 144/2^{10} = .141$.

**13.** Consider an experiment whose sample space consists of a countably infinite number of points. Show that not all points can be equally likely. Can all points have positive probability of occurring?

**14.** Consider Example 5k, which is concerned with the number of runs of wins obtained when $n$ wins and $m$ losses are randomly permuted. Now consider the total number of runs—that is, win runs plus loss runs—and show that

$$P\{2k \text{ runs}\} = 2\frac{\binom{m-1}{k-1}\binom{n-1}{k-1}}{\binom{m+n}{n}}$$

$$P\{2k+1 \text{ runs}\} = \frac{\binom{m-1}{k-1}\binom{n-1}{k} + \binom{m-1}{k}\binom{n-1}{k-1}}{\binom{m+n}{n}}$$

**15.** Starting with Boole's inequality for a finite number of events, show that for any infinite sequence of events $E_i$, $i \geq 1$,

$$P\left(\bigcup_1^\infty E_i\right) \leq \sum_1^\infty P(E_i)$$

**16.** Show that if $P(E_i) = 1$ for all $i \geq 1$, then $P\left(\bigcap_1^\infty E_i\right) = 1$.

**17.** For a sequence of events $E_i$, $i \geq 1$, we define the new event—called $\limsup_i E_i$—to consist of all outcomes that are contained in an infinite number of the $E_i$, $i \geq 1$. Show that

$$\limsup_i E_i = \bigcap_{n=1}^\infty \bigcup_{i=n}^\infty E_i$$

**18.** Show that if $\sum_{i=1}^\infty P(E_i) < \infty$, then $P\left(\limsup_i E_i\right) = 0$. This is an important result which states that if $\sum_1^\infty P(E_i) < \infty$, then the probability that an infinite number of the $E_i$ will occur is 0.

HINT: Use the inequality

$$\limsup_i E_i \equiv \bigcap_{n=1}^\infty \bigcup_{i=n}^\infty E_i \subset \bigcup_{i=n}^\infty E_i$$

## Problems

### Sections 1–2

**1.** A box contains 3 marbles, 1 red, 1 green, and 1 blue. Consider an experiment that consists of taking 1 marble from the box, then replacing it in the box and drawing a second marble from the box. Describe the sample space. Repeat when the second marble is drawn without first replacing the first marble.

**2.** A die is continually rolled until a 6 appears, at which point the experiment stops. What is the sample space of this experiment? Let $E_n$ denote the event that $n$ rolls are necessary to complete the experiment. What points of the sample space are contained in $E_n$? What is $\left(\bigcup_1^\infty E_n\right)^c$?

**3.** Two dice are thrown. Let $E$ be the event that the sum of the dice is odd; let $F$ be the event that at least one of the dice lands on 1; and let $G$ be the event that the sum is 5. Describe the events $EF$, $E \cup F$, $FG$, $EF^c$, and $EFG$.

**4.** $A$, $B$, and $C$ take turns in flipping a coin. The first one to get a head wins. The sample space of this experiment can be defined by

$$S = \begin{cases} 1, 01, 001, 0001, \ldots, \\ 0000 \cdots \end{cases}$$

(a) Interpret the sample space.
(b) Define the following events in terms of $S$:
  (i) $A$ wins $= A$.
  (ii) $B$ wins $= B$.
  (iii) $(A \cup B)^c$.

Assume that $A$ flips first, then $B$, then $C$, then $A$, and so on.

### Sections 3–6

**1.** A certain town of population size 100,000 has 3 newspapers: I, II, and III. The proportion of townspeople that read these papers are as follows:

| | | |
|---|---|---|
| I: 10 percent | I and II: 8 percent | I and II and III: 1 percent |
| II: 30 percent | I and III: 2 percent | |
| III: 5 percent | II and III: 4 percent | |

(The list tells us, for instance, that 8000 people read newpapers I and II.)

(a) Find the number of people reading only one newspaper.
(b) How many people read at least two newspapers?
(c) If I and III are morning papers and II is an evening paper, how many people read at least one morning paper plus an evening paper?

(d) How many people read only one morning paper and one evening paper?

**2.** The following data were given in a study of a group of 1000 subscribers to a certain magazine: In reference to sex, marital status, and education, there were 312 males, 470 married persons, 525 college graduates, 42 male college graduates, 147 married college graduates, 86 married males, and 25 married male college graduates. Show that the numbers reported in the study must be incorrect.

HINT: Let $M$, $W$, and $G$ denote, respectively, the set of males, married persons, and college graduates. Assume that one of the 1000 persons is chosen at random and use Proposition 4.4 to show that if the above numbers are correct, then $P(M \cup W \cup G) > 1$.

**3.** A deck of cards is dealt out. What is the probability that the fourteenth card dealt is an ace? What is the probability that the first ace occurs on the fourteenth card?

**4.** If it is assumed that all $\binom{52}{5}$ poker hands are equally likely, what is the probability of being dealt:

(a) A flush? (A hand is said to be a flush if all 5 cards are of the same suit.)
(b) One pair? (This occurs when the cards have denominations $a$, $a$, $b$, $c$, $d$, where $a$, $b$, $c$, and $d$ are all distinct.)
(c) Two pairs? (This occurs when the cards have denominations $a$, $a$, $b$, $b$, $c$, where $a$, $b$, and $c$ are all distinct.)
(d) Three of a kind? (This occurs when the cards have denominations $a$, $a$, $a$, $b$, $c$, where $a$, $b$, and $c$ are all distinct.)
(e) Four of a kind? (This occurs when the cards have denominations $a$, $a$, $a$, $a$, $b$.)

**5.** Poker dice is played by simultaneously rolling 5 dice. Show that

(a) $P\{\text{no two alike}\} = .0926$; (b) $P\{\text{one pair}\} = .4630$;
(c) $P\{\text{two pair}\} = .2315$; (d) $P\{\text{three alike}\} = .1543$;
(e) $P\{\text{full house}\} = .0386$; (f) $P\{\text{four alike}\} = .0193$;
(g) $P\{\text{five alike}\} = .0008$.

**6.** If 8 castles (that is, rooks) are randomly placed on a chessboard, compute the probability that none of the rooks can capture any of the others. That is, compute the probability that no row or file contains more than one rook.

**7.** Two cards are randomly selected from an ordinary playing deck. What is the probability that they form a blackjack? That is, what is the probability that one of the cards is an ace and the other one either a ten, jack, queen, or king?

**8.** If two dice are rolled, what is the probability that the sum of the upturned faces equals $i$? Do it for $i = 2, 3, \ldots, 11, 12$.

**9.** A pair of dice is rolled until a sum of either 5 or 7 appears. Find the probability that a 5 occurs first.

HINT: Let $E_n$ denote the event that a 5 occurs on the $n$th roll and no 5 or 7 occurs on the first $n - 1$ rolls. Compute $P(E_n)$ and argue that $\sum_{n=1}^{\infty} P(E_n)$ is the desired probability.

**10.** The game of craps is played as follows: A player rolls two dice. If the sum of the dice is either a 2, 3, or 12, the player loses; if the sum is either a 7 or an 11, he or she wins. If the outcome is anything else, the player continues to roll the dice until he or she rolls either the initial outcome or a 7. If the 7 comes first, the player loses; whereas if the initial outcome reoccurs before the 7, the player wins. Compute the probability of a player winning at craps.

HINT: Let $E_i$ denote the event that the initial outcome is $i$ and the player wins. The desired probability is $\sum_{i=2}^{12} P(E_i)$. To compute $P(E_i)$, define the events $E_{i,n}$ to be the event that the initial sum is $i$ and the player wins on the $n$th roll. Argue that $P(E_i) = \sum_{n=1}^{\infty} P(E_{i,n})$.

**11.** An urn contains 3 red and 7 black balls. Players $A$ and $B$ withdraw balls from the urn consecutively until a red ball is selected. Find the probability that $A$ selects the red ball. ($A$ draws the first ball, then $B$, and so on. There is no replacement of the drawn balls.)

**12.** An urn contains 5 red, 6 blue, and 8 green balls. If a set of 3 balls is randomly selected, what is the probability that each of the balls will be of the same color? Of different colors? Repeat under the assumption that whenever a ball is selected, its color is noted and it is then replaced in the urn before the next selection. This is known as sampling with replacement.

**13.** Urn $A$ contains 3 red and 3 black balls, whereas urn $B$ contains 4 red and 6 black balls. If a ball is randomly selected from each urn, what is the probability that the balls will be of the same color?

**14.** A 3-man basketball team consists of a guard, a forward, and a center. If a man is chosen at random from each of three different 3-men teams, what is the probability of selecting a complete team? What is the probability that all 3 players selected play the same position?

**15.** A group of individuals containing $b$ boys and $g$ girls is lined up in random order—that is, each of the $(b + g)!$ permutations is assumed to be equally likely. What is the probability that the person in the $i$th position, $1 \leq i \leq b + g$, is a girl?

**16.** A forest contains 20 elk, of which 5 are captured, tagged, and then released. A certain time later 4 of the 20 elk are captured. What is the probability that 2 of these 4 have been tagged? What assumptions are you making?

**17.** There are 5 hotels in a certain town. If 3 people check into hotels in a day, what is the probability they each check into a different hotel? What assumptions are you making?

**18.** A town contains 4 television repairmen. If 4 sets break down, what is the probability that exactly $i$ of the repairmen are called. Solve the problem for $i = 1, 2, 3, 4$. What assumptions are you making?

**19.** If a die is rolled 4 times, what is the probability of 6 coming up at least once?

**20.** Two dice are thrown $n$ times in succession. Compute the probability that double 6 appears at least once. How large need $n$ be to make this probability at least $\frac{1}{2}$?

**21.** If $N$ people, including $A$ and $B$, are randomly arranged in a line, what is the probability that $A$ and $B$ are next to each other? What would the probability be if the people were randomly arranged in a circle?

**22.** From a group of 3 freshmen, 4 sophomores, 4 juniors, and 3 seniors a committee of size 4 is randomly selected. Find the probability that the committee will consist of (a) 1 from each class, (b) 2 sophomores and 2 juniors, and (c) only sophomores and juniors.

**23.** A woman has $n$ keys, of which one will open her door. If she tries the keys at random, discarding those that do not work, what is the probability that she will open the door on her $k$th try? What if she does not discard previously tried keys?

**24.** Given 20 people, what is the probability that among the 12 months in the year there are 4 months containing exactly 2 birthdays and 4 containing exactly 3 birthdays?

**25.** A group of 6 men and 6 women is randomly divided into 2 groups of size 6 each. What is the probability that both groups will have the same number of men?

**26.** In a hand of bridge, find the probability that you have 5 spades and your partner has the remaining 8.

**27.** Suppose that $n$ balls are randomly distributed in $N$ compartments. Find the probability that $m$ balls will fall in the first compartment. Assume that all $N^n$ arrangements are equally likely.

**28.** A closet contains 10 pairs of shoes. If 8 shoes are randomly selected, what is the probability that there will be (a) no complete pair, and (b) exactly 1 complete pair?

**29.** A basketball team consists of 6 black and 4 white players. If players are divided into roommates at random, find the probability that there will be 2 black-white roommate pairs.

**30.** If 4 married couples are arranged in a row, find the probability that no husband sits next to his wife.

**31.** Compute the probability that a bridge hand is void in at least one suit. Note that the answer is not

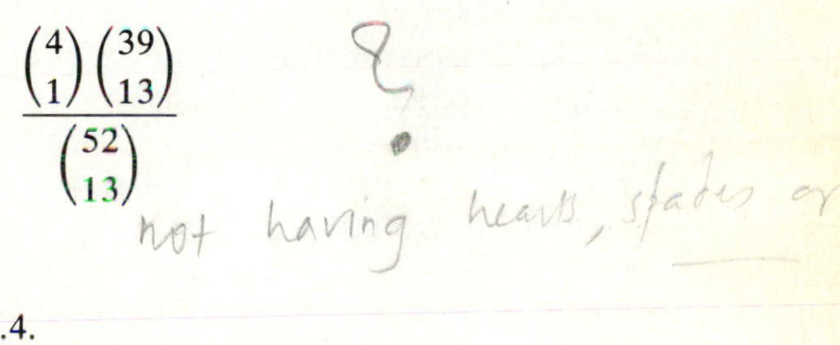

$$\frac{\binom{4}{1}\binom{39}{13}}{\binom{52}{13}}$$

Why not?

HINT: Use Proposition 4.4.

**32.** Compute the probability that a hand of 13 cards contains (a) the ace and king of some suit, (b) all 4 of at least 1 of the 13 denominations.

# 3

# Conditional Probability and Independence

## 1 Introduction

In this chapter we introduce one of the most important concepts in probability theory, that of conditional probability. The importance of this concept is twofold. In the first place, we are often interested in calculating probabilities when some partial information concerning the result of the experiment is available; in such a situation the desired probabilities are conditional ones. Second, even when no partial information is available, it is often useful to use conditional probabilities as a tool to enable us to compute the desired probabilities more easily.

## 2 Conditional Probabilities

Suppose that we toss 2 dice and suppose that each of the 36 possible outcomes is equally likely to occur and hence has probability $\frac{1}{36}$. Suppose further that we observe that the first die is a 3. Then, given this information, what is the probability that the sum of the 2 dice equals 8? To calculate this probability, we reason as follows: Given that the initial die is a 3, it follows that there can be at most 6 possible outcomes of our experiment, namely, $(3, 1)$, $(3, 2)$, $(3, 3)$, $(3, 4)$, $(3, 5)$, and $(3, 6)$. Since each of these outcomes originally had the same probability of occurring, the outcomes should still have equal probabilities. That is, given that the first die is a 3, the (conditional) probability of each of the outcomes $(3, 1)$, $(3, 2)$, $(3, 3)$, $(3, 4)$, $(3, 5)$, and $(3, 6)$ is $\frac{1}{6}$, whereas the (conditional) probability of the other 30 points in the sample space is 0. Hence the desired probability will be $\frac{1}{6}$.

If we let $E$ and $F$ denote, respectively, the event that the sum of the dice is 8 and the event that the first die is a 3, then the probability just obtained is called the conditional probability that $E$ occurs given that $F$

has occurred and is denoted by

$$P(E|F)$$

A general formula for $P(E|F)$ that is valid for all events $E$ and $F$ is derived in the same manner: If the event $F$ occurs, then in order for $E$ to occur it is necessary that the actual occurrence be a point in both $E$ and in $F$; that is, it must be in $EF$. Now, as we know that $F$ has occurred, it follows that $F$ becomes our new or reduced sample space; hence the probability that the event $EF$ occurs will equal the probability of $EF$ relative to the probability of $F$. That is, we have the following definition.

---

*Definition*

*If* $P(F) > 0$, *then*

$$P(E|F) = \frac{P(EF)}{P(F)} \tag{2.1}$$

---

**Example 2a.** A coin is flipped twice. If we assume that all four points in the sample space $S = \{(H, H), (H, T), (T, H), (T, T)\}$ are equally likely, what is the conditional probability that both flips result in heads, given that the first flip does?

***Solution:*** If $E = \{(H, H)\}$ denotes the event that both flips land heads, and $F = \{(H, H), (H, T)\}$ the event that the first flip lands heads, then the desired probability is given by

$$\begin{aligned} P(E|F) &= \frac{P(EF)}{P(F)} \\ &= \frac{P(\{(H, H)\})}{P(\{(H, H), (H, T)\})} \\ &= \frac{\frac{1}{4}}{\frac{2}{4}} = \frac{1}{2} \end{aligned}$$ ∎

**Example 2b.** An urn contains 10 white, 5 yellow, and 10 black marbles. A marble is chosen at random from the urn, and it is noted that it is not one of the black marbles. What is the probability that it is yellow?

***Solution:*** Let $Y$ denote the event that the marble selected is yellow, and let $B^c$ denote the event that it is not black. Now, from Equation (2.1)

$$P(Y|B^c) = \frac{P(YB^c)}{P(B^c)}$$

However, $YB^c = Y$, since the marble will be both yellow and not black if and only if it is yellow. Hence, assuming that each of the 25 marbles is equally likely to be chosen, we obtain that

$$P(Y|B^c) = \frac{\frac{5}{25}}{\frac{15}{25}} = \frac{1}{3}$$

It should be noted that we also could have derived this probability by working directly with the reduced sample space. That is, as we know that the chosen marble is not black, the problem reduces to computing the probability that a marble, chosen at random from an urn containing 10 white and 5 yellow marbles, is yellow. This is clearly equal to $\frac{5}{15} = \frac{1}{3}$. ∎

When all outcomes are assumed to be equally likely, it is often easier to compute a conditional probability by a consideration of the reduced sample space, as opposed to a direct application of (2.1).

**Example 2c.** In the card game bridge the 52 cards are dealt out equally to 4 players—called East, West, North, and South. If North and South have a total of 8 spades among them, what is the probability that East has 3 of the remaining 5 spades?

***Solution:*** Probably the easiest way to compute this is to work with the reduced sample space. That is, given that North-South have a total of 8 spades among their 26 cards, there remains a total of 26 cards, exactly 5 of them being spades, to be distributed among the East-West hands. As each distribution is equally likely, it follows that the conditional probability that East will have exactly 3 spades among his or her 13 cards is

$$\frac{\binom{5}{3}\binom{21}{10}}{\binom{26}{13}} = .339$$ ∎

**Example 2d.** The organization for which Mr. Jones works is running a father-son dinner for those employees having at least one son. Each of these employees is invited to attend along with his eldest son. If Jones is known to have two children, what is the conditional probability that they are both boys, given that he is invited to the dinner? Assume that the sample space $S$ is given by $S = \{(b, b), (b, g), (g, b), (g, g)\}$ and all outcomes are equally likely [$(b, g)$ means, for instance, that the older child is a boy and the younger child is a girl].

***Solution:*** The knowledge that Jones has been invited to the dinner is equivalent to knowing that he has at least one son. Hence, letting $E$ denote the event that both children are boys and $F$ the event that at least one of them is a boy, we have that the desired probability $P(E|F)$ is given by

$$\begin{aligned} P(E|F) &= \frac{P(EF)}{P(F)} \\ &= \frac{P(\{(b, b)\})}{P(\{(b, b), (b, g), (g, b)\})} = \frac{\frac{1}{4}}{\frac{3}{4}} = \frac{1}{3} \end{aligned}$$

Many readers incorrectly reason that the conditional probability of two boys given at least one is $\frac{1}{2}$, as opposed to the correct $\frac{1}{3}$, since they reason that the Jones child not attending the dinner is equally likely to be a boy or a girl. Their mistake, however, is in assuming that these two possibilities are equally likely. For, initially, there were 4 equally likely outcomes. Now the information that at least one child is a boy is equivalent to knowing that the outcome is not $(g, g)$. Hence we are left with the 3 equally likely outcomes $(b, b), (b, g), (g, b)$ thus showing that the Jones child not attending the dinner is twice as likely to be a girl as to be a boy. ∎

By multiplying both sides of Equation (2.1) by $P(F)$, we obtain that

$$P(EF) = P(F)P(E|F) \tag{2.2}$$

In words, Equation (2.2) states that the probability that both $E$ and $F$ occur is equal to the probability that $F$ occurs multiplied by the conditional probability of $E$ given that $F$ occurred. Equation (2.2) is often quite useful in computing the probability of the intersection of events.

**Example 2e.** Celine is undecided as to whether to take a French course or a chemistry course. Although she actually prefers chemistry, Celine estimates that her probability of receiving an A grade would be $\frac{1}{2}$ in a French course, whereas it would be only $\frac{1}{3}$ in a chemistry course. If Celine decides to base her decision on the flip of a fair coin, what is the probability that she gets an A in chemistry?

***Solution:*** If we let $C$ be the event that Celine takes chemistry and $A$ denote the event that she receives an A in whatever course she takes, then the desired probability is $P(CA)$. This is calculated by using Equation (2.2) as follows:

$$\begin{aligned} P(CA) &= P(C)P(A|C) \\ &= (\tfrac{1}{2})(\tfrac{1}{3}) = \tfrac{1}{6} \end{aligned}$$ ∎

**Example 2f.** Suppose an urn contains 8 red balls and 4 white balls. We draw 2 balls from the urn without replacement. If we assume that at each draw each ball in the urn is equally likely to be chosen, what is the probability that both drawn balls are red?

***Solution:*** Let $R_1$ and $R_2$ denote, respectively, the events that the first and second ball drawn is red. Now, given that the first ball selected is red, there are 7 remaining red balls and 4 white balls, and so $P(R_2|R_1) = \frac{7}{11}$. As $P(R_1)$ is clearly $\frac{8}{12}$, the desired probability is

$$\begin{aligned} P(R_1R_2) &= P(R_1)P(R_2|R_1) \\ &= (\tfrac{2}{3})(\tfrac{7}{11}) = \tfrac{14}{33} \end{aligned}$$

Of course, this probability could also have been computed by

$$P(R_1R_2) = \frac{\binom{8}{2}}{\binom{12}{2}}$$ ∎

## 3 Bayes' Formula

Let $E$ and $F$ be events. We may express $E$ as

$$E = EF \cup EF^c$$

for, in order for a point to be in $E$, it must either be in both $E$ and $F$ or be in $E$ but not in $F$. (See Figure 3.1.) As $EF$ and $EF^c$ are clearly mutually exclusive, we have by Axiom 3 that

$$\begin{aligned} P(E) &= P(EF) + P(EF^c) \\ &= P(E|F)P(F) + P(E|F^c)P(F^c) \\ &= P(E|F)P(F) + P(E|F^c)[1 - P(F)]. \end{aligned} \tag{3.1}$$

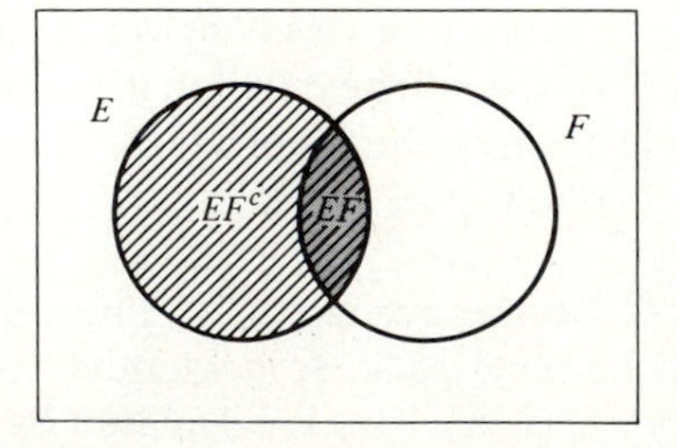


**Figure 3.1** $E = EF \cup EF^c$. $EF$ = shaded area; $EF^c$ = striped area

Equation (3.1) states that the probability of the event $E$ is a weighted average of the conditional probability of $E$ given that $F$ has occurred and the conditional probability of $E$ given that $F$ has not occurred—each conditional probability being given as much weight as the event that it is conditioned on has of occurring. This is an extremely useful formula because its use often enables us to determine the probability of an event by first "conditioning" upon whether or not some second event has occurred. That is, there are many instances where it is difficult to compute the probability of an event directly, but it is straightforward to compute it once we know whether or not some second event has occurred. We illustrate this with some examples.

**Example 3a** (part 1). An insurance company believes that people can be divided into two classes—those that are accident prone and those that are not. Their statistics show that an accident-prone person will have an accident at some time within a fixed 1-year period with probability .4, whereas this probability decreases to .2 for a non-accident-prone person. If we assume that 30 percent of the population is accident prone, what is the probability that a new policy holder will have an accident within a year of purchasing a policy?

***Solution:*** We shall obtain the desired probability by first conditioning upon whether or not the policy holder is accident prone. Let $A_1$ denote the event that the policy holder will have an accident within a year of purchase; and let $A$ denote the event that the policy holder is accident prone. Hence the desired probability, $P(A_1)$, is given by

$$\begin{aligned} P(A_1) &= P(A_1|A)P(A) + P(A_1|A^c)P(A^c) \\ &= (.4)(.3) + (.2)(.7) = .26 \end{aligned}$$

∎

**Example 3a** (part 2). Suppose a new policy holder has an accident within a year of purchasing a policy. What is the probability that he or she is accident prone?

***Solution:*** The desired probability is $P(A|A_1)$, which is given by

$$\begin{aligned} P(A|A_1) &= \frac{P(AA_1)}{P(A_1)} \\ &= \frac{P(A)P(A_1|A)}{P(A_1)} \\ &= \frac{(.3)(.4)}{.26} = \frac{6}{13} \end{aligned}$$

∎

**Example 3b.** In answering a question on a multiple-choice test, a student either knows the answer or guesses. Let $p$ be the probability that the student knows the answer and $1-p$ the probability that the student guesses. Assume that a student who guesses at the answer will be correct with probability $1/m$, where $m$ is the number of multiple-choice alternatives. What is the conditional probability that a student knew the answer to a question, given that he or she answered it correctly?

***Solution:*** Let $C$ and $K$ denote, respectively, the events that the student answers the question correctly and the event that he or she actually knows the answer. Now

$$
\begin{aligned}
P(K|C) &= \frac{P(KC)}{P(C)} \\
&= \frac{P(C|K)P(K)}{P(C|K)P(K) + P(C|K^c)P(K^c)} \\
&= \frac{p}{p + (1/m)(1-p)} \\
&= \frac{mp}{1 + (m-1)p}
\end{aligned}
$$

Thus, for example, if $m = 5$, $p = \frac{1}{2}$, then the probability that a student knew the answer to a question he or she correctly answered is $\frac{5}{6}$. ∎

**Example 3c.** A laboratory blood test is 95 percent effective in detecting a certain disease when it is, in fact, present. However, the test also yields a "false positive" result for 1 percent of the healthy persons tested. (That is, if a healthy person is tested, then, with probability .01, the test result will imply he or she has the disease.) If .5 percent of the population actually has the disease, what is the probability a person has the disease given that the test result is positive?

***Solution:*** Let $D$ be the event that the tested person has the disease and $E$ the event that the test result is positive. The desired probability $P(D|E)$ is obtained by

$$
\begin{aligned}
P(D|E) &= \frac{P(DE)}{P(E)} \\
&= \frac{P(E|D)P(D)}{P(E|D)P(D) + P(E|D^c)P(D^c)} \\
&= \frac{(.95)(.005)}{(.95)(.005) + (.01)(.995)} \\
&= \frac{95}{294} \approx .323
\end{aligned}
$$

Thus only 32 percent of those persons whose test results are positive actually have the disease. As many students are often surprised at this result (as they expected this figure to be much higher, since the blood test seems to be a good one), it is probably worthwhile to present a second argument that, although less rigorous than the preceding one, is probably more revealing. We now do so.

Since .5 percent of the population actually has the disease, it follows that, on the average, 1 person out of every 200 tested will have it. The test will correctly confirm that this person has the disease with probability .95. Thus on the average out of every 200 persons tested the test will correctly confirm that .95 persons have the disease. On the other hand, however, out of the (on the average) 199 healthy people, the test will incorrectly state that (199)(.01) of these people have the disease. Hence, for every .95 diseased persons that the test correctly states are ill, there are (on the average) (199)(.01) healthy persons that the test incorrectly states are ill. Hence the proportion of time that the test result is correct when it states that a person is ill is

$$\frac{.95}{.95 + (199)(.01)} = \frac{95}{294} \approx .323$$ ▮

Equation (3.1) is also useful when one has to reassess one's personal probabilities in the light of additional information. For instance, consider the following examples.

**Example 3d.** Consider a medical practitioner pondering the following dilemma: "If I'm at least 80 percent certain that my patient has this disease, then I always recommend surgery, whereas if I'm not quite as certain, then I recommend additional tests that are expensive and sometimes painful. Now, initially I was only 60 percent certain that Jones had the disease, so I ordered the series A test, which always gives a positive result when the patient has the disease and almost never does when he is healthy. The test result was positive, and I was all set to recommend surgery when Jones informed me, for the first time, that he is a diabetic. This information complicates matters because, although it doesn't change my original 60 percent estimate of his chances of having the disease, it does affect the interpretation of the results of the A test. This is so because the A test, while never yielding a positive result when the patient is healthy, does unfortunately yield a positive result 30 percent of the time in the case of *diabetic* patients not suffering from the disease. Now what do I do? More tests or immediate surgery?"

***Solution:*** In order to decide whether or not to recommend surgery, the doctor should first compute his updated probability that Jones has the disease given that the A test result was positive. Let $D$ denote the event that Jones has the disease, and $E$ the event of a positive A test result. The desired conditional probability $P(D|E)$ is obtained by

$$\begin{aligned}P(D|E) &= \frac{P(DE)}{P(E)}\\ &= \frac{P(D)P(E|D)}{P(E|D)P(D) + P(E|D^c)P(D^c)}\\ &= \frac{(.6)1}{1(.6) + (.3)(.4)}\\ &= .833\end{aligned}$$

Note that we have computed the probability of a positive test result by conditioning on whether or not Jones has the disease and then using the fact that because Jones is a diabetic his conditional probability of a positive result given he does not have the disease, $P(E|D^c)$, equals .3. Hence, as the doctor should now be over 80 percent certain that Jones has the disease, he should recommend surgery. ∎

**Example 3e.** At a certain stage of a criminal investigation the inspector in charge is 60 percent convinced of the guilt of a certain suspect. Suppose now that a *new* piece of evidence that shows that the criminal has a certain characteristic (such as left-handedness, baldness, or brown hair) is uncovered. If 20 percent of the population possesses this characteristic, how certain of the guilt of the suspect should the inspector now be if it turns out that the suspect has this characteristic?

***Solution:*** Letting $G$ denote the event that the suspect is guilty and $C$ the event that he posesses the characteristic of the criminal, we have

$$\begin{aligned}P(G|C) &= \frac{P(GC)}{P(C)}\\ &= \frac{P(C|G)P(G)}{P(C|G)P(G) + P(C|G^c)P(G^c)}\\ &= \frac{1(.6)}{1(.6) + (.2)(.4)}\\ &= .882\end{aligned}$$

where we have supposed that the probability of the suspect having the characteristic if he is, in fact, innocent is equal to .2, the proportion of the population possessing the characteristic. ∎

**Example 3f.** In the world bridge championships held in Buenos Aires in May 1965 the famous British bridge partnership of Terrence Reese and

Boris Schapiro was accused of cheating by using a system of finger signals that could indicate the number of hearts held by the players. Reese and Schapiro denied the accusation, and eventually a hearing was held by the British bridge league. The hearing was in the form of a legal proceeding with a prosecuting and defense team, both having the power to call and cross-examine witnesses. During the course of these proceedings the prosecutor examined specific hands played by Reese and Schapiro and claimed that their playing in these hands was consistent with the hypothesis that they were guilty of having illicit knowledge of the heart suit. At this point, the defense attorney pointed out that their play of these hands was also perfectly consistent with their standard line of play. However, the prosecution then argued that as long as their play was consistent with the hypothesis of guilt then it must be counted as evidence toward this hypothesis. What do you think of the reasoning of the prosecution?

***Solution:*** The problem is basically one of determining how the introduction of new evidence (in the above example, the playing of the hands) affects the probability of a particular hypothesis. Now, if we let $H$ denote a particular hypothesis (such as the guilt of Reese and Schapiro), and $E$ the new evidence, then

$$\begin{aligned} P(H|E) &= \frac{P(HE)}{P(E)} \\ &= \frac{P(E|H)P(H)}{P(E|H)P(H) + P(E|H^c)[1 - P(H)]} \end{aligned} \tag{3.2}$$

where $P(H)$ is our evaluation of the likelihood of the hypothesis before the introduction of the new evidence. The new evidence will be in support of the hypothesis whenever it makes the hypothesis more likely, that is, whenever $P(H|E) \geq P(H)$. From Equation (3.2), this will be the case whenever

$$P(E|H) \geq P(E|H)P(H) + P(E|H^c)[1 - P(H)]$$

or, equivalently, whenever

$$P(E|H) \geq P(E|H^c)$$

In other words, any new evidence can only be considered to be in support of a particular hypothesis if its occurrence is more likely when the hypothesis is true than when it is false. In fact, the new probability of the hypothesis depends on its initial probability and the ratio of these conditional probabilities, since from Equation (3.2)

$$P(H|E) = \frac{P(H)}{P(H) + [1 - P(H)]\dfrac{P(E|H^c)}{P(E|H)}}$$

Hence, in the problem under consideration, the play of the cards can only be considered to support the hypothesis of guilt if such playing would have been more likely if the partnership were cheating than if they were not. As the prosecutor never made this claim, his assertion that the evidence is in support of the guilt hypothesis is invalid. ∎

Equation (3.1) may be generalized in the following manner: Suppose that $F_1, F_2, \ldots, F_n$ are mutually exclusive events such that

$$\bigcup_{i=1}^{n} F_i = S$$

In other words, exactly one of the events $F_1, F_2, \ldots, F_n$ must occur. By writing

$$E = \bigcup_{i=1}^{n} EF_i$$

and using the fact that the events $EF_i$, $i = 1, \ldots, n$ are mutually exclusive, we obtain that

$$\begin{aligned} P(E) &= \sum_{i=1}^{n} P(EF_i) \\ &= \sum_{i=1}^{n} P(E \mid F_i)P(F_i) \end{aligned} \tag{3.3}$$

Thus Equation (3.3) shows how, for given events $F_1, F_2, \ldots, F_n$ of which one and only one must occur, we can compute $P(E)$ by first conditioning on which one of the $F_i$ occurs. That is, Equation (3.3) states that $P(E)$ is equal to a weighted average of $P(E \mid F_i)$, each term being weighted by the probability of the event on which it is conditioned.

Suppose now that $E$ has occurred and we are interested in determining which one of the $F_j$ also occurred. By Equation (3.3), we have the following proposition.

---

*Proposition 3.1*

$$\begin{aligned} P(F_j \mid E) &= \frac{P(EF_j)}{P(E)} \\ &= \frac{P(E \mid F_j)P(F_j)}{\sum_{i=1}^{n} P(E \mid F_i)P(F_i)} \end{aligned} \tag{3.4}$$

---

Equation (3.4) is known as Bayes' formula, after the English philosopher Thomas Bayes. If we think of the events $F_j$ as being possible "hypotheses" about some subject matter, then Bayes' formula may be interpreted as showing us how opinions about these hypotheses held before the experiment [that is, the $P(F_j)$] should be modified by the evidence of the experiment.

**Example 3g.** A plane is missing, and it is presumed that it was equally likely to have gone down in any of 3 possible regions. Let $1 - \alpha_i$ denote the probability that the plane will be found upon a search of the $i$th region when the plane is, in fact, in that region, $i = 1, 2, 3$. (The constants $\alpha_i$ are called overlook probabilities because they represent the probability of overlooking the plane; they are generally attributable to the geographical and environmental conditions of the regions.) What is the conditional probability that the plane is in the $i$th region, given that a search of region 1 is unsuccessful, $i = 1, 2, 3$?

***Solution:*** Let $R_i$, $i = 1, 2, 3$, be the event that the plane is in region $i$; and let $E$ be the event that a search of region 1 is unsuccessful. From Bayes' formula we obtain

$$
\begin{aligned}
P(R_1 | E) &= \frac{P(ER_1)}{P(E)} \\
&= \frac{P(E | R_1)P(R_1)}{\sum_{i=1}^{3} P(E | R_i)P(R_i)} \\
&= \frac{(\alpha_1)\frac{1}{3}}{(\alpha_1)\frac{1}{3} + (1)\frac{1}{3} + (1)\frac{1}{3}} \\
&= \frac{\alpha_1}{\alpha_1 + 2}
\end{aligned}
$$

For $j = 2, 3$,

$$
\begin{aligned}
P(R_j | E) &= \frac{P(E | R_j)P(R_j)}{P(E)} \\
&= \frac{(1)\frac{1}{3}}{(\alpha_1)\frac{1}{3} + \frac{1}{3} + \frac{1}{3}} \\
&= \frac{1}{\alpha_1 + 2}, \qquad j = 2, 3
\end{aligned}
$$

It should be noted that the updated (that is, the conditional) probability that the plane is in region $j$, given the information that a search of region 1 did not find it, is greater than the initial probability that it was in region

$j$ when $j \neq 1$ and is less than the initial probability when $j = 1$; which is certainly intuitive, since not finding it when searching region 1 would seem to decrease its chance of being in that region and increase its chance of being elsewhere. Also, the conditional probability that the plane is in region 1, given an unsuccessful search of that region, is an increasing function of the overlook probability $\alpha_1$, which is also intuitive since the larger $\alpha_1$ is, the more it is reasonable to attribute the unsuccessful search to "bad luck" as opposed to the plane not being there. Similarly, $P(R_j|E)$, $j \neq 1$ is a decreasing function of $\alpha$. ∎

The next example has often been used by unscrupulous probability students to win money from their less enlightened friends.

**Example 3h.** Suppose we have 3 cards identical in form except that both sides of the first card are colored red, both sides of the second card are colored black, and one side of the third card is colored red and the other side black. The 3 cards are mixed up in a hat, and 1 card is randomly selected and put down on the ground. If the upper side of the chosen card is colored red, what is the probability that the other side is colored black?

***Solution:*** Let $RR$, $BB$, and $RB$ denote, respectively, the events that the chosen card is the all red, all black, or the red-black card. Letting $R$ be the event that the upturned side of the chosen card is red, we have that the desired probability is obtained by

$$
\begin{aligned}
P(RB|R) &= \frac{P(RB \cap R)}{P(R)} \\
&= \frac{P(R|RB)P(RB)}{P(R|RR)P(RR) + P(R|RB)P(RB) + P(R|BB)P(BB)} \\
&= \frac{(\frac{1}{2})(\frac{1}{3})}{(1)(\frac{1}{3}) + (\frac{1}{2})(\frac{1}{3}) + 0(\frac{1}{3})} = \frac{1}{3}
\end{aligned}
$$

Hence the answer is $\frac{1}{3}$. Some students guess $\frac{1}{2}$ as the answer by incorrectly reasoning that given that a red side appears, there are two equally likely possibilities: that the card is the all red card or the red-black card. Their mistake, however, is in assuming that these two possibilities are equally likely. For, if we think of each card as consisting of two distinct sides, then there are 6 equally likely outcomes of the experiment—namely, $R_1$, $R_2$, $B_1$, $B_2$, $R_3$, $B_3$—where the outcome is $R_1$ if the first side of the all red card is turned face up, $R_2$ if the second side of the all red card is turned face up, $R_3$ if the red side of the red-black card is turned face up, and so on. Since the other side of the upturned red side will be black only if the outcome is $R_3$, we see that the desired probability is the conditional probability of $R_3$ given that either $R_1$ or $R_2$ or $R_3$ occurred, which obviously equals $\frac{1}{3}$. ∎

**Example 3i.** At a psychiatric clinic the social workers are so busy that, on the average, only 60 percent of potential new patients that telephone are able to talk immediately with a social worker when they call. The other 40 percent are asked to leave their phone numbers. About 75 percent of the time a social worker is able to return the call on the same day, and the other 25 percent of the time the caller is contacted on the following day. Experience at the clinic indicates that the probability a caller will actually visit the clinic for consultation is .8 if the caller was immediately able to speak to a social worker, whereas it is .6 and .4, respectively, if the patient's call was returned the same day or the following day.

1. What percentage of people that telephone visit the clinic for consultation?
2. What percentage of patients that visit the clinic did not have to have their telephone calls returned?

***Solution:*** Define the events $V$, $I$, $S$, $F$ by

$V$: caller visits the clinic for consultation;
$I$: caller immediately speaks to a social worker;
$S$: caller is contacted later on the same day;
$F$: caller is contacted on the following day.

Then

$$\begin{aligned} P(V) &= P(V|I)P(I) + P(V|S)P(S) + P(V|F)P(F) \\ &= (.8)(.6) + (.6)(.4)(.75) + (.4)(.4)(.25) \\ &= .70 \end{aligned}$$

where we have used the fact that $P(S) = (.4)(.75)$ and $P(F) = (.4)(.25)$. Hence (1) is answered. To answer (2), we note that

$$\begin{aligned} P(I|V) &= \frac{P(V|I)P(I)}{P(V)} \\ &= \frac{(.8)(.6)}{.7} \\ &= .686 \end{aligned}$$

Hence approximately 69 percent of the patients that visit the clinic had their phone call immediately answered by a social worker. ∎

## 4 Independent Events

The previous examples of this chapter show that $P(E|F)$, the conditional probability of $E$ given $F$, is not generally equal to $P(E)$, the unconditional probability of $E$. In other words, knowing that $F$ has occurred generally changes the chances of $E$'s occurrence. In the special cases where $P(E|F)$ does in fact equal $P(E)$, we say that $E$ is independent of $F$. That is, $E$ is independent of $F$ if knowledge that $F$ has occurred does not change the probability that $E$ occurs.

Since $P(E|F) = P(EF)/P(F)$, we see that $E$ is independent of $F$ if

$$P(EF) = P(E)P(F) \tag{4.1}$$

As Equation (4.1) is symmetric in $E$ and $F$, it shows that whenever $E$ is independent of $F$, $F$ is also independent of $E$. We thus have the following definition.

---

*Definition*

*Two events E and F are said to be* independent *if Equation* (4.1) *holds. Two events E and F that are not independent are said to be* dependent.

---

**Example 4a.** A card is selected at random from an ordinary deck of 52 playing cards. If $E$ is the event that the selected card is an ace and $F$ is the event that it is a spade, then $E$ and $F$ are independent. This follows because $P(EF) = \frac{1}{52}$, whereas $P(E) = \frac{4}{52}$ and $P(F) = \frac{13}{52}$. ∎

**Example 4b.** Two coins are flipped, and all 4 outcomes are assumed to be equally likely. If $E$ is the event that the first coin lands heads and $F$ the event that the second lands tails, then $E$ and $F$ are independent, since $P(EF) = P(\{(H, T)\}) = \frac{1}{4}$; whereas $P(E) = P(\{(H, H), (H, T)\}) = \frac{1}{2}$ and $P(F) = P(\{(H, T), (T, T)\}) = \frac{1}{2}$. ∎

**Example 4c.** Suppose that we toss 2 fair dice. Let $E_1$ denote the event that the sum of the dice is 6 and $F$ denote the event that the first die equals 4. Then

$$P(E_1F) = P(\{(4, 2)\}) = \tfrac{1}{36}$$

whereas

$$P(E_1)P(F) = (\tfrac{5}{36})(\tfrac{1}{6}) = \tfrac{5}{216}$$

Hence $E_1$ and $F$ are not independent. Intuitively, the reason for this is clear because if we are interested in the possibility of throwing a 6 (with 2 dice) we shall be quite happy if the first die lands 4 (or any of the numbers 1, 2, 3, 4, 5), for then we shall still have a possibility of getting a total of 6. On the other hand, if the first die landed 6, we would be unhappy because we would no longer have a chance of getting a total of 6. In other words, our chance of getting a total of six depends on the outcome of the first die; hence $E_1$ and $F$ cannot be independent.

Now, suppose that we let $E_2$ be the event that the sum of the dice equals 7. Is $E_2$ independent of $F$? The answer is yes, since

$$P(E_2F) = P(\{(4, 3)\}) = \tfrac{1}{36}$$

whereas

$$P(E_2)P(F) = (\tfrac{1}{6})(\tfrac{1}{6}) = \tfrac{1}{36}$$

We leave it for the reader to present the intuitive argument why the event that the sum of the dice equals seven is independent of the outcome on the first die. ∎

**Example 4d.** If we let $E$ denote the event that the next president is a Republican and $F$ the event that there will be a major earthquake within the next year, then most people would probably be willing to assume that $E$ and $F$ are independent. However, there would probably be some controversy over whether it is reasonable to assume that $E$ is independent of $G$, where $G$ is the event that there will be a major war within two years after the election. ∎

We now show that if $E$ is independent of $F$, then $E$ is also independent of $F^c$.

---

*Proposition 4.1*

*If $E$ and $F$ are independent, then so are $E$ and $F^c$.*

---

**Proof:** Assume that $E$ and $F$ are independent. Since $E = EF \cup EF^c$, and $EF$ and $EF^c$ are obviously mutually exclusive, we have that

$$\begin{aligned} P(E) &= P(EF) + P(EF^c) \\ &= P(E)P(F) + P(EF^c) \end{aligned}$$

or equivalently,

$$\begin{aligned} P(EF^c) &= P(E)[1 - P(F)] \\ &= P(E)P(F^c) \end{aligned}$$

and the result is proved. ∎

Thus, if $E$ is independent of $F$, then the probability of $E$'s occurrence is unchanged by information as to whether or not $F$ has occurred.

Suppose now that $E$ is independent of $F$ and is also independent of $G$. Is $E$ then necessarily independent of $FG$? The answer, somewhat surprisingly, is no. Consider the following example.

**Example 4e.** Two fair dice are thrown. Let $E$ denote the event that the sum of the dice is 7. Let $F$ denote the event that the first die equals 4 and let $G$ be the event that the second die equals 3. From Example 4c we know that $E$ is independent of $F$, and the same reasoning as applied there shows that $E$ is also independent of $G$; but clearly $E$ is not independent of $FG$ [since $P(E|FG) = 1$]. ∎

It would appear to follow from Example 4e that an appropriate definition of the independence of three events $E$, $F$, and $G$ would have to go further than merely assuming that all of the $\binom{3}{2}$ pairs of events are independent. We are thus led to the following definition.

---

*Definition*

*The three events E, F, and G are said to be independent if*

$$\begin{aligned} P(EFG) &= P(E)P(F)P(G) \\ P(EF) &= P(E)P(F) \\ P(EG) &= P(E)P(G) \\ P(FG) &= P(F)P(G) \end{aligned}$$

---

It should be noted that if $E$, $F$, and $G$ are independent, then $E$ will be independent of any event formed from $F$ and $G$. For instance, $E$ is independent of $F \cup G$, since

$$\begin{aligned} P[E(F \cup G)] &= P(EF \cup EG) \\ &= P(EF) + P(EG) - P(EFG) \\ &= P(E)P(F) + P(E)P(G) - P(E)P(FG) \\ &= P(E)[P(F) + P(G) - P(FG)] \\ &= P(E)P(F \cup G) \end{aligned}$$

Of course, we may also extend the definition of independence to more than three events. The events $E_1, E_2, \ldots, E_n$ are said to be independent if, for every subset $E_{1'}, E_{2'}, \ldots, E_{r'}$, $r \leq n$, of these events

$$P(E_{1'}E_{2'} \cdots E_{r'}) = P(E_{1'})P(E_{2'}) \cdots P(E_{r'})$$

Finally, we define an infinite set of events to be independent if every finite subset of these events is independent.

It is sometimes the case that the probability experiment under consideration consists of performing a sequence of subexperiments. For instance, if the experiment consists of continually tossing a coin, we may think of each toss as being a subexperiment. In many cases it is reasonable to assume that the outcomes of any group of the subexperiments have no effect on the probabilities of the outcomes of the other subexperiments. If such is the case, we say that the subexperiments are independent. More formally, we say that the subexperiments are independent if $E_1, E_2, \ldots, E_n, \ldots$ is necessarily an independent sequence of events whenever $E_i$ is an event whose occurrence is completely determined by the outcome of the $i$th subexperiment.

If each subexperiment is identical—that is, if each subexperiment has the same (sub) sample space and the same probability function on its events—then the subexperiments are called *trials.*

**Example 4f.** An infinite sequence of independent trials is to be performed. Each trial results in a success with probability $p$ and a failure with probability $1 - p$. What is the probability that

1. At least 1 success occurs in the first $n$ trials?
2. Exactly $k$ successes occur in the first $n$ trials?
3. All trials result in successes?

***Solution:*** In order to determine the probability of at least 1 success in the first $n$ trials, it is easiest to compute first the probability of the complementary event, that of no successes in the first $n$ trials. If we let $E_i$ denote the event of a failure on the $i$th trial, then the probability of no successes is, by independence,

$$P(E_1E_2 \cdots E_n) = P(E_1)P(E_2) \cdots P(E_n) = (1 - p)^n$$

Hence the answer to (1) is $1 - (1 - p)^n$.

To compute (2), consider any particular sequence of the first $n$ outcomes containing $k$ successes and $n - k$ failures. Each one of these sequences will, by the assumed independence of trials, occur with probability $p^k(1 - p)^{n-k}$. As there are $\binom{n}{k}$ such sequences (there are $n!/k!(n - k)!$ permutations of $k$ successes and $n - k$ failures), the desired probability in (2) is

$$P\{\text{exactly } k \text{ successes}\} = \binom{n}{k} p^k(1 - p)^{n-k}$$

To answer (3), we note by (1) that the probability of the first $n$ trials all resulting in successes is given by

$$P(E_1^c E_2^c \cdots E_n^c) = p^n$$

Hence, using the continuity property of probabilities (Section 6 of Chapter 2), we have that the desired probability $P\left(\bigcap_1^\infty E_i^c\right)$ is given by

$$\begin{aligned}
P\left(\bigcap_{i=1}^{\infty} E_i^c\right) &= P\left(\lim_{n\to\infty} \bigcap_{i=1}^{n} E_i^c\right) \\
&= \lim_{n\to\infty} P\left(\bigcap_{i=1}^{n} E_i^c\right) \\
&= \lim_n p^n = \begin{cases} 0 & \text{if } p < 1 \\ 1 & \text{if } p = 1 \end{cases}
\end{aligned}$$

▮

**Example 4g.** A system composed of $n$ separate components is said to be a parallel system if it functions when at least one of the components functions. (See Figure 3.2.) For such a system, if component $i$, independent of other components, functions with probability $p_i$, $i = 1, \ldots, n$, what is the probability the system functions?

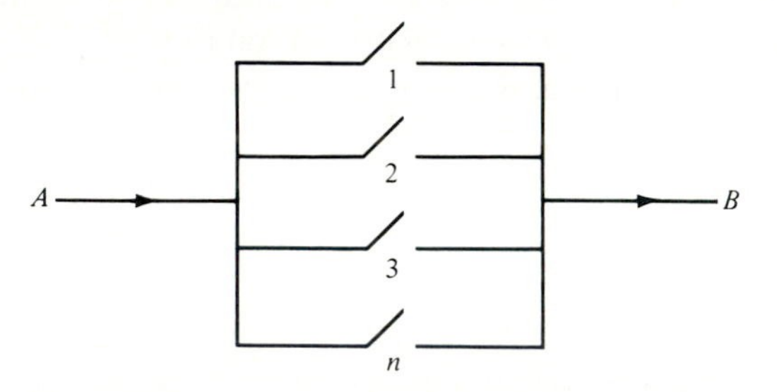


**Figure 3.2** Parallel system: functions if current flows from $A$ to $B$

***Solution:*** Let $A_i$ denote the event that component $i$ functions. Then

$$\begin{aligned}
P\{\text{system functions}\} &= 1 - P\{\text{system does not function}\} \\
&= 1 - P\{\text{all components do not function}\} \\
&= 1 - P\left(\bigcap_i A_i^c\right) \\
&= 1 - \prod_{i=1}^{n} (1 - p_i) \qquad \text{by independence}
\end{aligned}$$

▮

**Example 4h.** Independent trials, consisting of rolling a pair of fair dice, are performed. What is the probability that an outcome of 5 appears before an outcome of 7 when the outcome of a roll is the sum of the dice?

***Solution:*** If we let $E_n$ denote the event that no 5 or 7 appears on the first $n - 1$ trials and a 5 appears on the $n$th trial, then the desired probability is

$$P\left(\bigcup_{n=1}^{\infty} E_n\right) = \sum_{n=1}^{\infty} P(E_n)$$

Now, since $P\{5 \text{ on any trial}\} = \frac{4}{36}$ and $P\{7 \text{ on any trial}\} = \frac{6}{36}$, we obtain, by the independence of trials

$$P(E_n) = (1 - \tfrac{10}{36})^{n-1}\tfrac{4}{36}$$

and thus

$$\begin{aligned} P\left(\bigcup_{n=1}^{\infty} E_n\right) &= \left(\frac{1}{9}\right)\sum_{n=1}^{\infty}\left(\frac{13}{18}\right)^{n-1} \\ &= \frac{1}{9}\frac{1}{1 - \frac{13}{18}} \\ &= \frac{2}{5} \end{aligned}$$

This result may also have been obtained by using conditional probabilities. If we let $E$ be the event that a 5 occurs before a 7, then we can obtain the desired probability, $P(E)$, by conditioning on the outcome of the first trial, as follows: Let $F$ be the event that the first trial results in a 5; let $G$ be the event that it results in a 7; and let $H$ be the event that the first trial results in neither a 5 nor a 7. Since $E = EF \cup EG \cup EH$, we have

$$\begin{aligned} P(E) &= P(EF) + P(EG) + P(EH) \\ &= P(E|F)P(F) + P(E|G)P(G) + P(E|H)P(H) \end{aligned}$$

However,

$$\begin{aligned} P(E|F) &= 1 \\ P(E|G) &= 0 \\ P(E|H) &= P(E) \end{aligned}$$

The first two equalities are obvious. The third follows because, if the first outcome results in neither a 5 nor a 7, then at that point the situation is exactly as when the problem first started; namely, the experimenter will continually roll a pair of fair dice until either a 5 or 7 appears. Furthermore, the trials are independent; therefore, the outcome of the first trial will have no effect on subsequent rolls of the dice. Since $P(F) = \frac{4}{36}$, $P(G) = \frac{6}{36}$,

$P(H) = \frac{26}{36}$ we see that

$$P(E) = \tfrac{1}{9} + P(E)\tfrac{13}{18}$$

or

$$P(E) = \tfrac{2}{5}$$

The reader should note that the answer is quite intuitive. That is, since a 5 occurs on any roll with probability $\frac{4}{36}$ and a 7 with probability $\frac{6}{36}$, it seems intuitive that the odds that a 5 appears before a 7 should be 6 to 4 against. The probability should be $\frac{4}{10}$, as indeed it is.

The same argument shows that if $E$ and $F$ are mutually exclusive events of an experiment, then, when independent trials of this experiment are performed, the event $E$ will occur before the event $F$ with probability

$$\frac{P(E)}{P(E) + P(F)}$$ ▮

The next example presents a problem that occupies an honored place in the history of probability theory. This is the famous *problem of the points.* In general terms, the problem is this: Two players put up stakes and play some game, with the stakes to go to the winner of the game. An interruption requires them to stop before either has won, and when each has some sort of a "partial score." How should the stakes be divided?

This problem was posed to the French mathematician Pascal in 1654 by the Chevalier de Méré, who was a professional gambler at that time. In attacking the problem, Pascal introduced the important idea that the proportion of the prize deserved by the competitors should depend on their respective probabilities of winning if the game were to be continued at that point. Pascal worked out some special cases, and, more important, initiated a correspondence with the famous Frenchman Fermat, who had a great reputation as a mathematician. The resulting exchange of letters led not only to a complete solution of the problem of the points, but also laid the framework for the solution of many other problems connected with games of chance. This celebrated correspondence, dated by some as the birth date of probability theory, was also important in stimulating interest in probability among the mathematicians in Europe, for Pascal and Fermat were both recognized as being among the foremost mathematicians of the time. For instance, within a short time of their correspondence, the young Dutch genius Huygens came to Paris to discuss these problems and solutions; and interest and activity in this new field grew rapidly.

**Example 4i.** The Problem of the Points. Independent trials, resulting in a success with probability $p$ and a failure with probability $1 - p$, are performed. What is the probability that $n$ successes occur before $m$

failures? If we think of $A$ and $B$ as playing a game such that $A$ gains 1 point when a success occurs and $B$ gains 1 point when a failure occurs, then the desired probability is the probability that $A$ would win if the game were to be continued in a position where $A$ needed $n$ and $B$ needed $m$ more points to win.

***Solution:*** We shall present two solutions. The first is due to Fermat and the second to Pascal.

Let us denote by $P_{n,m}$ the probability than $n$ successes occur before $m$ failures. By conditioning on the outcome of the first trial we obtain: (Why? Reason it out.)

$$P_{n,m} = pP_{n-1,m} + (1-p)P_{n,m-1} \qquad n \geq 1, m \geq 1$$

By using the obvious boundary conditions $P_{n,0} = 0$, $P_{0,m} = 1$, these equations can be solved for $P_{n,m}$. Rather than go through the tedious details, let us instead consider Pascal's solution.

Pascal argued that in order for $n$ successes to occur before $m$ failures, it is necessary and sufficient that there be at least $n$ successes in the first $m + n - 1$ trials. (Even if the game were to end before a total of $m + n - 1$ trials were completed, we could still imagine that the necessary additional trials were performed.) This is true, for if there are at least $n$ successes in the first $m + n - 1$ trials, there could be at most $m - 1$ failures in those $m + n - 1$ trials; thus $n$ successes would occur before $m$ failures. On the other hand, if there were fewer than $n$ successes in the first $m + n - 1$ trials, there would have to be at least $m$ failures in that same number of trials; thus $n$ successes would not occur before $m$ failures.

Hence, as the probability of exactly $k$ successes in $m + n - 1$ trials is, as shown in Example 4f, $\binom{m+n-1}{k}p^k(1-p)^{m+n-1-k}$, we see that the desired probability of $n$ successes before $m$ failures is

$$P_{n,m} = \sum_{k=n}^{m+n-1} \binom{m+n-1}{k} p^k (1-p)^{m+n-1-k}$$

Another solution of the problem of the points is presented in Theoretical Exercise 12. As an illustration of the problem of the points suppose that 2 players each put up $\$A$ and each of them has an equal chance of winning each point ($p = \frac{1}{2}$). *If* $n$ points are required to win and the first player has 1 point and the second none, then the first player is entitled to

$$2AP_{n-1,n} = 2A \sum_{k=n-1}^{2n-2} \binom{2n-2}{k} \left(\frac{1}{2}\right)^{2n-2}$$

Now,

$$\sum_{k=n-1}^{2n-2} \binom{2n-2}{k} = \sum_{k=n-1}^{2n-2} \binom{2n-2}{2n-2-k}$$
$$= \sum_{i=n-1}^{0} \binom{2n-2}{i}$$

where the last identity follows from the substitution $i = 2n - 2 - k$. Thus

$$2 \sum_{k=n-1}^{2n-2} \binom{2n-2}{k} = \sum_{k=0}^{2n-2} \binom{2n-2}{k} + \binom{2n-2}{n-1}$$
$$= (1+1)^{2n-2} + \binom{2n-2}{n-1}$$

and so the first player is entitled to

$$A\left[1 + \left(\frac{1}{2}\right)^{2n-2} \binom{2n-2}{n-1}\right]$$ ∎

The next example deals with a famous problem known as the gambler's ruin problem.

**Example 4j.** The Gambler's Ruin Problem. Two gamblers, $A$ and $B$, bet on the outcomes of successive flips of a coin. On each flip, if the coin comes up heads, $A$ collects from $B$ 1 unit, whereas, if it comes up tails, $A$ pays to $B$ 1 unit. They continue to do this until one of them runs out of money. If it is assumed that the successive flips of the coin are independent and each flip results in a head with probability $p$, what is the probability that $A$ ends up with all the money if he starts with $i$ units and $B$ starts with $N - i$ units?

***Solution:*** Let $E$ denote the event that $A$ ends up with all the money when he starts with $i$ and $B$ with $N - i$, and to make clear the dependence on the initial fortune of $A$, let $P_i = P(E)$. We shall obtain an expression for $P(E)$ by conditioning on the outcome of the first flip as follows: Let $H$ denote the event that the first flip lands heads, then

$$P_i = P(E) = P(E \mid H)P(H) + P(E \mid H^c)P(H^c)$$
$$= pP(E \mid H) + (1 - p)P(E \mid H^c)$$

Now, given that the first flip lands heads, the situation after the first bet is that $A$ has $i + 1$ units and $B$ has $N - (i + 1)$. Since the successive flips are assumed to be independent with a common probability $p$ of heads, it follows that, from that point on, $A$'s probability of winning all the money is exactly the same as if the game were just starting with $A$ having an initial fortune of $i+1$ and $B$ having an initial fortune of $N-(i+1)$. Therefore,

$$P(E|H) = P_{i+1}$$

and similarly,

$$P(E|H^c) = P_{i-1}$$

Hence, letting $q = 1-p$, we obtain

$$P_i = pP_{i+1} + qP_{i-1} \qquad i = 1, 2, \ldots, N-1 \tag{4.2}$$

By making use of the obvious boundary conditions $P_0 = 0$ and $P_N = 1$, we shall now solve Equation (4.2). Since $p+q=1$, these equations are equivalent to

$$pP_i + qP_i = pP_{i+1} + qP_{i-1}$$

or

$$P_{i+1} - P_i = \frac{q}{p}(P_i - P_{i-1}) \qquad i = 1, 2, \ldots, N-1 \tag{4.3}$$

Hence, since $P_0 = 0$, we obtain from Equation (4.3):

$$\begin{aligned}
P_2 - P_1 &= \frac{q}{p}(P_1 - P_0) = \frac{q}{p}P_1 \\
P_3 - P_2 &= \frac{q}{p}(P_2 - P_1) = \left(\frac{q}{p}\right)^2 P_1 \\
&\vdots \\
P_i - P_{i-1} &= \frac{q}{p}(P_{i-1} - P_{i-2}) = \left(\frac{q}{p}\right)^{i-1} P_1 \\
&\vdots \\
P_N - P_{N-1} &= \frac{q}{p}(P_{N-1} - P_{N-2}) = \left(\frac{q}{p}\right)^{N-1} P_1
\end{aligned} \tag{4.4}$$

Adding the first $i-1$ of Equation (4.4) yields

$$P_i - P_1 = P_1\left[\left(\frac{q}{p}\right) + \left(\frac{q}{p}\right)^2 + \cdots + \left(\frac{q}{p}\right)^{i-1}\right]$$

or

$$P_i = \begin{cases} \dfrac{1-(q/p)^i}{1-(q/p)} P_1 & \text{if } \dfrac{q}{p} \neq 1 \\[2ex] iP_1 & \text{if } \dfrac{q}{p} = 1 \end{cases}$$

Now, using the fact that $P_N = 1$, we obtain

$$P_1 = \begin{cases} \dfrac{1-(q/p)}{1-(q/p)^N} & \text{if } p \neq \frac{1}{2} \\[2ex] \dfrac{1}{N} & \text{if } p = \frac{1}{2} \end{cases}$$

and hence,

$$P_i = \begin{cases} \dfrac{1-(q/p)^i}{1-(q/p)^N} & \text{if } p \neq \frac{1}{2} \\[2ex] \dfrac{i}{N} & \text{if } p = \frac{1}{2} \end{cases} \tag{4.5}$$

Let $Q_i$ denote the probability that $B$ winds up with all the money when $A$ starts with $i$ and $B$ with $N - i$. Then, by symmetry with the situation described and on replacing $p$ by $q$ and $i$ by $N - i$, we see that

$$Q_i = \begin{cases} \dfrac{1-(p/q)^{N-i}}{1-(p/q)^N} & \text{if } q \neq \frac{1}{2} \\[2ex] \dfrac{N-i}{N} & \text{if } q = \frac{1}{2} \end{cases}$$

Moreover, since $q = \frac{1}{2}$ is equivalent to $p = \frac{1}{2}$, we have when $q \neq \frac{1}{2}$,

$$\begin{aligned} P_i + Q_i &= \frac{1-(q/p)^i}{1-(q/p)^N} + \frac{1-(p/q)^{N-i}}{1-(p/q)^N} \\ &= \frac{p^N - p^N(q/p)^i}{p^N - q^N} + \frac{q^N - q^N(p/q)^{N-i}}{q^N - p^N} \\ &= \frac{p^N - p^{N-i}q^i - q^N + q^i p^{N-i}}{p^N - q^N} \\ &= 1 \end{aligned}$$

As this result also holds when $p = q = \frac{1}{2}$, we see that

$$P_i + Q_i = 1$$

In words, this equation states that with probability 1 either $A$ or $B$ will wind up with all of the money; or, in other words, the probability that the game continues indefinitely with $A$'s fortune always being between 1 and $N - 1$ is zero. (The reader must be careful because, a priori, there are three possible outcomes of this gambling game, not two. Either $A$ wins or $B$ wins or it goes on forever with nobody winning. We have just shown that this last event has probability 0.)

As a numerical illustration of the above result, if $A$ were to start with 5 units and $B$ with 10, then the probability of $A$'s winning would be $\frac{1}{3}$ if $p$ were $\frac{1}{2}$, whereas it would jump to

$$\frac{1-(\frac{2}{3})^5}{1-(\frac{2}{3})^{15}} \approx .87$$

if $p$ were .6. ∎

A special case of the gambler's ruin problem, which is also known as the problem of *duration of play*, was proposed to the Dutch mathematician Christian Huygens by the Frenchman Fermat in 1657. The version he proposed, which was solved by Huygens, was that $A$ and $B$ each have 12 coins. They play for these coins in a game with 3 dice as follows. Whenever 11 is thrown (by either—it makes no difference who rolls the dice), then $A$ gives a coin to $B$. Whenever 14 is thrown, $B$ gives a coin to $A$. The person who first wins all the coins wins the game. Since $P\{\text{roll } 11\} = \frac{27}{216}$ and $P\{\text{roll } 14\} = \frac{15}{216}$, we see from Example 4h that for $A$ this is just the gambler's ruin problem (Example 4j) with $p = \frac{15}{42}$, $i = 12$, $N = 24$. The general form of the gambler's ruin problem was solved by the mathematician James Bernoulli and published 8 years after his death in 1713.

For an application of the gambler's ruin problem to drug testing, suppose that two new drugs have been developed for treating a certain disease. Drug $i$ has a cure rate $P_i$, $i = 1, 2$, in the sense that each patient treated with drug $i$ will be cured with probability $P_i$. These cure rates are, however, not known, and we are interested in a method for deciding whether $P_1 > P_2$ or $P_2 > P_1$. To decide on one of these alternatives, consider the following test: Pairs of patients are to be treated sequentially, with one member of the pair receiving drug 1 and the other drug 2. The results for each pair are determined, and the testing stops when the cumulative number of cures from one of the drugs exceeds the cumulative number of cures from the other by some fixed, predetermined number. More formally, let

$$X_j = \begin{cases} 1 & \text{if the patient in the } j\text{th pair that receives drug 1 is cured} \\ 0 & \text{otherwise} \end{cases}$$

$$Y_j = \begin{cases} 1 & \text{if the patient in the } j\text{th pair that receives drug 2 is cured} \\ 0 & \text{otherwise} \end{cases}$$

For a predetermined positive integer $M$, the test stops after pair $N$ where $N$ is the first value of $n$ such that either

$$X_1 + \cdots + X_n - (Y_1 + \cdots + Y_n) = M$$

or

$$X_1 + \cdots + X_n - (Y_1 + \cdots + Y_n) = -M$$

In the former case we then assert that $P_1 > P_2$ and in the latter that $P_2 > P_1$.

In order to help ascertain whether the above is a good test, one thing we would like to know is the probability that it leads to an incorrect decision. That is, for given $P_1$ and $P_2$, where $P_1 > P_2$, what is the probability that the test will incorrectly assert that $P_2 > P_1$? To determine this probability, note that, after each pair is checked, the cumulative difference of cures using drug 1 versus drug 2 will go up by 1 with probability $P_1(1 - P_2)$—since this is the probability that drug 1 leads to a cure and drug 2 does not—or go down by 1 with probability $(1 - P_1)P_2$ or remain the same with probability $P_1P_2 + (1 - P_1)(1 - P_2)$. Hence, if we only consider those pairs in which the cumulative difference changes, then the difference will go up 1 with probability

$$\begin{aligned} P &= P\{\text{up } 1 \mid \text{up } 1 \text{ or down } 1\} \\ &= \frac{P_1(1 - P_2)}{P_1(1 - P_2) + (1 - P_1)P_2} \end{aligned}$$

and down 1 with probability

$$1 - P = \frac{P_2(1 - P_1)}{P_1(1 - P_2) + (1 - P_1)P_2}$$

Hence the probability that the test will assert that $P_2 > P_1$ is equal to the probability that a gambler who wins each (one unit) bet with probability $P$ will go down $M$ before going up $M$. But Equation (4.5), with $i = M$, $N = 2M$, shows that this probability is given by

$$\begin{aligned} &P\{\text{test asserts that } P_2 > P_1\} \\ &= 1 - \frac{1 - \left(\dfrac{1 - P}{P}\right)^M}{1 - \left(\dfrac{1 - P}{P}\right)^{2M}} \\ &= 1 - \frac{1}{1 + \left(\dfrac{1 - P}{P}\right)^M} \\ &= \frac{1}{1 + \gamma^M} \end{aligned}$$

where

$$\begin{aligned} \gamma &= \frac{P}{1 - P} \\ &= \frac{P_1(1 - P_2)}{P_2(1 - P_1)} \end{aligned}$$

Thus, for instance, if $P_1 = .6$ and $P_2 = .4$, then the probability of an incorrect decision is .017 when $M = 5$ and reduces to .0003 when $M = 10$.

## 5 $P(\cdot|F)$ Is a Probability

Conditional probabilities satisfy all of the properties of ordinary probabilities. This is proved by Proposition 5.1, which shows that $P(E|F)$ satisfies the three axioms of a probability.

---

*Proposition 5.1*

(a) $0 \le P(E|F) \le 1$.
(b) $P(S|F) = 1$.
(c) *If* $E_i$, $i = 1, 2, \ldots$ *are mutually exclusive events, then*

$$P\left(\bigcup_1^\infty E_i|F\right) = \sum_1^\infty P(E_i|F).$$

---

**Proof:** To prove part (a), we must show that $0 \le P(EF)/P(F) \le 1$. The left-side inequality is obvious, whereas the right side follows because $EF \subset F$, which implies that $P(EF) \le P(F)$.

Part (b) follows because

$$P(S|F) = \frac{P(SF)}{P(F)} = \frac{P(F)}{P(F)} = 1$$

Part (c) follows since

$$\begin{aligned}
P\left(\bigcup_{i=1}^\infty E_i|F\right) &= \frac{P\left(\left(\bigcup_{i=1}^\infty E_i\right)F\right)}{P(F)} \\
&= \frac{P\left(\bigcup_1^\infty E_iF\right)}{P(F)} \qquad \text{since } \left(\bigcup_1^\infty E_i\right)F = \bigcup_1^\infty E_iF \\
&= \frac{\sum_1^\infty P(E_iF)}{P(F)} \\
&= \sum_1^\infty P(E_i|F)
\end{aligned}$$

where the next to last equality follows because $E_iE_j = \varnothing$ implies that $E_iFE_jF = \varnothing$. ∎

If we define $Q(E) = P(E|F)$, then it follows from Proposition 5.1 that $Q(E)$ may be regarded as a probability function on the events of $S$. Hence all of the propositions previously proved for probabilities apply to it. For instance, we have

$$Q(E_1 \cup E_2) = Q(E_1) + Q(E_2) - Q(E_1E_2)$$

or, equivalently,

$$P(E_1 \cup E_2|F) = P(E_1|F) + P(E_2|F) - P(E_1E_2|F)$$

Also, if we define the conditional probability $Q(E_1|E_2)$ by $Q(E_1|E_2) = Q(E_1E_2)/Q(E_2)$, then from Equation (3.1) we see that

$$Q(E_1) = Q(E_1|E_2)Q(E_2) + Q(E_1|E_2^c)Q(E_2^c) \tag{5.1}$$

Since

$$\begin{aligned} Q(E_1|E_2) &= \frac{Q(E_1E_2)}{Q(E_2)} \\ &= \frac{P(E_1E_2|F)}{P(E_2|F)} \\ &= \frac{\dfrac{P(E_1E_2F)}{P(F)}}{\dfrac{P(E_2F)}{P(F)}} \\ &= P(E_1|E_2F) \end{aligned}$$

We see that Equation (5.1) is equivalent to

$$P(E_1|F) = P(E_1|E_2F)P(E_2|F) + P(E_1|E_2^cF)P(E_2^c|F)$$

**Example 5a.** Consider Example 3a, which is concerned with an insurance company that believes that people can be divided into two distinct classes—those that are accident prone and those that are not. During any given year an accident-prone person will have an accident with probability .4, whereas the corresponding figure for a non-accident-prone person is .2. What is the conditional probability that a new policy holder will have an accident in his or her second year of policy ownership, given that the policy holder has had an accident in the first year?

***Solution:*** If we let $A$ be the event that the policy holder is accident prone and we let $A_i$, $i = 1, 2$, be the event that he or she has had an accident in the $i$th year, then the desired probability $P(A_2|A_1)$ may be obtained by conditioning on whether or not the policy holder is accident prone, as follows:

$$P(A_2|A_1) = P(A_2|AA_1)P(A|A_1) + P(A_2|A^cA_1)P(A^c|A_1)$$

Now,

$$P(A|A_1) = \frac{P(A_1A)}{P(A_1)} = \frac{P(A_1|A)P(A)}{P(A_1)}$$

However, $P(A)$ is assumed to equal $\frac{3}{10}$, and it was shown in Example 3a that $P(A_1) = .26$. Hence

$$P(A|A_1) = \frac{(.4)(.3)}{.26} = \frac{6}{13}$$

and thus

$$P(A^c|A_1) = 1 - P(A|A_1) = \tfrac{7}{13}$$

Since $P(A_2|AA_1) = .4$ and $P(A_2|A^cA_1) = .2$, we see that

$$P(A_2|A_1) = (.4)\tfrac{6}{13} + (.2)\tfrac{7}{13} \approx .29$$ ▮

The next example deals with a problem in the theory of runs.

**Example 5b.** Independent trials, each resulting in a success with probability $p$ or a failure with probability $q = 1 - p$ are performed. We are interested in computing the probability that a run of $n$ consecutive successes occurs before a run of $m$ consecutive failures.

***Solution:*** Let $E$ be the event that a run of $n$ consecutive successes occurs before a run of $m$ consecutive failures. To obtain $P(E)$, we start by conditioning on the outcome of the first trial. That is, letting $H$ denote the event that the first trial results in a success, we obtain

$$P(E) = pP(E|H) + qP(E|H^c) \tag{5.2}$$

Now, given that the first trial was successful, one way we can get a run of $n$ successes before a run of $m$ failures would be to have the next $n-1$ trials all result in successes. So, let us condition on whether or not that occurs. That is, letting $F$ be the event that trials 2 through $n$ all are successes, we obtain

$$P(E|H) = P(E|FH)P(F|H) + P(E|F^cH)P(F^c|H) \tag{5.3}$$

Clearly, $P(E|FH) = 1$; on the other hand, if the event $F^cH$ occurs, then the first trial would result in a success, but there would be a failure some time during the next $n-1$ trials. However, when this failure occurs, it would wipe out all of the previous successes, and the situation would be exactly as if we started out with a failure. Hence

$$P(E|F^cH) = P(E|H^c)$$

As the independence of trials implies that $F$ and $H$ are independent and as $P(F) = p^{n-1}$, we obtain from (5.3)

$$P(E|H) = p^{n-1} + (1 - p^{n-1})P(E|H^c) \tag{5.4}$$

We now obtain an expression for $P(E|H)^c)$ in a similar manner. That is, we let $G$ denote the event that trials 2 through $m$ are all *failures.* Then

$$P(E|H^c) = P(E|GH^c)P(G|H^c) + P(E|G^cH^c)P(G^c|H^c) \tag{5.5}$$

Now, $GH^c$ is the event that the first $m$ trials all result in failures, so $P(E|GH^c) = 0$. Also, if $G^cH^c$ occurs, the first trial is a failure, but there is at least one success in the next $m - 1$ trials. Hence, as this success wipes out all previous failures, we see that

$$P(E|G^cH^c) = P(E|H)$$

Thus, because $P(G^c|H^c) = P(G^c) = 1 - q^{m-1}$, we obtain from (5.5)

$$P(E|H^c) = (1 - q^{m-1})P(E|H) \tag{5.6}$$

Solving Equations (5.4) and (5.6) yields

$$P(E|H) = \frac{p^{n-1}}{p^{n-1} + q^{m-1} - p^{n-1}q^{m-1}}$$

and

$$P(E|H^c) = \frac{(1 - q^{m-1})p^{n-1}}{p^{n-1} + q^{m-1} - p^{n-1}q^{m-1}}$$

and thus

$$\begin{aligned} P(E) &= pP(E|H) + qP(E|H^c) \\ &= \frac{p^n + qp^{n-1}(1 - q^{m-1})}{p^{n-1} + q^{m-1} - p^{n-1}q^{m-1}} \\ &= \frac{p^{n-1}(1 - q^m)}{p^{n-1} + q^{m-1} - p^{n-1}q^{m-1}} \end{aligned} \tag{5.7}$$

It is interesting to note that, by the symmetry of the problem, the probability of obtaining a run of $m$ failures before a run of $n$ successes would be given by Equation (5.7) with $p$ and $q$ interchanged and $n$ and $m$

interchanged. Hence this probability would equal

$$P\{\text{run of } m \text{ failures before a run of } n \text{ successes}\}$$
$$= \frac{q^{m-1}(1-p^n)}{q^{m-1}+p^{n-1}-q^{m-1}p^{n-1}} \tag{5.8}$$

Since Equations (5.7) and (5.8) sum to 1, it follows that, with probability 1, either a run of $n$ successes or a run of $m$ failures will eventually occur.

As an example of Equation (5.7) we note that in tossing a fair coin the probability that a run of 2 heads will precede a run of 3 tails is $\frac{7}{10}$; for 2 consecutive heads before 4 consecutive tails the probability rises to $\frac{5}{6}$. ∎

In our next example we return to Montmort's matching problem (Example 5i, Chapter 2), and this time obtain a solution by using conditional probabilities.

**Example 5c.** At a party $n$ men take off their hats. The hats are then mixed up, and each man randomly selects one. We say that a match occurs if a man selects his own hat.

1. What is the probability of no matches?
2. What is the probability of exactly $k$ matches?

***Solution:*** Let $E$ denote the event that no matches occur, and to make explicit the dependence on $n$ write $P_n = P(E)$. We start by conditioning on whether or not the first man selects his own hat—call these events $M$ and $M^c$. Then

$$P_n = P(E) = P(E\,|\,M)P(M) + P(E\,|\,M^c)P(M^c)$$

Clearly, $P(E\,|\,M) = 0$, and so

$$P_n = P(E\,|\,M^c)\frac{n-1}{n} \tag{5.9}$$

Now, $P(E\,|\,M^c)$ is the probability of no matches when $n-1$ men select from a set of $n-1$ hats that does not contain the hat of one of these men. This can happen in either of two mutually exclusive ways. Either there are no matches and the extra man does not select the extra hat (this being the hat of the man that chose first), or there are no matches and the extra man does select the extra hat. The probability of the first of these events is just $P_{n-1}$, which is seen by regarding the extra hat as "belonging" to the extra man. As the second event has probability $[1/(n-1)]P_{n-2}$, we have

$$P(E\,|\,M^c) = P_{n-1} + \frac{1}{n-1}P_{n-2}$$

and thus, from Equation (5.9),

$$P_n = \frac{n-1}{n} P_{n-1} + \frac{1}{n} P_{n-2}$$

or, equivalently,

$$P_n - P_{n-1} = -\frac{1}{n}(P_{n-1} - P_{n-2}) \tag{5.10}$$

However, as $P_n$ is the probability of no matches when $n$ men select among their own hats, we have

$$P_1 = 0 \qquad P_2 = \tfrac{1}{2}$$

and so, from Equation (5.10),

$$P_3 - P_2 = -\frac{(P_2 - P_1)}{3} = -\frac{1}{3!} \quad \text{or} \quad P_3 = \frac{1}{2!} - \frac{1}{3!}$$

$$P_4 - P_3 = -\frac{(P_3 - P_2)}{4} = \frac{1}{4!} \quad \text{or} \quad P_4 = \frac{1}{2!} - \frac{1}{3!} + \frac{1}{4!}$$

and, in general, we see that

$$P_n = \frac{1}{2!} - \frac{1}{3!} + \frac{1}{4!} - \cdots + \frac{(-1)^n}{n!}$$

To obtain the probability of exactly $k$ matches, we consider any fixed group of $k$ men. The probability that they, and only they, select their own hats is

$$\frac{1}{n}\frac{1}{n-1}\cdots\frac{1}{n-(k-1)}P_{n-k} = \frac{(n-k)!}{n!}P_{n-k}$$

where $P_{n-k}$ is the conditional probability that the other $n-k$ men, selecting among their own hats, have no matches. As there are $\binom{n}{k}$ choices of a set of $k$ men, the desired probability of exactly $k$ matches is

$$\frac{P_{n-k}}{k!} = \frac{\dfrac{1}{2!} - \dfrac{1}{3!} + \cdots + \dfrac{(-1)^{n-k}}{(n-k)!}}{k!}$$ ∎

An important concept in probability theory is that of the conditional independence of events. We say that the events $E_1$ and $E_2$ are *conditionally independent* given $F$ if, given that $F$ occurs, the conditional probability of $E_1$ occurring is unchanged by information as to whether or not $E_2$ occurs. More formally, $E_1$ and $E_2$ are said to be conditionally independent given $F$ if

$$P(E_1 | E_2 F) = P(E_1 | F) \tag{5.11}$$

or equivalently

$$P(E_1E_2|F) = P(E_1|F)P(E_2|F) \tag{5.12}$$

The notion of conditional independence can easily be extended to more than two events and is left as an exercise.

The reader should note that the concept of conditional independence was implicitly employed in Example 5a, where it was implicitly assumed that the events that a policy holder had an accident in his or her $i$th year, $i = 1, 2, \ldots$, were conditionally independent given whether or not the individual was accident prone. [This was used to evaluate $P(A_2|AA_1)$ and $P(A_2|A^cA_1)$ as, respectively, .4 and .2.] The following example, sometimes referred to as Laplace's rule of succession, further illustrates the concept of conditional independence.

**Example 5d.** Laplace's Rule of Succession. There are $k + 1$ coins in a box. The $i$th coin will, when flipped, turn up heads with probability $i/k$, $i = 0, 1, \ldots, k$. A coin is randomly selected from the box and is then repeatedly flipped. If the first $n$ flips all result in heads, what is the conditional probability that the $(n + 1)$st flip will do likewise?

***Solution:*** Let $E_i$ denote the event that the $i$th coin is initially selected, $i = 0, 1, \ldots, k$; let $F_n$ denote the event that the first $n$ flips all result in heads; and let $F$ be the event that the $(n + 1)$st flip is a head. The desired probability, $P(F|F_n)$, is now obtained as follows:

$$P(F|F_n) = \sum_{i=0}^{k} P(F|F_nE_i)P(E_i|F_n)$$

Now, given that the $i$th coin is selected, it is reasonable to assume that the outcomes will be conditionally independent, with each one resulting in a head with probability $i/k$. Hence

$$P(F|F_nE_i) = P(F|E_i) = \frac{i}{k}$$

Also,

$$\begin{aligned} P(E_i|F_n) &= \frac{P(E_iF_n)}{P(F_n)} \\ &= \frac{P(F_n|E_i)P(E_i)}{\sum_{j=0}^{k} P(F_n|E_j)P(E_j)} \\ &= \frac{(i/k)^n[1/(k+1)]}{\sum_{j=0}^{k} (j/k)^n[1/(k+1)]} \end{aligned}$$

Hence

$$P(F|F_n) = \frac{\sum_{i=0}^{k} (i/k)^{n+1}}{\sum_{j=0}^{k} (j/k)^n}$$

But if $k$ is large, we can use the integral approximations

$$\frac{1}{k}\sum_{i=0}^{k}\left(\frac{i}{k}\right)^{n+1} \approx \int_0^1 x^{n+1}\,dx = \frac{1}{n+2}$$

$$\frac{1}{k}\sum_{j=0}^{k}\left(\frac{j}{k}\right)^{n} \approx \int_0^1 x^{n}\,dx = \frac{1}{n+1}$$

and so, for $k$ large,

$$P(F|F_n) \approx \frac{n+1}{n+2}$$ ∎

## *Theoretical Exercises*

**1.** Assume the truth of Bernoulli's law of large numbers; this states that if an experiment is continually repeated under the exact same conditions, the proportion of time that the event $E$ occurs will, with probability 1, equal $P(E)$. Consider now those repetitions of the experiment in which the outcome is a point in $F$. Show that the proportion of these repetitions in which the outcome is also in $E$ is, with probability 1, $P(E|F)$.

**2.** Prove that if $P(E_i|E_1 \cdots E_{i-1}) > 0$, $i = 1, \ldots, n$, then

$$P(E_1E_2 \cdots E_n) = P(E_1)P(E_2|E_1)P(E_3|E_1E_2) \cdots P(E_n|E_1E_2 \cdots E_{n-1})$$

**3.** A ball is in any one of $n$ boxes. It is in the $i$th box with probability $P_i$. If the ball is in box $i$, a search of that box will uncover it with probability $\alpha_i$. Show that the conditional probability that the ball is in box $j$, given that a search of box $i$ did not uncover it, is

$$\frac{P_j}{1-\alpha_i P_i} \quad \text{if } j \neq i$$

$$\frac{(1-\alpha_i)P_i}{1-\alpha_i P_i} \quad \text{if } j = i$$

**4.** Prove or give counterexamples to the following statements:
(a) If $E$ is independent of $F$ and $E$ is independent of $G$, then $E$ is independent of $F \cup G$.
(b) If $E$ is independent of $F$, and $E$ is independent of $G$, and $FG = \emptyset$, then $E$ is independent of $F \cup G$.
(c) If $E$ is independent of $F$, and $F$ is independent of $G$, and $E$ is independent of $FG$, then $G$ is independent of $EF$.

**5.** An event $F$ is said to carry negative information about an event $E$, and we write $F \searrow E$ if

$$P(E|F) \leq P(E)$$

Prove or give counterexamples to the following assertions:
(a) If $F \searrow E$, then $E \searrow F$.
(b) If $F \searrow E$ and $E \searrow G$, then $F \searrow G$.
(c) If $F \searrow E$ and $G \searrow E$, then $FG \searrow E$.
Repeat (a), (b), and (c) when $\searrow$ is replaced by $\nearrow$, where we say that $F$ carries positive information about $E$, written $F \nearrow E$, when $P(E|F) \geq P(E)$.

**6.** Suppose that $\{E_n, n \geq 1\}$ and $\{F_n, n \geq 1\}$ are increasing sequences of events having limits $E$ and $F$. Show that if $E_n$ is independent of $F_n$ for all $n$, then $E$ is independent of $F$.

**7.** Prove that if $E_1, E_2, \ldots, E_n$ are independent events, then

$$P(E_1 \cup E_2 \cup \cdots \cup E_n) = 1 - [1 - P(E_1)][1 - P(E_2)] \cdots [1 - P(E_n)]$$

**8.** (a) An urn contains $n$ white and $m$ black balls. The balls are withdrawn one at a time until only those of the same color are left. Show that with probability $n/(n + m)$ they are all white.
HINT: Imagine that the experiment continues until all the balls are removed and consider the last ball withdrawn.
(b) A pond contains 3 distinct species of fish, which we will call the Red, Blue, and Green fish. There are $r$ Red, $b$ Blue, and $g$ Green fish. Suppose that the fish are removed from the pond in a random order (that is, each selection is equally likely to be any of the remaining fish). What is the probability that the Red fish are the first species to become extinct in the pond?
HINT: Write $P\{R\} = P\{RBG\} + P\{RGB\}$, and compute the probabilities on the right by first conditioning on the last species to be removed.

**9.** If $0 \leq a_i \leq 1$, $i = 1, 2, \ldots$, show that

$$\sum_{i=1}^{\infty} \left[ a_i \prod_{j=1}^{i-1} (1 - a_j) \right] + \prod_{i=1}^{\infty} (1 - a_i) = 1$$

HINT: Suppose an infinite number of coins are to be flipped. Let $a_i$ be the probability that the $i$th coin lands heads.

**10.** The probability of getting a head on a single toss of a coin is $p$. Consider that $A$ starts and continues to flip the coin until a tail shows up, at which point $B$ starts flipping. Then $B$ continues to flip until a tail comes up, at which point $A$ takes over, and so on. Let $P_{n,m}$ denote the probability that $A$ accumulates a total of $n$ heads before $B$ accumulates $m$. Show that

$$P_{n,m} = pP_{n-1,m} + (1 - p)(1 - P_{m,n})$$

**11.** Suppose that you are gambling against an infinitely rich adversary and at each stage you either win or lose 1 unit with respective probabilities $p$ and $1 - p$. Show that the probability that you eventually go broke is

$$\begin{array}{ll} 1 & \text{if } p \leq \frac{1}{2} \\ (q/p)^i & \text{if } p > \frac{1}{2} \text{ where } q = 1 - p \end{array}$$

where $i$ is your initial fortune.

**12.** Independent trials that result in a success with probability $p$ are successively performed until a total of $r$ successes is obtained. Show that the probability that exactly $n$ trials are required is

$$\binom{n-1}{r-1} p^r (1 - p)^{n-r}$$

Use this result to solve the problem of the points (Example 4i).

**13.** Independent trials that result in a success with probability $p$ and a failure with probability $1 - p$ are called Bernoulli trials. Let $P_n$ denote the probability that $n$ Bernoulli trials result in an even number of successes (0 being considered as an even number). Show that

$$P_n = p(1 - P_{n-1}) + (1 - p)P_{n-1} \qquad n \geq 1$$

and use this to prove (by induction) that

$$P_n = \frac{1 + (1 - 2p)^n}{2}$$

**14.** Let $Q_n$ denote the probability that in $n$ tosses of a fair coin no run of 3 consecutive heads appears. Show that

$$Q_n = \tfrac{1}{2}Q_{n-1} + \tfrac{1}{4}Q_{n-2} + \tfrac{1}{8}Q_{n-3}$$
$$Q_0 = Q_1 = Q_2 = 1$$

Find $Q_8$.

**15.** If $A$ has $n + 1$ and $B$ has $n$ fair coins, which they flip, show that the probability that $A$ gets more heads than $B$ is $\frac{1}{2}$.

HINT: Condition on which player has more heads after each has flipped $n$ coins. (There are three possibilities.)

**16.** Consider the gamblers ruin problem with the exception that $A$ and $B$ agree to play no more than $n$ games. Let $P_{n,i}$ denote the probability that $A$ winds up with all the money when $A$ starts with $i$ and $B$ with $N - i$. Derive an equation for $P_{n,i}$ in terms of $P_{n-1,i+1}$ and $P_{n-1,i-1}$ and compute $P_{7,3}$, $N = 5$.

**17.** Consider two urns, each containing both white and black balls. The probabilities of drawing white balls from the first and second urns are, respectively, $p$ and $p'$. Balls are sequentially selected with replacement as follows: With probability $\alpha$ a ball is initially chosen from the first urn, and with probability $1 - \alpha$ it is chosen from the second urn. The subsequent selections are then made according to the rule that whenever a white ball is drawn (and replaced), the next ball is drawn from the same urn; but when a black ball is drawn, the next ball is taken from the other urn. Let $\alpha_n$ denote the probability that the $n$th ball is chosen from the first urn. Show that

$$\alpha_{n+1} = \alpha_n(p + p' - 1) + 1 - p' \qquad n \geq 1$$

and use this to prove that

$$\alpha_n = \frac{1 - p'}{2 - p - p'} + \left(\alpha - \frac{1 - p'}{2 - p - p'}\right)(p + p' - 1)^{n-1}$$

Let $P_n$ denote the probability that the $n$th ball selected is white. Find $P_n$. Also compute $\lim_{n\to\infty} \alpha_n$ and $\lim_{n\to\infty} P_n$.

**18.** The Ballot Problem. In an election candidate $A$ receives $n$ votes, and candidate $B$ receives $m$ votes, where $n > m$. Assuming that all of the $(n + m)!/n!\,m!$ orderings of the votes are equally likely, let $P_{n,m}$ denote the probability that $A$ is always ahead in the counting of the votes.

(a) Compute $P_{2,1}$, $P_{3,1}$, $P_{3,2}$, $P_{4,1}$, $P_{4,2}$, $P_{4,3}$.
(b) Based on your results in (a), conjecture the value of $P_{n,m}$.
(c) Derive a recursion for $P_{n,m}$ in terms of $P_{n-1,m}$ and $P_{n,m-1}$ by conditioning on who receives the ____ vote. (Fill in the missing word.)
(d) Use (c) to verify your conjecture in (b) by an induction proof on $n + m$.

**19.** As a simplified model for weather forecasting, suppose that the weather (either wet or dry) tomorrow will be the same as the weather today with probability $p$. If the weather is dry on January 1, show that $P_n$, the probability that it will be dry $n$ days later, satisfies

$$P_n = (2p - 1)P_{n-1} + (1 - p) \qquad n \geq 1$$
$$P_0 = 1$$

Prove that

$$P_n = \tfrac{1}{2} + \tfrac{1}{2}(2p - 1)^n \qquad n \geq 0$$

**20.** A bag contains $a$ white and $b$ black balls. Balls are chosen from the bag according to the following method:
(a) A ball is chosen at random and is discarded.
(b) A second ball is then chosen. If its color is different from that of the preceding ball, it is replaced in the bag, and the process is repeated from the beginning. If its color is the same, it is discarded, and we start from (b).

In other words, balls are sampled and discarded until a change in color occurs, at which point the last ball is returned to the urn and the process starts anew. Let $P_{a,b}$ denote the probability that the last ball in the bag is white. Prove that

$$P_{a,b} = \tfrac{1}{2}$$

HINT: Use induction on $k \equiv a + b$.

**21.** Prove directly that

$$P(E|F) = P(E|FG)P(G|F) + P(E|FG^c)P(G^c|F)$$

**22.** Prove the equivalence of Equations (5.11) and (5.12).

**23.** Extend the definition of conditional independence to more than 2 events.

**24.** Prove or give a counterexample. If $E_1$ and $E_2$ are independent, then they are conditionally independent given $F$.

**25.** In Laplace's rule of succession (Example 5d) show that if the first $n$ flips all result in heads, then the conditional probability that the next $m$ flips also result in all heads is $(n + 1)/(n + m + 1)$.

**26.** In Laplace's rule of succession, suppose that the first $n$ flips resulted in $r$ heads and $n - r$ tails. Show that the probability that the $(n + 1)$st flip turns up heads is $(r + 1)/(n + 2)$. To do so, you will have to prove and use the identity

$$\int_0^1 y^n(1 - y)^m dy = \frac{n!\, m!}{(n + m + 1)!}$$

HINT: To prove the identity, let $C(n, m) = \int_0^1 y^n(1-y)^m\,dy$. Integrating by parts yields that

$$C(n, m) = \frac{m}{n+1}C(n+1, m-1)$$

Starting with $C(n, 0) = 1/(n+1)$, prove the identity by induction on $m$.

**27.** Suppose that a nonmathematical but philosophically minded friend of yours claims that Laplace's rule of succession must be incorrect because it can lead to ridiculous conclusions. "For instance," says he, "if a boy is 10 years old, the rule states that, having lived 10 years, the boy has probability $\frac{11}{12}$ of living another year. On the other hand, if the boy has an 80-year-old grandfather, then by Laplace's rule the grandfather has probability $\frac{81}{82}$ of surviving another year. However, this is ridiculous. Clearly, the boy is more likely to survive an additional year than is the grandfather." How would you answer your friend?

## *Problems*

**1.** Two fair dice are rolled. What is the conditional probability that at least one lands on 6 given that the dice land on different numbers?

**2.** If two fair dice are rolled, what is the conditional probability that the first one lands on 6 given that the sum of the dice is $i$? Compute for all values of $i$ between 2 and 12.

**3.** Use Equation (2.1) to compute, in a hand of bridge, the conditional probability that East has 3 spades given that North-South have a combined total of 8 spades.

**4.** What is the probability that at least one of a pair of fair dice lands on 6, given that the sum of the dice is $i$, $i = 2, 3, \ldots, 12$?

**5.** An urn contains 6 white and 9 black balls. If 4 balls are to be randomly selected without replacement, what is the probability that the first 2 selected are white and the last 2 black?

**6.** Consider an urn containing 12 balls of which 8 are white. A sample of size 4 is to be drawn with replacement (without replacement). What is the conditional probability (in each case) that the first and third balls drawn will be white, given that the sample drawn contains exactly 3 white balls?

**7.** The king comes from a family of 2 children. What is the probability that the other child is his sister?

**8.** A couple has 2 children. What is the probability that both are girls if the eldest is a girl?

**9.** Consider 3 urns. Urn $A$ contains 2 white and 4 red balls; urn B contains 8 white and 4 red balls; and urn C contains 1 white and 3 red balls. If 1 ball is selected from each urn, what is the probability that the ball chosen from urn $A$ was white, given that exactly 2 white balls were selected?

**10.** In a game of bridge, West has no aces. What is the probability of his partner's having (a) no aces, and (b) 2 or more aces? What would the probabilities be if West had exactly 1 ace?

**11.** Three cards are randomly selected, without replacement, from an ordinary deck of 52 playing cards. Compute the conditional probability that the first card selected is a spade, given that the second and third cards are spades.

**12.** An urn initially contains 5 white and 7 black balls. Each time a ball is selected, its color is noted and it is replaced in the urn along with 2 other balls of the same color.
(a) Compute the probability that the first 2 balls selected are black and the next 2 white.
(b) Compute the probability that, of the first 4 balls selected, exactly 2 are black.

**13.** Urn I contains 2 white and 4 red balls, whereas urn II contains 1 white and 1 red ball. A ball is randomly chosen from urn I and put into urn II, and a ball is then randomly selected from urn II.
(a) What is the probability that the ball selected from urn II is white?
(b) What is the conditional probability that the transferred ball was white, given that a white ball is selected from urn II?

**14.** How can 20 balls, 10 white and 10 black, be put into two urns so as to maximize the probability of drawing a white ball if an urn is selected at random and a ball is drawn at random from it?

**15.** Each of 2 balls is painted either black or gold and then placed in an urn. Suppose that each ball is colored black, with probability $\frac{1}{2}$, and that these events are independent.
(a) Suppose that you obtain information that the gold paint has been used (and thus at least one of the balls is painted gold). Compute the conditional probability that both balls are painted gold.
(b) Suppose, now, that the urn tips over and 1 ball falls out. It is painted gold. What is the probability that both balls are gold in this case? Explain.

**16.** The following method was proposed to estimate the number of people over the age of 50 that reside in a town of known population 100,000. "As you walk along the streets, keep a running count of the percentage of people that you encounter that are over 50. Do this for a few days; then multiply the obtained percentage by 100,000 to obtain the estimate." Comment on this method.

HINT: Let $p$ denote the proportion of people in this town that are over 50. Furthermore, let $\alpha_1$ denote the proportion of time that a person under the age of 50 spends in the streets, and let $\alpha_2$ be the corresponding value for those over 50. What quantity does the method suggested estimate? When is it approximately equal to $p$?

**17.** Suppose that 5 percent of men and .25 percent of women are color blind. A color blind person is chosen at random. What is the probability of this person's being male? Assume that there are an equal number of males and females. What if the population consisted of twice as many males as females?

**18.** Consider two boxes, one containing 1 black and 1 white marble, the other 2 black and 1 white marble. A box is selected at random, and a marble is drawn at random from the selected box. What is the probability that the marble is black? What is the probability that the first box was the one selected, given that the marble is white?

**19.** English and American spellings are *rigour* and *rigor*, respectively. A man staying at a Parisian hotel writes this word, and a letter taken at random from his spelling is found to be a vowel. If 40 percent of the English-speaking men at the hotel are English and 60 percent are Americans, what is the probability that the writer is an Englishman?

**20.** Urn $A$ contains 2 white balls and 1 black ball, whereas urn $B$ contains 1 white ball and 5 black balls. A ball is drawn at random from urn $A$ and placed in urn $B$. A ball is then drawn from urn $B$. It happens to be white. What is the probability that the transferred ball was white?

**21.** In Example 3e suppose that the new evidence is subject to different possible interpretations and in fact only shows that it is 90 percent likely that the criminal possesses this certain characteristic. In this case how likely would it be that the suspect is guilty (assuming, as before, that he has this characteristic)?

**22.** One probability class of 30 students contains 15 that are good, 10 that are average, and 5 that are of poor quality. A second probability class, also of 30 students, contains 5 that are good, 10 that are fair, and 15 that are poor. You (the expert) are aware of these numbers, but you have no idea which class is which. If you examine one student selected at random from each class and find that the student from class $A$ is a fair student whereas the student from class $B$ is a poor student, what is the probability that class $A$ is the superior class?

**23.** Stores $A$, $B$, and $C$ have 50, 75, and 100 employees and, respectively, 50, 60, and 70 percent of these are women. Resignations are equally likely among all employees, regardless of sex. One employee resigns, and this is a woman. What is the probability that she works in store $C$?

**24.** (a) A gambler has in his pocket a fair coin and a two-headed coin. He selects one of the coins at random; when he flips it, it shows heads. What is the probability that it is the fair coin?
(b) Suppose that he flips the same coin a second time and again it shows heads. What is now the probability that it is the fair coin?
(c) Suppose that he flips the same coin a third time and it shows tails. What is now the probability that it is the fair coin?

**25.** Urn $A$ has 5 white and 7 black balls. Urn $B$ has 3 white and 12 black balls. We flip a fair coin. If the outcome is heads, then a ball from urn $A$ is selected, whereas if the outcome is tails, then a ball from urn $B$ is selected. Suppose that a white ball is selected. What is the probability that the coin landed tails?

**26.** In Example 3a what is the probability that someone has an accident in the second year, given that he or she has had no accidents in the first year?

**27.** Consider a sample of size 3 drawn in the following manner: We start with an urn containing 5 white and 7 red balls. At each stage a ball is drawn and its color is noted. The ball is then returned to the urn along with an additional ball of the same color. Find the probability that the sample will contain exactly (a) 0 white balls, (b) 1 white ball, (c) 3 white balls, and (d) 2 white balls.

**28.** An urn contains $b$ black balls and $r$ red balls. One of the balls is drawn at random, but when it is put back in the urn, $c$ additional balls of the same color are put in with it. Now, suppose that we draw another ball. Show that the probability that the first ball was black, given that the second ball drawn was red, is $b/(b + r + c)$.

**29.** A deck of cards is shuffled and then divided into two halves of 26 cards each. A card is drawn from one of the halves; it turns out to be an ace. The ace is then placed in the second half-deck. The half is then shuffled, and a card is drawn from it. Compute the probability that this drawn card is an ace.

HINT: Condition on whether or not the interchanged card is selected.

**30.** Three cooks, $A$, $B$, and $C$, bake a special kind of cake, and with respective probabilities .02, .03, and .05 it fails to rise. In the restaurant where they work, $A$ bakes 50 percent of these cakes, $B$ 30 percent, and $C$ 20 percent. What proportion of "failures" is caused by $A$?

**31.** There are 3 coins in a box. One is a two-headed coin; another is a fair coin; and the third is a biased coin that comes up heads 75 percent of the time. When one of the 3 coins is selected at random and flipped, it shows heads. What is the probability that it was the two-headed coin?

**32.** Three prisoners are informed by their jailer that one of them has been chosen at random to be executed, and the other two are to be freed. Prisoner $A$ asks the jailer to tell him privately which of his fellow prisoners will be set free, claiming that there would be no harm in divulging this information because he already knows that at least one of the two will go free. The jailer refuses to answer this question, pointing out that if $A$ knew which of his fellow prisoners were to be set free, then his own probability of being executed would rise from $\frac{1}{3}$ to $\frac{1}{2}$ because he would then be one of two prisoners. What do you think of the jailer's reasoning?

**33.** Suppose we have 10 coins such that if the $i$th coin is flipped, heads will appear with probability $i/10$, $i = 1, 2, \ldots, 10$. When one of the coins is randomly selected and flipped, it shows heads. What is the conditional probability that it was the fifth coin?

**34.** An urn contains 5 white and 10 black balls. A fair die is rolled and that number of balls are chosen from the urn. What is the probability that all of the balls selected are white? What is the conditional probability that the die landed on 3 if all the balls selected are white?

**35.** Each of 2 cabinets identical in appearance has 2 drawers. Cabinet $A$ contains a silver coin in each drawer, and cabinet $B$ contains a silver coin in one of its drawers and a gold coin in the other. A cabinet is randomly selected, one of its drawers is opened, and a silver coin is found. What is the probability that there is a silver coin in the other drawer?

**36.** Suppose that there was a cancer diagnostic test that was 95 percent accurate both on those that do and those that do not have the disease. If .4 percent of the population have cancer, compute the probability that a tested person has cancer, given that his or her test result indicates so.

**37.** Suppose that an insurance company classifies people into one of three classes—good risks, average risks, and bad risks. Their records indicate that the probabilities that good, average, and bad risk persons will be involved in an accident over a 1-year span are, respectively, .05, .15, and .30. If 20 percent of the population are "good risks," 50 percent are "average risks," and 30 percent are "bad risks," what proportion of people have accidents in a fixed year? If policy holder $A$ had no accidents in 1972, what is the probability that he or she is a good (average) risk?

**38.** If you had to construct a mathematical model for events $E$ and $F$, as described in parts (a) through (e), would you assume that they were independent events? Explain your reasoning.
(a) $E$ is the event that a businesswoman has blue eyes, and $F$ is the event that her secretary has blue eyes.
(b) $E$ is the event that a professor owns a car, and $F$ is the event that he is listed in the telephone book.
(c) $E$ is the event that a man is under 6 feet tall, and $F$ is the event that he weighs over 200 pounds.
(d) $E$ is the event that a woman lives in the United States, and $F$ is the event that she lives in the Western Hemisphere.
(e) $E$ is the event that it will rain tomorrow, and $F$ is the event that it will rain the day after tomorrow.

**39.** In a class there are 4 freshman boys, 6 freshman girls, and 6 sophomore boys. How many sophomore girls must be present if sex and class are to be independent when a student is selected at random?

**40.** Mr. Jones has devised a gambling system for winning at roulette. When he bets, he bets on red, and places a bet only when the 10 previous spins of the roulette have landed on a black number. He reasons that his chance of winning is quite large because the probability of 11 consecutive spins resulting in black is quite small. What do you think of this system?

**41.** The probability of the closing of the $i$th relay in the circuits shown is given by $p_i$, $i = 1, 2, 3, 4, 5$. If all relays function independently, what is the probability that a current flows between $A$ and $B$ for the respective circuits?

(a)

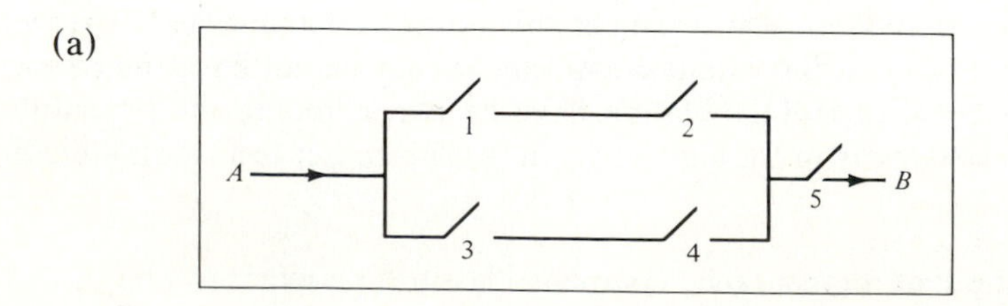

(b)

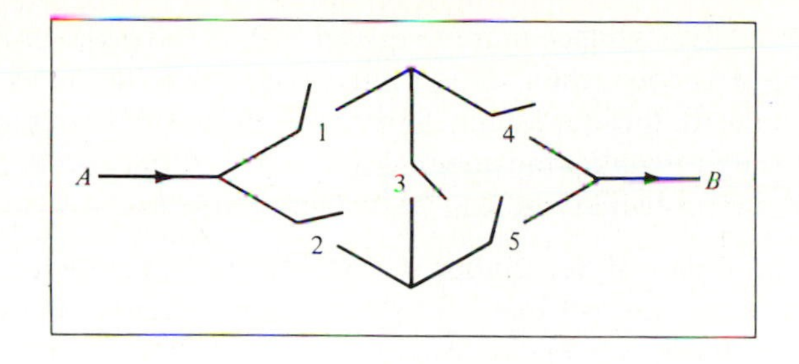


**42.** An engineering system consisting of $n$ components is said to be a $k$-out-of-$n$ system ($k \leq n$) if the system functions if and only if at least $k$ of the $n$ components function. Suppose that all components function independently of each other.

(a) If the $i$th component functions with probability $P_i$, $i = 1, 2, 3, 4$, compute the probability that a 2-out-of-4 system functions.

(b) Repeat (a) for a 3-out-of-5-system.

(c) Repeat for a $k$-out-of-$n$ system when all the $P_i$ equal $p$ (that is, $P_i = p$, $i = 1, 2, \ldots, n$).

**43.** A certain organism possesses a pair of each of 5 different genes (which we will designate by the first 5 letters of the English alphabet). Each gene appears in 2 forms (which we designate by lowercase and capital letters). The capital letter will be assumed to be the dominant gene in the sense that if an organism possesses the gene pair $xX$, then it will outwardly have the appearance of the $X$ gene. For instance, if $X$ stands for brown eyes and $x$ for blue eyes, then an individual having either gene pair $XX$ or $xX$ will have brown eyes, whereas one having gene pair $xx$ will be blue eyed. The characteristic appearance of an organism is called its phenotype, whereas its genetic constitution is called its genotype. (Thus 2 organisms with respective genotypes $aA$, $bB$, $cc$, $dD$, $ee$ and $AA$, $BB$, $cc$, $DD$, $ee$ would have different genotypes but the same phenotype.) In a mating between 2 organisms each one contributes, at random, one of its gene pairs of each type. The 5 contributions of an organism (one of each of the 5 types) are assumed to be independent and are also independent of the contributions of its mate. In a mating between organisms having genotypes $aA$, $bB$, $cC$, $dD$, $eE$ and $aa$, $bB$, $cc$, $Dd$, $ee$ what is the probability that the progeny will (1) phenotypically, (2) genotypically resemble

(a) the first parent;

(b) the second parent;

(c) either parent;

(d) neither parent?

**44.** There is a 50–50 chance that the queen carries the gene for hemophilia. If she is a carrier, then each prince has a 50–50 chance of having hemophilia. If the queen has had three princes without the disease, what is the probability the queen is a carrier? If there is a fourth prince, what is the probability he will have hemophilia?

**45.** On the morning of September 31, 1982, the won-lost records of the three leading baseball teams in the western division of the National League of the U.S. were as follows:

| *Team* | *Won* | *Lost* |
|---|---|---|
| Atlanta Braves | 87 | 72 |
| San Francisco Giants | 86 | 73 |
| Los Angeles Dodgers | 86 | 73 |

Each team had 3 games remaining to be played. All 3 of the Giants games were with the Dodgers, and the Braves remaining 3 games were against the San Diego Padres. Suppose that the outcomes of all remaining games are independent and each game is equally likely to be won by either participant. What are the probabilities that each of the teams wins the division? If two teams tie for first place, they have a playoff game, which each team has an equal chance of winning.

**46.** A town council of 7 members contains a steering committee of size 3. New ideas for legislation go first to the steering committee and then on to the council as a whole if at least 2 of the 3 committee members approve the legislation. Once at the full council, the legislation requires a majority vote (of at least 4) to pass. Consider now a new piece of legislation and suppose that each town council member will approve it, independently, with probability $p$. What is the probability that a given steering committee member's vote is decisive in the sense that if that person's vote were reversed, then the final fate of the legislation would be reversed? What is the corresponding probability for a given council member not on the steering committee?

**47.** Suppose that each child born to a couple is equally likely to be a boy as a girl independent of the sex distribution of the other children in the family. For a couple having 5 children, compute the probabilities of the following events:

(a) All children are of the same sex.
(b) The 3 eldest are boys and the others girls.
(c) Exactly 3 are boys.
(d) The 2 oldest are boys.
(e) There is at least 1 girl.

**48.** The probability of winning on a single toss of the dice is $p$. $A$ starts, and if he fails, he passes the dice to $B$, who then attempts to win on his roll. They continue tossing the dice back and forth until one of them wins. What are their respective probabilities of winning? Repeat if there are $k$ players.

**49.** Repeat Problem 48 under the assumption that when $A$ rolls the dice, she wins with probability $P_1$, and, when $B$ rolls, then $B$ wins with probability $P_2$.

**50.** Three players simultaneously toss coins. The coin tossed by $A$ ($B$) [$C$] turns up heads with probability $P_1$ ($P_2$) [$P_3$]. If one person gets a different outcome than the other two, then he is the odd man out. If there is no odd man out, the players flip again and continue to do so until they get an odd man out. What is the probability that $A$ will be the odd man?

**51.** Suppose that $E$ and $F$ are mutually exclusive events of an experiment. Show that if independent trials of this experiment are performed, then $E$ will occur before $F$ with probability $P(E)/[P(E) + P(F)]$.

**52.** When $A$ and $B$ flip coins, the one coming closest to a given line wins 1 penny from the other. If $A$ starts with 3 and $B$ with 7 pennies, what is the probability that $A$ winds up with all of the money if both players are equally skilled? What if $A$ were a better player who won 60 percent of the time?

**53.** In successive rolls of a pair of fair dice, what is the probability of getting 2 sevens before 6 even numbers?

**54.** Players are of equal skill, and in a contest the probability is $\frac{1}{2}$ that a specified one of the two contestants will be the victor. A group of $2^n$ players are paired off against each other at random. The $2^{n-1}$ winners are again paired off randomly, and so on, until a single winner remains. Consider two specified contestants, $A$ and $B$, and define the events $A_i$, $i \leq n$, $E$ by

$A_i$: $A$ plays in exactly $i$ contests;
$E$: $A$ and $B$ ever play each other.

Find

(a) $P(A_i)$, $i = 1, \ldots, n$.
(b) $P(E)$.
(c) Let $P_n = P(E)$. Show that

$$P_n = \frac{1}{2^n - 1} + \frac{2^n - 2}{2^n - 1}\left(\frac{1}{2}\right)^2 P_{n-1}$$

and use this to check your answer obtained in part (b).

**55.** A stock market investor owns shares in a stock whose present value is 25. She has decided that she must sell her stock if it either goes down to 10 or goes up to 40. If each change of price is either up 1 point with probability .55 or down 1 point with probability .45, and the successive changes are independent, what is the probability that the investor retires a winner?

**56.** $A$ and $B$ flip coins. $A$ starts and continues flipping until a tail occurs. At this point $B$ starts flipping and continues until there is a tail, then $A$ takes over, and so on. Let $P_1$ be the probability of the coin's landing heads when $A$ flips, and let it be $P_2$ when $B$ flips. The winner of the game is the first one to get:
(a) 2 heads in a row;
(b) a total of 2 heads;
(c) 3 heads in a row;
(d) a total of 3 heads.
Find the probability, in each case, that $A$ wins.

**57.** Die $A$ has 4 red and 2 white faces, whereas die $B$ has 2 red and 4 white faces. A fair coin is flipped once. If it lands on heads, the game continues with die $A$; if it lands tails, then die $B$ is to be used.
(a) Show that the probability of red at any throw is $\frac{1}{2}$.
(b) If the first two throws result in red, what is the probability of red at the third throw?
(c) If red turns up at the first two throws, what is the probability that it is die $A$ that is being used?

**58.** There are 12 balls, of which 4 are white, in an urn. Three players—$A$, $B$, $C$—successively draw from the urn, $A$ first, then $B$, then $C$, then $A$, and so on. The winner is the first one to draw a white ball. Find the win probabilities for each player if (a) each ball is replaced after it is drawn, and (b) the withdrawn balls are not replaced.

**59.** Repeat Problem 58 if each of the 3 players selects from his other own urn. That is, suppose that there are 3 different urns of 12 balls with 4 white in each.

**60.** In Example 5d what is the conditional probability that the $i$th coin was selected, given that the first $n$ trials all result in heads?

**61.** In Laplace's rule of succession, Example 5d, are the outcomes of the successive flips independent? Explain.

**62.** An individual tried by a 3-judge panel is declared guilty if at least 2 judges cast votes of guilty. Suppose that when the defendant is, in fact, guilty, each judge will independently vote guilty with probability .7, whereas when the defendant is, in fact, innocent, this probability drops

to .2. If 70 percent of defendants are guilty, compute the conditional probability that judge number 3 votes guilty given that:
(a) judges 1 and 2 vote guilty;
(b) judges 1 and 2 cast 1 guilty and 1 innocent vote;
(c) judges 1 and 2 both case innocent votes.
Let $E_i$, $i = 1, 2, 3$ denote the event that judge $i$ casts a guilty vote. Are these events independent? Are they conditionally independent? Explain.

# 4

# Random Variables

## 1 Random Variables

It is frequently the case when an experiment is performed that we are mainly interested in some function of the outcome as opposed to the actual outcome itself. For instance, in tossing dice we are often interested in the sum of the two dice and are not really concerned about the separate values of the dice. That is, we may be interested in knowing that the sum is 7 and not be concerned over whether the actual outcome was $(1, 6)$ or $(2, 5)$ or $(3, 4)$ or $(4, 3)$ or $(5, 2)$ or $(6, 1)$. Also, in coin-flipping, we may be interested in the total number of heads that occur and not care at all about the actual head-tail sequence that results. These quantities of interest, or more formally, these real-valued functions defined on the sample space, are known as *random variables.*

Because the value of a random variable is determined by the outcome of the experiment, we may assign probabilities to the possible values of the random variable.

**Example 1a.** Suppose that our experiment consists of tossing 3 fair coins. If we let $Y$ denote the number of heads appearing, then $Y$ is a random variable taking on one of the values 0, 1, 2, 3 with respective probabilities

$$
\begin{aligned}
&P\{Y = 0\} = P\{(T, T, T)\} = \tfrac{1}{8} \\
&P\{Y = 1\} = P\{(T, T, H), (T, H, T), (H, T, T)\} = \tfrac{3}{8} \\
&P\{Y = 2\} = P\{(T, H, H), (H, T, H), (H, H, T)\} = \tfrac{3}{8} \\
&P\{Y = 3\} = P\{(H, H, H)\} = \tfrac{1}{8}
\end{aligned}
$$

Since $Y$ must take on one of the values 0 through 3, we must have

$$
1 = P\left(\bigcup_{i=0}^{3} \{Y = i\}\right) = \sum_{i=0}^{3} P\{Y = i\}
$$

which, of course, is in accord with the above probabilities. ∎

**Example 1b.** Three balls are to be randomly selected without replacement from an urn containing 20 balls numbered 1 through 20. If we bet that at least one of the drawn balls has a number as large as or larger than 17, what is the probability that we win the bet?

***Solution:*** Let $X$ denote the largest number selected. Then $X$ is a random variable taking on one of the values $3, 4, \ldots, 20$. Furthermore, if we suppose that each of the $\binom{20}{3}$ possible selections are equally likely to occur, then

$$P\{X = i\} = \frac{\binom{i-1}{2}}{\binom{20}{3}} \qquad i = 3, \ldots, 20 \tag{1.1}$$

Equation (1.1) follows because the number of selections that result in the event $\{X = i\}$ is just the number of selections that result in ball numbered $i$ and two of the balls numbered 1 through $i - 1$ being chosen. As there are clearly $\binom{1}{1}\binom{i-1}{2}$ such selections, we obtain the probabilities expressed in Equation (1.1). From this equation we see that

$$P\{X = 20\} = \frac{\binom{19}{2}}{\binom{20}{3}} = \frac{3}{20} = .150$$

$$P\{X = 19\} = \frac{\binom{18}{2}}{\binom{20}{3}} = \frac{51}{380} = .134$$

$$P\{X = 18\} = \frac{\binom{17}{2}}{\binom{20}{3}} = \frac{34}{285} = .119$$

$$P\{X = 17\} = \frac{\binom{16}{2}}{\binom{20}{3}} = \frac{2}{19} = .105$$

Hence, as the event $\{X \geq 17\}$ is the union of the disjoint events $\{X = i\}$, $i = 17, 18, 19, 20$, it follows that the probability of our winning the bet is given by

$$P\{X \geq 17\} = .105 + .119 + .134 + .150 = .508 \qquad \blacksquare$$

**Example 1c.** Independent trials, consisting of the flipping of a coin having probability $p$ of coming up heads, are continually performed until either a head occurs or a total of $n$ flips are made. If we let $X$ denote the number of times the coin is flipped, then $X$ is a random variable taking on one of the values $1, 2, 3, \ldots, n$ with respective probabilities

$$P\{X = 1\} = P\{H\} = p$$

$$P\{X = 2\} = P\{(T, H)\} = (1 - p)p$$

$$P\{X = 3\} = P\{(T, T, H)\} = (1 - p)^2 p$$

$$\vdots$$

$$P\{X = n - 1\} = P\{(\underbrace{T, T, \ldots, T}_{n-2}, H)\} = (1 - p)^{n-2} p$$

$$P\{X = n\} = P\{(\underbrace{T, T, \ldots, T}_{n-1}, T), (\underbrace{T, T, \ldots, T}_{n-1}, H)\} = (1 - p)^{n-1}$$

As a check, note that

$$\begin{aligned} P\left(\bigcup_{i=1}^{n} \{X = i\}\right) &= \sum_{i=1}^{n} P\{X = i\} \\ &= \sum_{i=1}^{n-1} p(1 - p)^{i-1} + (1 - p)^{n-1} \\ &= p\left[\frac{1 - (1 - p)^{n-1}}{1 - (1 - p)}\right] + (1 - p)^{n-1} \\ &= 1 - (1 - p)^{n-1} + (1 - p)^{n-1} \\ &= 1 \end{aligned} \qquad \blacksquare$$

**Example 1d.** Three balls are randomly chosen from an urn containing 3 white, 3 red, and 5 black balls. Suppose that we win \$1 for each white ball selected and lose \$1 for each red selected. If we let $X$ denote our total winnings from the experiment, then $X$ is a random variable taking on the possible values $0, \pm 1, \pm 2, \pm 3$ with respective probabilities

$$P\{X = 0\} = \frac{\binom{5}{3} + \binom{3}{1}\binom{3}{1}\binom{5}{1}}{\binom{11}{3}} = \frac{55}{165}$$

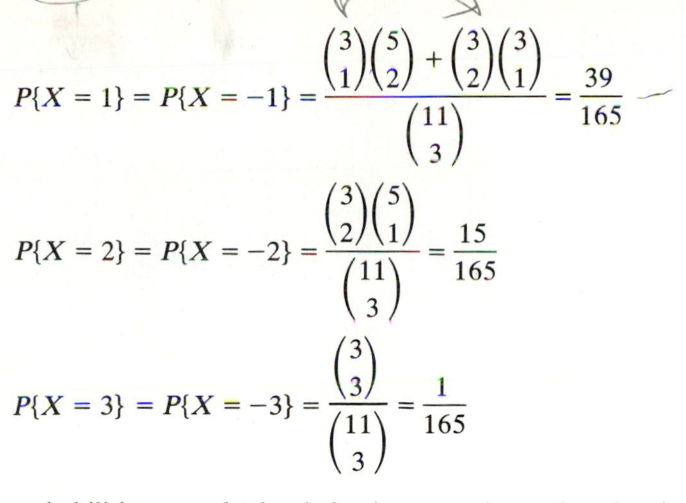

$$P\{X = 1\} = P\{X = -1\} = \frac{\binom{3}{1}\binom{5}{2} + \binom{3}{2}\binom{3}{1}}{\binom{11}{3}} = \frac{39}{165}$$

$$P\{X = 2\} = P\{X = -2\} = \frac{\binom{3}{2}\binom{5}{1}}{\binom{11}{3}} = \frac{15}{165}$$

$$P\{X = 3\} = P\{X = -3\} = \frac{\binom{3}{3}}{\binom{11}{3}} = \frac{1}{165}$$

These probabilities are obtained, for instance, by noting that in order for $X$ to equal 0, either all 3 balls selected must be black or 1 ball of each color must be selected. Similarly, the event $\{X = 1\}$ occurs either if 1 white and 2 black balls are selected or if 2 white and 1 red is selected. As a check we note that

$$\sum_{i=0}^{3} P\{X = i\} + \sum_{i=1}^{3} P\{X = -i\} = \frac{55 + 39 + 15 + 1 + 39 + 15 + 1}{165} = 1$$

The probability that we win money is given by

$$\sum_{i=1}^{3} P\{X = i\} = \tfrac{55}{165} = \tfrac{1}{3}$$ ∎

**Example 1e.** Suppose that there are $N$ distinct types of coupons and each time one obtains a coupon it is, independent of prior selections, equally likely to be any one of the $N$ types. One random variable of interest is $T$, the number of coupons that need be collected until one obtains a complete set of at least one of each type. Rather than derive $P\{T = n\}$ directly, let us start by considering the probability that $T$ is greater than $n$. To do so, fix $n$ and define the events $A_1, A_2, \ldots, A_N$ as follows: $A_j$ is the event that no type $j$ coupon is contained among the first $n$, $j = 1, \ldots, N$. Hence

$$\begin{aligned} P\{T > n\} &= P\left(\bigcup_{j=1}^{N} A_j\right) \\ &= \sum_j P(A_j) - \sum\sum_{j_1 < j_2} P(A_{j_1} A_{j_2}) + \cdots \\ &\quad + (-1)^{k+1} \sum\sum_{j_1 < j_2 < \cdots < j_k}\sum P(A_{j_1} A_{j_2} \cdots A_{j_k}) \cdots \\ &\quad + (-1)^{N+1} P(A_1 A_2 \cdots A_N) \end{aligned}$$

Now $A_j$ will occur if each of the $n$ coupons is not of type $j$. As each of the coupons will not be of type $j$ with probability $(N-1)/N$, we have, by the assumed independence of the types of successive coupons, that

$$P(A_j) = \left(\frac{N-1}{N}\right)^n$$

Also, the event $A_{j_1}A_{j_2}$ will occur if none of the first $n$ are of either type $j_1$ or $j_2$. Thus, again using independence, we see that

$$P(A_{j_1}A_{j_2}) = \left(\frac{N-2}{N}\right)^n$$

The same reasoning gives that

$$P(A_{j_1}A_{j_2}\cdots A_{j_k}) = \left(\frac{N-k}{N}\right)^n$$

and we see that, for $n > 0$,

$$\begin{aligned} P\{T > n\} &= N\left(\frac{N-1}{N}\right)^n - \binom{N}{2}\left(\frac{N-2}{N}\right)^n + \binom{N}{3}\left(\frac{N-3}{N}\right)^n - \cdots \\ &\quad + (-1)^N \binom{N}{N-1}\left(\frac{1}{N}\right)^n \\ &= \sum_{i=1}^{N-1} \binom{N}{i}\left(\frac{N-i}{N}\right)^n (-1)^{i+1} \end{aligned} \tag{1.2}$$

The probability that $T$ equals $n$ can now be obtained from the above by using

$$P\{T > n-1\} = P\{T = n\} + P\{T > n\}$$

or, equivalently,

$$P\{T = n\} = P\{T > n-1\} - P\{T > n\}$$

Another random variable of interest is the number of distinct types of coupons that are contained in the first $n$ selections—call this random variable $D_n$. To compute $P\{D_n = k\}$, let us start by fixing attention on a particular set of $k$ distinct types, and let us then determine the probability that this set constitutes the set of distinct types obtained in the first $n$ selections. Now, in order for this to be the situation, it is necessary and sufficient that of the first $n$ coupons obtained

$A$: each is one of these $k$ types;
$B$: each of these $k$ types is represented.

Now each coupon selected will be one of the $k$ types with probability $k/N$, and so the probability that $A$ will be valid is $(k/N)^n$. Also, given that a coupon is of one of the $k$ types under consideration, it is easy to see that it is equally likely to be of any one of these $k$ types. Hence the conditional probability of $B$ given that $A$ occurs is the same as the probability that a set of $n$ coupons, each equally likely to be any of $k$ possible types, contains a complete set of all $k$ types. But this is just the probability that the number needed to amass a complete set, when choosing among $k$ types, is less than or equal to $n$ and is thus obtainable from Equation (1.2) with $k$ replacing $N$. Hence we see that

$$P(A) = \left(\frac{k}{N}\right)^n$$

$$P(B|A) = 1 - \sum_{i=1}^{k-1} \binom{k}{i} \left(\frac{k-i}{k}\right)^n (-1)^{i+1}$$

Finally, as there are $\binom{N}{k}$ possible choices for the set of $k$ types, we arrive at

$$\begin{aligned} P\{D_n = k\} &= \binom{N}{k} P(AB) \\ &= \binom{N}{k} \left(\frac{k}{N}\right)^n \left[1 - \sum_{i=1}^{k-1} \binom{k}{i} \left(\frac{k-i}{k}\right)^n (-1)^{i+1}\right] \end{aligned}$$ ∎

## 2 Distribution Functions

The cumulative distribution function (c.d.f.), or more simply the distribution function $F$ of the random variable $X$, is defined for all real numbers $b$, $-\infty < b < \infty$, by

$$F(b) = P\{X \le b\}$$

In words, $F(b)$ denotes the probability that the random variable $X$ takes on a value that is less than or equal to $b$. Some properties of the c.d.f. $F$ are

1. $F$ is a nondecreasing function; that is, if $a < b$, then $F(a) \le F(b)$.
2. $\lim_{b \to \infty} F(b) = 1$.
3. $\lim_{b \to -\infty} F(b) = 0$.
4. $F$ is right continuous. That is, for any $b$ and any decreasing sequence $b_n$, $n \ge 1$, that converges to $b$, $\lim_{n \to \infty} F(b_n) = F(b)$.

Property 1 follows because for $a < b$ the event $\{X \le a\}$ is contained in the event $\{X \le b\}$ and so cannot have a larger probability. Properties 2, 3, and 4 all follow from the continuity property of probabilities (Section 6 of Chapter 2). For instance, to prove Property 2 we note that if $b_n$ increases to $\infty$, then the events $\{X \le b_n\}$, $n \ge 1$, are increasing events whose union is the event $\{X < \infty\}$. Hence, by the continuity property of probabilities,

$$\lim_{n \to \infty} P\{X \le b_n\} = P\{X < \infty\} = 1$$

which proves Property 2.

The proof of Property 3 is similar and is left as an exercise. To prove Property 4, we note that if $b_n$ decreases to $b$, then $\{X \le b_n\}$, $n \ge 1$ are decreasing events whose intersection is $\{X \le b\}$. Hence the continuity property yields that

$$\lim_{n} P\{X \le b_n\} = P\{X \le b\}$$

which verifies Property 4.

All probability questions about $X$ can be answered in terms of the c.d.f. $F$. For example,

$$P\{a < X \le b\} = F(b) - F(a) \qquad \text{for all } a < b \tag{2.1}$$

This can best be seen by writing the event $\{X \le b\}$ as the union of the mutually exclusive events $\{X \le a\}$ and $\{a < X \le b\}$. That is,

$$\{X \le b\} = \{X \le a\} \cup \{a < X \le b\}$$

and so

$$P\{X \le b\} = P\{X \le a\} + P\{a < X \le b\}$$

which establishes Equation (2.1).

If we want to compute the probability that $X$ is strictly less than $b$, we can again apply the continuity property to obtain

$$\begin{aligned} P\{X < b\} &= P\left(\lim_{n \to \infty} \left\{X \le b - \frac{1}{n}\right\}\right) \\ &= \lim_{n \to \infty} P\left(X \le b - \frac{1}{n}\right) \\ &= \lim_{n \to \infty} F\left(b - \frac{1}{n}\right) \end{aligned}$$

Note that $P\{X < b\}$ does not necessarily equal $F(b)$, since $F(b)$ also includes the probability that $X$ equals $b$.

**Example 2a.** The distribution function of the random variable $X$ is given by

$$F(x) = \begin{cases} 0 & x < 0 \\ \dfrac{x}{2} & 0 \le x < 1 \\ \dfrac{2}{3} & 1 \le x < 2 \\ \dfrac{11}{12} & 2 \le x < 3 \\ 1 & 3 \le x \end{cases}$$

A graph of $F(x)$ is presented in Figure 4.1.

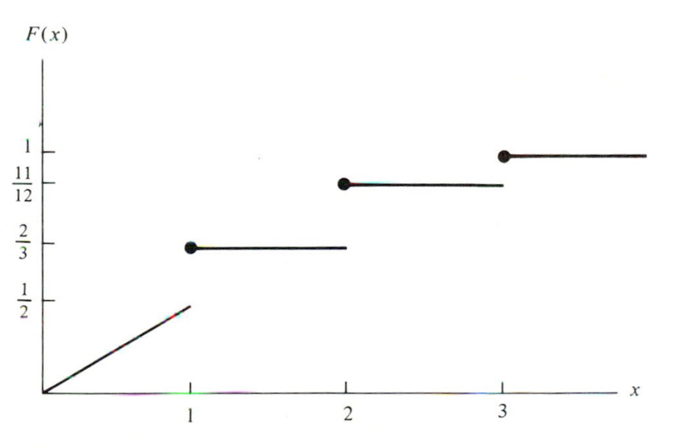


**Figure 4.1** Graph of $F(x)$

Compute (a) $P\{X < 3\}$, (b) $P\{X = 1\}$, (c) $P\{X > \frac{1}{2}\}$, and (d) $P\{2 < X \le 4\}$.

***Solution***

1. $$P\{X < 3\} = \lim_n P\left\{X \le 3 - \frac{1}{n}\right\} = \lim_n F\left(3 - \frac{1}{n}\right) = \frac{11}{12}$$

2. $$P\{X = 1\} = P\{X \le 1\} - P\{X < 1\}$$

$$= F(1) - \lim_n F\left(1 - \frac{1}{n}\right) = \frac{2}{3} - \frac{1}{2} = \frac{1}{6}$$

3. $$P\left\{X > \frac{1}{2}\right\} = 1 - P\left\{X \leq \frac{1}{2}\right\}$$
$$= 1 - F\left(\frac{1}{2}\right) = \frac{3}{4}$$

4. $$P\{2 < X \leq 4\} = F(4) - F(2)$$
$$= \frac{1}{12}$$ ∎

## 3 Discrete Random Variables

A random variable that can take on at most a countable number of possible values is said to be discrete. For a discrete random variable $X$, we define the probability mass function $p(a)$ of $X$ by

$$p(a) = P\{X = a\}$$

The probability mass function $p(a)$ is positive for at most a countable number of values of $a$. That is, if $X$ must assume one of the values $x_1, x_2, \ldots,$ then

$$p(x_i) \geq 0 \qquad i = 1, 2, \ldots$$
$$p(x) = 0 \qquad \text{all other values of } x$$

Since $X$ must take on one of the values $x_i$, we have

$$\sum_{i=1}^{\infty} p(x_i) = 1$$

It is often instructive to present the probability mass function in a graphical format by plotting $p(x_i)$ on the $y$-axis against $x_i$ on the $x$-axis. For instance, if the probability mass function of $X$ is

$$p(0) = \tfrac{1}{4} \qquad p(1) = \tfrac{1}{2} \qquad p(2) = \tfrac{1}{4}$$

we can represent this graphically as shown in Figure 4.2. Similarly, a graph of the probability mass function of the random variable representing the sum when two dice are rolled looks like the one shown in Figure 4.3.

**Example 3a.** The probability mass function of a random variable $X$ is given by $p(i) = c\lambda^i/i!$, $i = 0, 1, 2, \ldots,$ where $\lambda$ is some positive value.

1. Find $P\{X = 0\}$.
2. Find $P\{X > 2\}$.

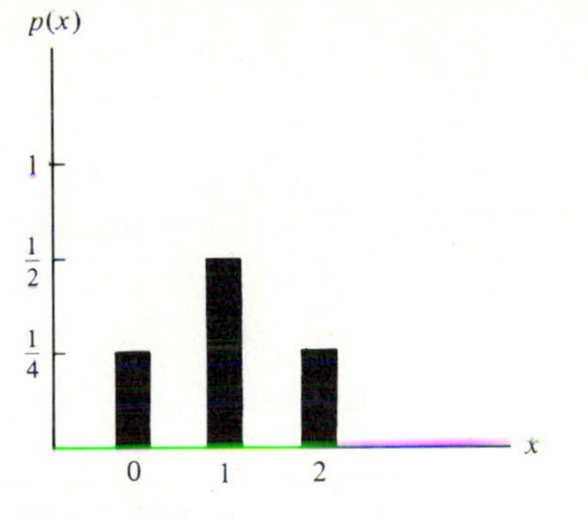

**Figure 4.2**

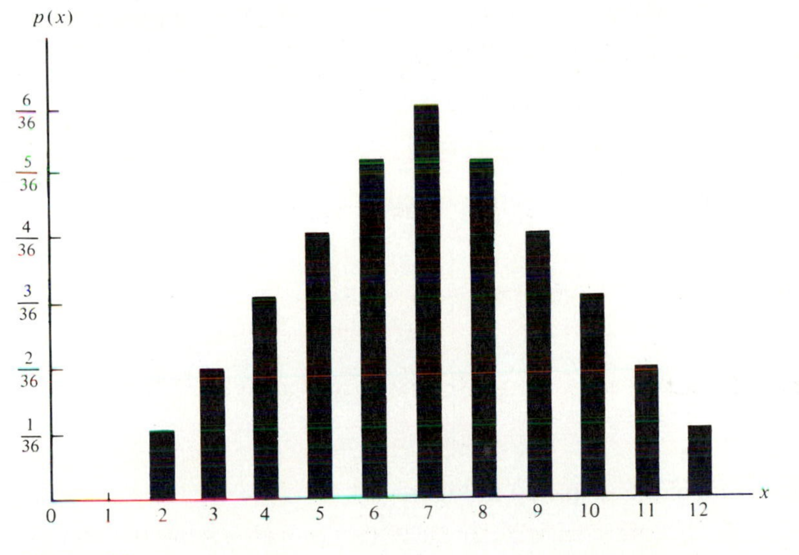


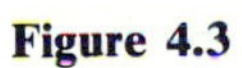
**Figure 4.3**

***Solution:*** Since $\sum_{i=0}^{\infty} p(i) = 1$, we have that

$$c \sum_{i=0}^{\infty} \frac{\lambda^i}{i!} = 1$$

implying, because $e^x = \sum_{i=0}^{\infty} x^i/i!$, that

$$ce^{\lambda} = 1 \qquad \text{or} \qquad c = e^{-\lambda}$$

Hence

$$P\{X = 0\} = e^{-\lambda}\lambda^0/0! = e^{-\lambda}$$

$$P\{X > 2\} = 1 - P\{X \le 2\} = 1 - P\{X = 0\} - P\{X = 1\} - P\{X = 2\}$$

$$= 1 - e^{-\lambda} - \lambda e^{-\lambda} - \frac{\lambda^2 e^{-\lambda}}{2}$$ ∎

The cumulative distribution function $F$ can be expressed in terms of $p(a)$ by

$$F(a) = \sum_{\text{all } x \le a} p(x)$$

If $X$ is a discrete random variable whose set of possible values are $x_1, x_2, x_3, \ldots$, where $x_1 < x_2 < x_3 < \cdots$, then its distribution function $F$ is a step function. That is, the value of $F$ is constant in the intervals $[x_{i-1}, x_i)$ and then takes a step (or jump) of size $p(x_i)$ at $x_i$. For instance, if $X$ has a probability mass function given by

$$p(1) = \tfrac{1}{4} \qquad p(2) = \tfrac{1}{2} \qquad p(3) = \tfrac{1}{8} \qquad p(4) = \tfrac{1}{8}$$

then its cumulative distribution function is given by

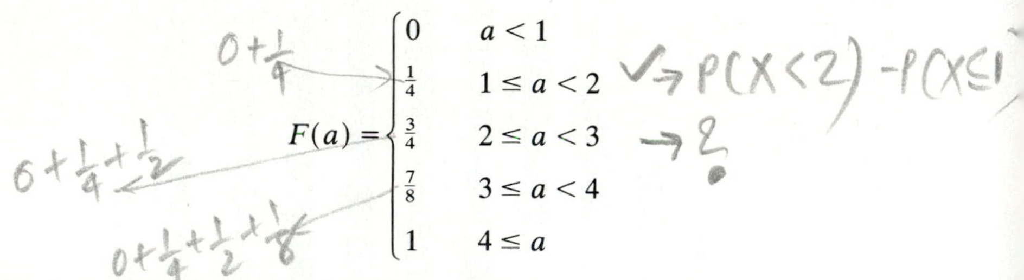

$$F(a) = \begin{cases} 0 & a < 1 \\ \frac{1}{4} & 1 \le a < 2 \\ \frac{3}{4} & 2 \le a < 3 \\ \frac{7}{8} & 3 \le a < 4 \\ 1 & 4 \le a \end{cases}$$

This is graphically depicted in Figure 4.4.

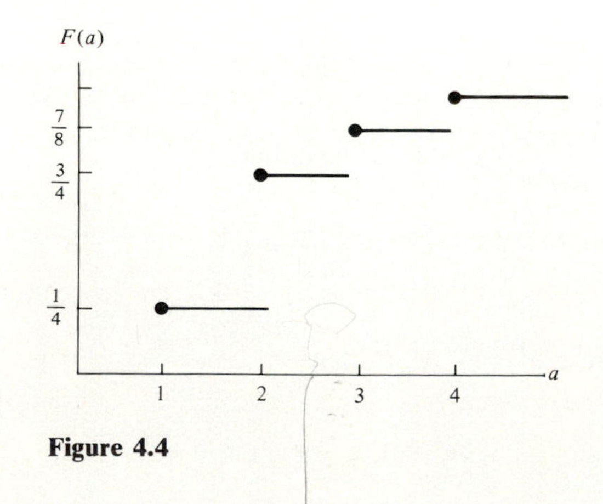


**Figure 4.4**

The reader should note that the size of the step at any of the values 1, 2, 3, 4 is equal to the probability that $X$ assumes that particular value.

Discrete random variables are often classified according to their probability mass function. In the next few sections we consider some of these.

## 4 The Bernoulli and Binomial Random Variables

Suppose that a trial, or an experiment, whose outcome can be classified as either a "success" or as a "failure" is performed. If we let $X = 1$ when the outcome is a success and $X = 0$ when it is a failure, then the probability mass function of $X$ is given by

$$\begin{aligned} p(0) &= P\{X = 0\} = 1 - p \\ p(1) &= P\{X = 1\} = p \end{aligned} \tag{4.1}$$

where $p, 0 \leq p \leq 1$, is the probability that the trial is a "success."

A random variable $X$ is said to be a Bernoulli random variable (after the Swiss mathematician James Bernoulli) if its probability mass function is given by Equations (4.1) for some $p \in (0, 1)$.

Suppose now that $n$ independent trials, each of which results in a "success" with probability $p$ and in a "failure" with probability $1 - p$, are to be performed. If $X$ represents the number of successes that occur in the $n$ trials, then $X$ is said to be a *binomial* random variable with parameters $(n, p)$. Thus a Bernoulli random variable is just a binomial random variable with parameters $(1, p)$.

The probability mass function of a binomial random variable having parameters $(n, p)$ is given by

$$p(i) = \binom{n}{i} p^i (1 - p)^{n-i} \qquad i = 0, 1, \ldots, n \tag{4.2}$$

The validity of Equation (4.2) may be verified by first noting that the probability of any particular sequence of $n$ outcomes containing $i$ successes and $n - i$ failures is, by the assumed independence of trials, $p^i (1 - p)^{n-i}$. Equation (4.2) then follows, since there are $\binom{n}{i}$ different sequences of the $n$ outcomes leading to $i$ successes and $n - i$ failures. This perhaps can most easily be seen by noting that there are $\binom{n}{i}$ different choices of the $i$ trials that result in successes. For instance, if $n = 4$, $i = 2$, then there are $\binom{4}{2} = 6$ ways in which the four trials can result in two successes, namely, any of the outcomes $(s, s, f, f)$, $(s, f, s, f)$, $(s, f, f, s)$, $(f, s, s, f)$, $(f, s, f,$

$s$), or ($f$, $f$, $s$, $s$), where the outcome ($s$, $s$, $f$, $f$) means, for instance, that the first two trials are successes and the last two failures.
Since each of these outcomes has probability $p^2(1-p)^2$ of occurring, the desired probability of 2 successes in the 4 trials is thus $\binom{4}{2}p^2(1-p)^2$.

Note that by the binomial theorem, the probabilities sum to one; that is,

$$\sum_{i=0}^{\infty} p(i) = \sum_{i=0}^{n} \binom{n}{i} p^i(1-p)^{n-i} = [p + (1-p)]^n = 1$$

**Example 4a.** Five fair coins are flipped. If the outcomes are assumed independent, find the probability mass function of the number of heads obtained.

***Solution:*** If we let $X$ equal the number of heads ("successes") that appear, then $X$ is a binomial random variable with parameters ($n = 5$, $p = \frac{1}{2}$). Hence, by Equation (4.2),

$$P\{X = 0\} = \binom{5}{0}\left(\frac{1}{2}\right)^0\left(\frac{1}{2}\right)^5 = \frac{1}{32}$$

$$P\{X = 1\} = \binom{5}{1}\left(\frac{1}{2}\right)^1\left(\frac{1}{2}\right)^4 = \frac{5}{32}$$

$$P\{X = 2\} = \binom{5}{2}\left(\frac{1}{2}\right)^2\left(\frac{1}{2}\right)^3 = \frac{10}{32}$$

$$P\{X = 3\} = \binom{5}{3}\left(\frac{1}{2}\right)^3\left(\frac{1}{2}\right)^2 = \frac{10}{32}$$

$$P\{X = 4\} = \binom{5}{4}\left(\frac{1}{2}\right)^4\left(\frac{1}{2}\right)^1 = \frac{5}{32}$$

$$P\{X = 5\} = \binom{5}{5}\left(\frac{1}{2}\right)^5\left(\frac{1}{2}\right)^0 = \frac{1}{32}$$ ∎

**Example 4b.** It is known that screws produced by a certain company will be defective with probability .01 independently of each other. The company sells the screws in packages of 10 and offers a money-back guarantee that at most 1 of the 10 screws is defective. What proportion of packages sold must the company replace?

***Solution:*** If $X$ is the number of defective screws in a package, then $X$ is a binomial random variable with parameters (10, .01). Hence the probability that a package will have to be replaced is

$$1 - P\{X = 0\} - P\{X = 1\} = 1 - \binom{10}{0}(.01)^0(.99)^{10} - \binom{10}{1}(.01)^1(.99)^9$$
$$\approx .004$$

Hence only .4 percent of the packages will have to be replaced. ∎

**Example 4c.** The following gambling game, known as the wheel of fortune (or chuck-a-luck), is quite popular at many carnivals and gambling casinos: A player bets on one of the numbers 1 through 6. Three dice are then rolled, and if the number bet by the player appears $i$ times, $i = 1, 2, 3$, then the player wins $i$ units; on the other hand, if the number bet by the player does not appear on any of the dice, then the player loses 1 unit. Is this game fair to the player? (Actually, the game is played by spinning a wheel that comes to rest on a slot labeled by three of the numbers 1 through 6, but it is mathematically equivalent to the dice version.)

***Solution:*** If we assume that the dice are fair and act independently of each other, then the number of times that the number bet appears is a binomial random variable with parameters $(3, \frac{1}{6})$. Hence, letting $X$ denote the players winnings in the game, we have

$$\begin{aligned}
P\{X = -1\} &= \binom{3}{0}\left(\frac{1}{6}\right)^0\left(\frac{5}{6}\right)^3 = \frac{125}{216} \\
P\{X = 1\} &= \binom{3}{1}\left(\frac{1}{6}\right)^1\left(\frac{5}{6}\right)^2 = \frac{75}{216} \\
P\{X = 2\} &= \binom{3}{2}\left(\frac{1}{6}\right)^2\left(\frac{5}{6}\right)^1 = \frac{15}{216} \\
P\{X = 3\} &= \binom{3}{3}\left(\frac{1}{6}\right)^3\left(\frac{5}{6}\right)^0 = \frac{1}{216}
\end{aligned} \tag{4.3}$$

In order to determine whether or not this is a fair game for the player, we shall take a long-run point of view. It follows from Equations (4.3) that if the player continues to play indefinitely, then in the long run, out of every 216 games he plays he will

Win −1 (that is, lose 1) 125 times.
Win 1 75 times.
Win 2 15 times.
Win 3 1 time.

Hence, in the long run, out of every 216 games he will win $-1(125) + 1(75) + 2(15) + 3(1) = -17$. That is, on the average, he will lose 17 units per every 216 games he plays.

From this point of view, the game is clearly not fair. (Although the above argument is somewhat heuristic, it is, nevertheless, completely valid and will be made rigorous by the strong law of large numbers, which will be presented in Chapter 8.) The correct interpretation of the somewhat imprecise statement that in the long run the players will lose 17 units per every 216 games is that the player's total winnings during the first $n$ games divided by $n$ will, with probability 1, converge to $-17/216$ as $n \to \infty$. ▮

In the next example we consider the simplest form of the theory of inheritance as developed by G. Mendel (1822–1884).

**Example 4d.** Suppose that a particular trait (such as eye color or left handedness) of a person is classified on the basis of one pair of genes and suppose that $d$ represents a dominant gene and $r$ a recessive gene. Thus a person with $dd$ genes is pure dominance, one with $rr$ is pure recessive, and one with $rd$ is hybrid. The pure dominance and the hybrid are alike in appearance. Children receive 1 gene from each parent. If, with respect to a particular trait, 2 hybrid parents have a total of 4 children, what is the probability that 3 of the 4 children have the outward appearance of the dominant gene?

***Solution:*** If we assume that each child is equally likely to inherit either of 2 genes from each parent, the probabilities that the child of 2 hybrid parents will have $dd$, $rr$, or $rd$ pairs of genes are, respectively, $\frac{1}{4}, \frac{1}{4}, \frac{1}{2}$. Hence, as an offspring will have the outward appearance of the dominant gene if its gene pair is either $dd$ or $rd$, it follows that the number of such children is binomially distributed with parameters $(4, \frac{3}{4})$. Thus the desired probability is

$$\binom{4}{3}\left(\frac{3}{4}\right)^3\left(\frac{1}{4}\right)^1 = \frac{27}{64}$$ ▮

**Example 4e.** Consider a jury trial in which it takes 8 of the 12 jurors to convict; that is, in order for the defendant to be convicted, at least 8 of the jurors must vote him guilty. If we assume that jurors act independently and each makes the right decision with probability $\theta$, what is the probability that the jury renders a correct decision?

***Solution:*** The problem, as stated, is incapable of solution, for there is not yet enough information. For instance, if the defendant is innocent, the probability of the jury's rendering a correct decision is

$$\sum_{i=5}^{12} \binom{12}{i} \theta^i (1 - \theta)^{12-i}$$

whereas, if he is guilty, the probability of a correct decision is

$$\sum_{i=8}^{12} \binom{12}{i} \theta^i (1 - \theta)^{12-i}$$

Therefore, if $\alpha$ represents the probability that the defendant is guilty, then, by conditioning on whether or not he is guilty, we obtain that the probability that the jury renders a correct decision is

$$\alpha \sum_{i=8}^{12} \binom{12}{i} \theta^i (1 - \theta)^{12-i} + (1 - \alpha) \sum_{i=5}^{12} \binom{12}{i} \theta^i (1 - \theta)^{12-i}$$ ∎

**Example 4f.** A communication system consists of $n$ components each of which will, independently, function with probability $p$. The total system will be able to operate effectively if at least one-half of its components function.

(a) For what values of $p$ is a 5-component system more likely to operate effectively than a 3-component system?
(b) In general, when is a $2k + 1$ component system better than a $2k - 1$ component system?

***Solution:*** (a) As the number of functioning components is a binomial random variable with parameters $(n,p)$, it follows that the probability that a 5-component system will be effective is

$$\binom{5}{3} p^3 (1 - p)^2 + \binom{5}{4} p^4 (1 - p) + p^5$$

whereas the corresponding probability for a 3-component system is

$$\binom{3}{2} p^2 (1 - p) + p^3.$$

Hence, the 5-component system is better if

$$10p^3 (1 - p)^2 + 5p^4 (1 - p) + p^5 \geq 3p^2 (1 - p) + p^3$$

which reduces to

$$3(p - 1)^2 (2p - 1) \geq 0$$

or

$$p \geq 1/2.$$

(b) In general, a system with $2k + 1$ components will be better than one with $2k - 1$ components if (and only if) $p \geq 1/2$. To prove this, consider a system of $2k + 1$ components and let $X$ denote the number of the first $2k - 1$ that function. Then

$$\begin{aligned} P_{2k+1}\text{ (effective)} &= P\{X \geq k + 1\} + P\{X = k\}(1 - (1 - p)^2) \\ &\quad + P\{X = k - 1\}p^2 \end{aligned}$$

which follows since the $2k + 1$ component system will be effective if either

(i) $X \geq k + 1$
(ii) $X = k$ and at least one of the remaining 2 components function

or

(iii) $X = k - 1$ and both of the next 2 functions.

As

$$\begin{aligned} P_{2k-1}\text{ (effective)} &= P\{X \geq k\} \\ &= P\{X = k\} + P\{X \geq k + 1\} \end{aligned}$$

we obtain that

$$\begin{aligned} &P_{2k+1}\text{ (effective)} - P_{2k-1}\text{ (effective)} \\ &\quad = P\{X = k - 1\}\, p^2 - (1 - p)^2\, P\{X = k\} \\ &\quad = \binom{2k-1}{k-1} p^{k-1}(1 - p)^k p^2 - (1 - p)^2 \binom{2k-1}{k} p^k(1 - p)^{k-1} \\ &\quad = \binom{2k-1}{k} p^k(1 - p)^k[p - (1 - p)] \text{since} \binom{2k-1}{k-1} = \binom{2k-1}{k} \\ &\quad \geq 0 \Leftrightarrow p \geq \tfrac{1}{2}. \end{aligned}$$

▮

We shall now examine the properties of the probability mass functions of a binomial random variable. Specifically, we prove the following proposition.

---

*Proposition 4.1*

*If $X$ is a binomial random variable with parameters $(n, p)$, where $0 < p < 1$, then as $k$ goes from 0 to $n$, $P\{X = k\}$ first increases monotonically and then decreases monotonically, reaching its largest value when $k$ is the largest integer less than or equal to $(n + 1)p$.*

---

**Proof:** We prove the proposition by considering $P\{X = k\}/P\{X = k - 1\}$ and determining for what values of $k$ it is greater or less than 1. Now,

$$\frac{P\{X = k\}}{P\{X = k-1\}} = \frac{\dfrac{n!}{(n-k)!\,k!}p^k(1-p)^{n-k}}{\dfrac{n!}{(n-k+1)!\,(k-1)!}p^{k-1}(1-p)^{n-k+1}}$$

$$= \frac{(n-k+1)p}{k(1-p)}$$

Hence $P\{X = k\} \geq P\{X = k - 1\}$ if and only if

$$(n-k+1)p \geq k(1-p)$$

or, equivalently, if and only if

$$k \leq (n+1)p$$

and the proposition is proved. ∎

As an illustration of Proposition 4.1 consider Figure 4.5, the graph of the probability mass function of a binomial random variable with parameters $(10, \frac{1}{2})$.

**Example 4g.** In a United States presidential election the candidate that gains the maximum number of votes in a state is awarded the total

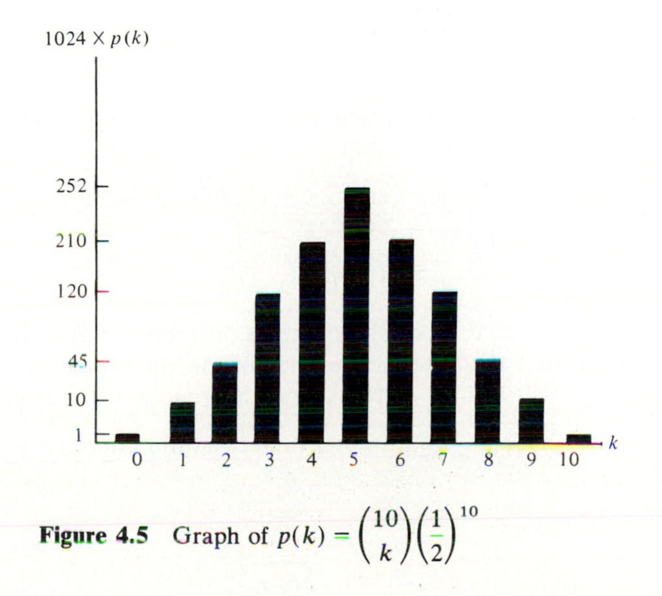


**Figure 4.5** Graph of $p(k) = \binom{10}{k}\left(\frac{1}{2}\right)^{10}$

number of electoral college votes allocated to that state. The number of electoral college votes of a given state is roughly proportional to the population of that state—that is, a state of population size $n$ has roughly $nc$ electoral votes. (Actually, it is closer to $nc + 2$ as a state is given an electoral vote for each member of the House of Representatives, the number of such representatives being roughly proportional to its population, and one electoral college vote for each of its two senators.) Let us determine the average power in a close presidential election of a citizen in a state of size $n$, where by *average power* in a close election we mean the following: A voter in a state of size $n = 2k + 1$ will be decisive if the other $n - 1$ voters split their votes evenly between the two candidates. (We are assuming here that $n$ is odd, but the case where $n$ is even is quite similar.) As the election is close, we shall suppose that each of the other $n - 1 = 2k$ voters acts independently and is equally likely to vote for either candidate. Hence the probability that a voter in a state of size $n = 2k + 1$ will make a difference to the outcome is the same as the probability that $2k$ tosses of a fair coin lands heads and tails an equal number of times. That is,

$$
\begin{aligned}
&P\{\text{voter in state of size } 2k+1 \text{ makes a difference}\}\\
&\quad = \binom{2k}{k}\left(\frac{1}{2}\right)^k\left(\frac{1}{2}\right)^k\\
&\quad = \frac{(2k)!}{k!\,k!\,2^{2k}}
\end{aligned}
$$

To approximate the above, we make use of Stirling's approximation, which says that for $k$ large

$$k! \sim k^{k+1/2}e^{-k}\sqrt{2\pi}$$

where we say that $a_k \sim b_k$ when the ratio $a_k/b_k$ approaches 1 as $k$ approaches $\infty$. Hence we see that

$$
\begin{aligned}
&P\{\text{voter in state of size } 2k+1 \text{ makes a difference}\}\\
&\quad \sim \frac{(2k)^{2k+1/2}e^{-2k}\sqrt{2\pi}}{k^{2k+1}e^{-2k}(2\pi)2^{2k}} = \frac{1}{\sqrt{k\pi}}
\end{aligned}
$$

As such a voter will, if he or she makes a difference, affect $nc$ electoral votes, we see that the average number of electoral votes a voter in a state of size $n$ will affect—or the voter's average power—is given by

$$
\begin{aligned}
\text{average power} &= ncP\{\text{makes a difference}\}\\
&\sim \frac{nc}{\sqrt{n\pi/2}}\\
&= c\sqrt{2n/\pi}
\end{aligned}
$$

Hence the average power of a voter in a state of size $n$ is proportional to the square root of $n$, thus showing that voters in large states have more power than those in smaller ones. ∎

## 4.1 Computing the Binomial Distribution Function

Suppose that $X$ is binomial with parameters $(n, p)$. The key to computing its distribution function

$$P\{X \le i\} = \sum_{k=0}^{i} \binom{n}{k} p^k (1-p)^{n-k}, \qquad i = 0, 1, \ldots, n$$

is to utilize the following relationship between $P\{X = k + 1\}$ and $P\{X = k\}$, which was established in the proof of proposition 4.1:

$$P\{X = k + 1\} = \frac{p}{1-p} \frac{n-k}{k+1} P\{X = k\} \tag{4.4}$$

**Example 4h.** Let $X$ be a binomial random variable with parameters $n = 6, p = .4$. Then, starting with $P\{X = 0\} = (.6)^6$ and recursively employing Equation 4.4, we obtain

$$\begin{aligned}
P\{X = 0\} &= (.6)^6 = .0467 \\
P\{X = 1\} &= \tfrac{4}{6}\tfrac{6}{1}P\{X = 0\} = .1866 \\
P\{X = 2\} &= \tfrac{4}{6}\tfrac{5}{2}P\{X = 1\} = .3110 \\
P\{X = 3\} &= \tfrac{4}{6}\tfrac{4}{3}P\{X = 2\} = .2765 \\
P\{X = 4\} &= \tfrac{4}{6}\tfrac{3}{4}P\{X = 3\} = .1382 \\
P\{X = 5\} &= \tfrac{4}{6}\tfrac{2}{5}P\{X = 4\} = .0369 \\
P\{X = 6\} &= \tfrac{4}{6}\tfrac{1}{6}P\{X = 5\} = .0041.
\end{aligned}$$ ∎

A computer program, written in Basic, which utilizes the recursion (4.4) to compute the binomial distribution function is presented at the end of this chapter. This program attempts to first compute $P\{X = 0\} = (1 - p)^n$ and then utilize Equation 4.4 to successively compute $P\{X = 1\}, \ldots, P\{X = i\}$. However, this will be successful only for moderate values of $n$, since in the case of large $n$, due to computer round-off error, $P\{X = 0\} = (1 - p)^n$ will be computed to equal 0. If this error is made, then all subsequent terms $P\{X = k\}, k = 1, \ldots, i$ will also be taken to equal 0 and so the program would incorrectly conclude that $P\{X \le i\} = 0$. To guard against this possibility, in cases where $(1 - p)^n$ is computed to equal 0, the basic program is instructed to begin calculations not with $P\{X = 0\}$ but with $P\{X = J\}$ where

$$J = \begin{cases} i & \text{if } i \le (n+1)p \\ [(n+1)p] & \text{if } i > (n+1)p \end{cases}$$

where $[x]$—called Int $(x)$ in the program—is the largest integer less than or equal to $x$. (Of all the probabilities $P\{X = k\}$, $k = 0, 1, \ldots, i$ that need be computed, $P\{X = J\}$ will be the largest). The program then recursively computes $P\{X = J - 1\}, P\{X = J - 2\}, \ldots, P\{X = 0\}$. In addition if $J < i$, it also computes $P\{X = J + 1\}, \ldots, P\{X = i\}$.

The computation of

$$P\{X = J\} = \binom{n}{J} p^J(1 - p)^{n-J}$$
$$= \frac{n(n - 1) \cdots (n - J + 1)}{J(J - 1) \cdots 1} p^J(1 - p)^{n-J}$$

is accomplished by first taking logarithms to compute

$$\log P\{X = J\} = \sum_{k=1}^{J} \log (n + 1 - k)$$
$$- \sum_{k=1}^{J} \log (k) + J \log p + (n - J) \log (1 - p)$$

and then taking

$$P\{X = J\} = \exp \{\log P\{X = J\}\}$$

**Example 4i.**

(i) Determine $P\{X \leq 145\}$ when $X$ is binomial with parameters (250, .5).

(ii) Determine $P\{X \leq 90\}$ when $X$ is binomial with parameters (1000, .1).

***Solution:*** Run the binomial distribution program:

```
RUN
THE DISTRIBUTION FUNCTION OF A BINOMIAL(n,p) RANDOM VARIABLE
ENTER n
? 250
ENTER p
? .5
ENTER i
? 145
THE PROBABILITY IS .995255
Ok

RUN
THE DISTRIBUTION FUNCTION OF A BINOMIAL(n,p) RANDOM VARIABLE
ENTER n
? 1000
ENTER p
? .1
ENTER i
? 90
THE PROBABILITY IS .1582189
Ok
```

∎

## 5 The Poisson Random Variable

A random variable $X$, taking on one of the values $0, 1, 2, \ldots$, is said to be a *Poisson* random variable with parameter $\lambda$ if for some $\lambda > 0$,

$$p(i) = P\{X = i\} = e^{-\lambda}\frac{\lambda^i}{i!} \qquad i = 0, 1, 2, \ldots \tag{5.1}$$

Equation (5.1) defines a probability mass function, since

$$\sum_{i=0}^{\infty} p(i) = e^{-\lambda} \sum_{i=0}^{\infty} \frac{\lambda^i}{i!} = e^{-\lambda}e^{\lambda} = 1$$

The Poisson probability distribution was introduced by S. D. Poisson in a book he wrote dealing with the application of probability theory to lawsuits, criminal trials, and the like. This book, published in 1837, was entitled *Recherches sur la probabilité des jugements en matière criminelle et en matière civile.*

The Poisson random variable has a tremendous range of applications in diverse areas because it may be used as an approximation for a binomial random variable with parameters $(n, p)$ when $n$ is large and $p$ is small enough so that $np$ is of a moderate size. To see this, suppose that $X$ is a binomial random variable with parameters $(n, p)$ and let $\lambda = np$. Then

$$\begin{aligned} P\{X = i\} &= \frac{n!}{(n-i)!\,i!}p^i(1-p)^{n-i} \\ &= \frac{n!}{(n-i)!\,i!}\left(\frac{\lambda}{n}\right)^i\left(1-\frac{\lambda}{n}\right)^{n-i} \\ &= \frac{n(n-1)\cdots(n-i+1)}{n^i}\frac{\lambda^i}{i!}\frac{(1-\lambda/n)^n}{(1-\lambda/n)^i} \end{aligned}$$

Now, for $n$ large and $\lambda$ moderate,

$$\left(1-\frac{\lambda}{n}\right)^n \approx e^{-\lambda} \qquad \frac{n(n-1)\cdots(n-i+1)}{n^i} \approx 1 \qquad \left(1-\frac{\lambda}{n}\right)^i \approx 1$$

Hence, for $n$ large and $\lambda$ moderate,

$$P\{X = i\} \approx e^{-\lambda}\frac{\lambda^i}{i!}$$

In other words, if $n$ independent trials, each of which results in a "success" with probability $p$, are performed, then, when $n$ is large and $p$ small enough to make $np$ moderate, the number of successes occurring is approximately a Poisson random variable with parameter $\lambda = np$. This value $\lambda$ (which will later be shown to equal the expected number of successes) will usually be determined empirically.

Some examples of random variables that usually obey the Poisson probability law [that is, they obey Equation (5.1)] follow:

1. The number of misprints on a page (or a group of pages) of a book.
2. The number of people in a community living to 100 years of age.
3. The number of wrong telephone numbers that are dialed in a day.
4. The number of packages of dog biscuits sold in a particular store each day.
5. The number of customers entering a post office on a given day.
6. The number of vacancies occurring during a year in the Supreme Court.
7. The number of $\alpha$-particles discharged in a fixed period of time from some radioactive material.

Each of the above, and numerous other random variables, are approximately Poisson for the same reason—namely, because of the Poisson approximation to the binomial. For instance, we can suppose that there is a probability $p$ that each letter typed on a page will be misprinted. Hence the number of misprints on a page will be approximately Poisson with $\lambda = np$, where $n$ is the number of letters on a page. Similarly, we can suppose that each person in a community has some probability of reaching age 100. Also, each person entering a store may be thought of as having some probability of buying a package of dog biscuits, and so on.

**Example 5a.** Suppose that the number of typographical errors on a single page of this book has a Poisson distribution with parameter $\lambda = \frac{1}{2}$. Calculate the probability that there is at least one error on this page.

***Solution:*** Letting $X$ denote the number of errors on this page, we have

$$P\{X \geq 1\} = 1 - P\{X = 0\} = 1 - e^{-1/2} \approx .393$$ ∎

**Example 5b.** Suppose that the probability that an item produced by a certain machine will be defective is .1. Find the probability that a sample of 10 items will contain at most 1 defective item.

***Solution:*** The desired probability is $\binom{10}{0}(.1)^0(.9)^{10} + \binom{10}{1}(.1)^1(.9)^9 =$ .7361, whereas the Poisson approximation yields the value $e^{-1} + e^{-1} \approx$ .7358. ∎

**Example 5c.** Consider an experiment that consists of counting the number of $\alpha$-particles given off in a 1-second interval by 1 gram of radioactive material. If we know from past experience that, on the average, 3.2 such $\alpha$-particles are given off, what is a good approximation to the probability that no more than 2 $\alpha$-particles will appear?

***Solution:*** If we think of the gram of radioactive material as consisting of a large number $n$ of atoms, each of which has probability $3.2/n$ of disintegrating and sending off an $\alpha$-particle during the second considered, then we see that, to a very close approximation, the number of $\alpha$-particles given off will be a Poisson random variable with parameter $\lambda = 3.2$. Hence the desired probability is

$$P\{X \leq 2\} = e^{-3.2} + 3.2e^{-3.2} + \frac{(3.2)^2}{2}e^{-3.2}$$

$$\approx .382$$ ∎

We have shown that the Poisson with parameter $np$ is a very good approximation to the distribution of the number of successes in $n$ independent trials when each trial has probability $p$ of being a success, provided that $n$ is large and $p$ small. In fact, it remains a good approximation even when the trials are not independent, provided that their dependence is "weak." For instance, recall the matching problem (Example 5i of Chapter 2) where $n$ men randomly select hats from a set consisting of one hat from each person. From the point of view of the number of men that select their own hat, we may regard the random selection as the result of $n$ trials where we say that trial $i$ is a success if person $i$ selects his own hat, $i = 1, \ldots, n$. Defining the events $E_i$, $i = 1, \ldots, n$, by

$$E_i = \{\text{trial } i \text{ is a success}\}$$

then it is easy to see that

$$P\{E_i\} = 1/n \qquad \text{and} \qquad P\{E_i \mid E_j\} = 1/(n-1), j \neq i$$

Thus, we see that while the events $E_i$, $i = 1, \ldots, n$ are not independent, their dependence, for large $n$, appears to be weak. Based on this, it seems reasonable to expect that the number of successes will approximately have a Poisson distribution with parameter $n \times 1/n = 1$, and indeed this is verified in Example 5i of Chapter 2.

For a second illustration of the strength of the Poisson approximation when the trials are weakly dependent, let us reconsider the birthday problem presented in Example 5g of Chapter 2. In this example we suppose that each of $n$ people is equally likely to have any of the 365 days of the year as their birthday, and the problem is to determine the probability that a set of $n$ independent people all have different birthdays. A combinatorial argument was used to determine this probability and it was then computed that when $n = 23$ this probability was less than $\frac{1}{2}$.

We can approximate the above probability by using the Poisson approximation as follows. Imagine that we have a trial for each of the $\binom{n}{2}$ pairs of individuals $i$ and $j$, $i \neq j$, and say that trial $i-j$ is a success if persons $i$ and $j$ have the same birthday. If we let $E_{ij}$ denote the event that trial $i-j$ is a success, then whereas the $\binom{n}{2}$ events $E_{ij}$, $1 \leq i < j \leq n$ are not independent (see Theoretical Exercise 14) their dependence appears to be rather weak. (Indeed, these events are even "pairwise independent" in that any 2 of the events $E_{ij}$ and $E_{kl}$ are independent—again see Exercise 14). As $P(E_{ij}) = 1/365$ it is thus reasonable to suppose that the number of successes should approximately have a Poisson distribution with parameter $\binom{n}{2}/365 = n(n-1)/730$. Therefore,

$$\begin{aligned} P\{\text{no 2 people have the same birthday}\} &= P\{0 \text{ successes}\} \\ &\approx \exp\{-n(n-1)/730\} \end{aligned}$$

To determine the smallest integer $n$ for which this probability is less than $\frac{1}{2}$ note that

$$\exp\{-n(n-1)/730\} \leq \tfrac{1}{2}$$

is equivalent to

$$\exp\{n(n-1)/730\} \geq 2$$

or, taking logarithms of both sides

$$\begin{aligned} n(n-1) &\geq 730 \log 2 \\ &\approx 505.997 \end{aligned}$$

which yields the solution $n = 23$, in agreement with the result of Example 5i.

Suppose now that we wanted the probability that among the $n$ people no 3 of them have their birthday on the same day. Whereas this now becomes a difficult combinatorial problem, it is a simple matter to obtain a good approximation. To begin, imagine that we have a trial for each of the $\binom{n}{3}$ triplets $i, j, k$ where $1 \leq i < j < k \leq n$, and call the $i-j-k$ trial a success if persons $i$, $j$, and $k$ all have their birthday on the same day. As above, we can then conclude that the number of successes is approximately a Poisson random variable with parameter

$$\begin{aligned} \binom{n}{3} P\{i, j, k \text{ have the same birthday}\} &= \binom{n}{3}\left(\frac{1}{365}\right)^2 \\ &= \frac{n(n-1)(n-2)}{6 \times (365)^2} \end{aligned}$$

Hence,

$$P\{\text{no 3 have the same birthday}\} \approx \exp\{-n(n-1)(n-2)/799350)\}$$

This probability will be less than $\frac{1}{2}$ when $n$ is such that

$$n(n-1)(n-2) \geq 799350\log 2 \approx 554067.1$$

which is equivalent to $n \geq 84$. Thus, the probability that at least 3 people in a group of size 84 or larger will have the same birthday exceeds $\frac{1}{2}$. ∎

Another use of the Poisson probability distribution arises in situations where "events" occur at certain points in time. One example of this is when an event is the occurrence of an earthquake; another possibility would be for an event to correspond to an individual entering a particular establishment (bank, post office, gas station, and so on); and a third possibility is for an event to occur whenever a war starts. Let us suppose that events are indeed occurring at certain (random) points of time, and let us assume that for some positive constant $\lambda$ the following assumptions hold true:

1. The probability that exactly 1 event occurs in a given interval of length $h$ is equal to $\lambda h + o(h)$, where $o(h)$ stands for any function $f(h)$ that is such that $\lim_{h\to 0} f(h)/h = 0$. [For instance, $f(h) = h^2$ is $o(h)$, whereas $f(h) = h$ is not.]
2. The probability that 2 or more events occur in an interval of length $h$ is equal to $o(h)$.
3. For any integers $n, j_1, j_2, \ldots, j_n$, and any set of $n$ nonoverlapping intervals, if we define $E_i$ to be the event that exactly $j_i$ of the events under consideration occur in the $i$th of these intervals, then events $E_1, E_2, \ldots, E_n$ are independent.

Loosely put, Assumptions 1 and 2 state that for small values of $h$, the probability of exactly 1 event occurring in an interval of size $h$ equals $\lambda h$ plus something that is small compared to $h$, whereas the probability of 2 or more events occurring is small compared to $h$. Assumption 3 states that whatever occurs in one interval has no (probability) effect on what will occur in other nonoverlapping intervals.

Under Assumptions 1, 2, and 3, we shall now show that the number of events occurring in any interval of length $t$ is a Poisson random variable with parameter $\lambda t$. To be precise, let us call the interval $[0, t]$ and denote by $N(t)$ the number of events occurring in that interval. To obtain an expression for $P\{N(t) = k\}$, we start by breaking the interval $[0, t]$ into $n$ nonoverlapping subintervals each of length $t/n$ (Figure 4.6).

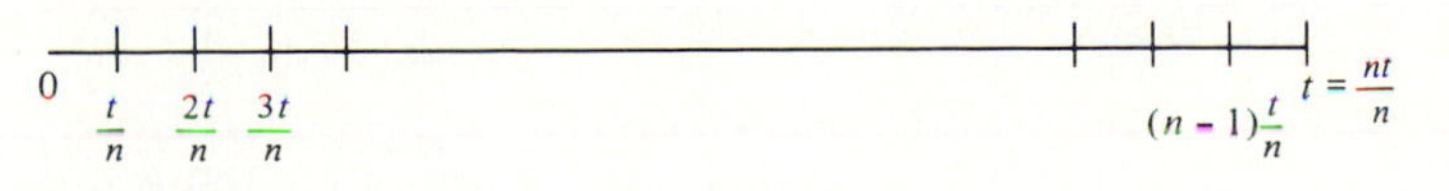


**Figure 4.6**

Now,

$$P\{N(t)=k\} = P\{k \text{ of the } n \text{ subintervals contain exactly 1 event and the other } n-k \text{ contain 0 events}\} + P\{N(t)=k \text{ and at least 1 subinterval contains 2 or more events}\} \tag{5.2}$$

This follows because the event on the left side of Equation (5.2), that is, $\{N(t) = k\}$, is clearly equal to the union of the two mutually exclusive events on the right side of the equation. Letting $A$ and $B$ denote the two mutually exclusive events on the right side of Equation (5.2), we have

$$\begin{aligned}
P(B) &\le P\{\text{at least one subinterval contains 2 or more events}\} \\
&= P\Big(\bigcup_{i=1}^{n}\{i\text{th subinterval contains 2 or more events}\}\Big) \\
&\le \sum_{i=1}^{n} P\{i\text{th subinterval contains 2 or more events}\} && \text{by Boole's inequality} \\
&= \sum_{i=1}^{n} o\left(\frac{t}{n}\right) && \text{by Assumption 2} \\
&= no\left(\frac{t}{n}\right) \\
&= t\left[\frac{o(t/n)}{t/n}\right]
\end{aligned}$$

Now, for any $t$, $t/n \to 0$ as $n \to \infty$ and so $o(t/n)/(t/n) \to 0$ as $n \to \infty$ by the definition of $o(h)$. Hence

$$P(B) \to 0 \text{ as } n \to \infty \tag{5.3}$$

On the other hand, since Assumptions 1 and 2 imply that

$$P\{0 \text{ events occur in an interval of length } h\} = 1 - [\lambda h + o(h) + o(h)] = 1 - \lambda h - o(h)$$[1]

we see from the independence assumption, number 3, that

$$\begin{aligned}
P(A) &= P\{k \text{ of the subintervals contain exactly 1 event and the other } n-k \text{ contain 0 events}\} \\
&= \binom{n}{k}\left[\frac{\lambda t}{n} + o\left(\frac{t}{n}\right)\right]^k \left[1 - \left(\frac{\lambda t}{n}\right) - o\left(\frac{t}{n}\right)\right]^{n-k}
\end{aligned}$$

[1] The sum of two functions both of which are $o(h)$ is also $o(h)$. This is so because if $\lim_{h\to 0} f(h)/h = \lim_{h\to 0} g(h)/h = 0$, then $\lim_{h\to 0} [f(h) + g(h)]/h = 0$.

However, since

$$n\left[\frac{\lambda t}{n} + o\left(\frac{t}{n}\right)\right] = \lambda t + t\left[\frac{o(t/n)}{t/n}\right] \to \lambda t \text{ as } n \to \infty$$

it follows, by the same argument that verified the Poisson approximation to the binomial, that

$$P(A) \to e^{-\lambda t}\frac{(\lambda t)^k}{k!} \qquad \text{as } n \to \infty \tag{5.4}$$

Thus, from Equations (5.2), (5.3), and (5.4), we obtain, by letting $n \to \infty$,

$$P\{N(t) = k\} = e^{-\lambda t}\frac{(\lambda t)^k}{k!} \qquad k = 0, 1, \ldots \tag{5.5}$$

Hence, if Assumptions 1, 2, and 3 are satisfied, the number of events occurring in any fixed interval of length $t$ is a Poisson random variable with parameter $\lambda t$; and we say that the events occur in accordance with a Poisson process having rate $\lambda$. The value $\lambda$, which can be shown to equal the rate per unit time at which events occur, is a constant that must be empirically determined.

The preceding discussion explains why a Poisson random variable is usually a good approximation for such diverse phenomena as the following:

1. The number of earthquakes occurring during some fixed time span.
2. The number of wars per year.
3. The number of electrons emitted from a heated cathode during a fixed time period.
4. The number of deaths in a given period of time of the policyholders of a life insurance company.

**Example 5d.** Suppose that earthquakes occur in the western portion of the United States in accordance with Assumptions 1, 2, and 3 with $\lambda = 2$ and with 1 week as the unit of time. (That is, earthquakes occur in accordance with the three assumptions at a rate of 2 per week.) (1) Find the probability that at least 3 earthquakes occur during the next 2 weeks. (2) Find the probability distribution of the time, starting from now, until the next earthquake.

***Solution***

1. From Equation (5.5) we have

$$\begin{aligned} P\{N(2) \geq 3\} &= 1 - P\{N(2) = 0\} - P\{N(2) = 1\} - P\{N(2) = 2\} \\ &= 1 - e^{-4} - 4e^{-4} - \frac{4^2}{2}e^{-4} \\ &= 1 - 13e^{-4} \end{aligned}$$

2. Let $X$ denote the amount of time (in weeks) until the next earthquake. Because $X$ will be greater than $t$ if and only if no events occur within the next $t$ units of time, we have from Equation (5.5) that

$$P\{X > t\} = P\{N(t) = 0\} = e^{-\lambda t}$$

and so, the probability distribution function $F$ of the random variable $X$ is given by

$$F(t) = P\{X \leq t\} = 1 - P\{X > t\} = 1 - e^{-\lambda t}$$
$$= 1 - e^{-2t}$$

■

## 5.1 *Computing the Poisson Distribution Function*

If $X$ is Poisson with parameter $\lambda$, then

$$\frac{P\{X = i + 1\}}{P\{X = i\}} = \frac{e^{-\lambda}\lambda^{i+1}/(i + 1)!}{e^{-\lambda}\lambda^{i}/i!} = \frac{\lambda}{i + 1}. \tag{5.6}$$

Starting with $P\{X = 0\} = e^{-\lambda}$, we can use (5.6) to successively compute

$$P\{X = 1\} = \lambda P\{X = 0\}$$

$$P\{X = 2\} = \frac{\lambda}{2} P\{X = 1\}$$

$$\vdots$$

$$P\{X = i + 1\} = \frac{\lambda}{i + 1} P\{X = i\}.$$

A Basic program, which uses Equation 5.6 to compute the Poisson distribution function is presented at the end of this chapter. The program starts the computation with $P\{X = 0\} = e^{-\lambda}$. However, if, as will be the case when $\lambda$ is large, the computer returns the value 0 for $e^{-\lambda}$, the program calls for beginning with $P\{X = J\}$ where

$$J = \begin{cases} \text{i} & \text{if} \quad i \leq \lambda \\ \text{Int}(\lambda) & \text{if} \quad i > \lambda \end{cases}$$

where $\text{Int}(\lambda)$ is the largest integer less than or equal to $\lambda$. The reason for this choice is that of all the values $P\{X = k\}$, $k = 0, 1, \ldots, i$ that we must compute the largest one is $P\{X = J\}$ (see Theoretical Exercise 11). The program computes $P\{X = J\}$ by first computing

$$\log P\{X = J\} = -\lambda + J \log(\lambda) - \sum_{k=1}^{J} \log k$$

and then takes $P\{X = J\} = \exp\{\log P\{X = J\}\}$.

Once $P\{X = J\}$ has been computed, the program uses (5.6) to recursively compute $P\{X = J - 1\}, P\{X = J - 2\}, \ldots, P\{X = 0\}$. If $J < i$ it again uses (5.6) to recursively compute $P\{X = J + 1\}, \ldots, P\{X = i\}$. Summing all of these quantities yields $P\{X \le i\}$.

**Example 5e.**
(i) Determine $P\{X \le 100\}$ when $X$ is Poisson with mean 90.
(ii) $P\{X \le 1075\}$ when $X$ is Poisson with mean 1000.

***Solution:*** Run the Poisson distribution program:

```
RUN
THIS PROGRAM COMPUTES THE PROBABILITY THAT A POISSON RANDOM VARIABLE
IS LESS THAN OR EQUAL TO i
ENTER THE MEAN OF THE RANDOM VARIABLE
? 100
ENTER THE DESIRED VALUE OF i
? 90
THE PROBABILITY THAT A POISSON RANDOM VARIABLE WITH MEAN  100
IS LESS THAN OR EQUAL TO 90 IS .1713914
Ok
RUN
THIS PROGRAM COMPUTES THE PROBABILITY THAT A POISSON RANDOM VARIABLE
IS LESS THAN OR EQUAL TO i
ENTER THE MEAN OF THE RANDOM VARIABLE
? 1000
ENTER THE DESIRED VALUE OF i
? 1075
THE PROBABILITY THAT A POISSON RANDOM VARIABLE WITH MEAN  1000
IS LESS THAN OR EQUAL TO 1075 IS .989354
Ok
```

## 6 Other Discrete Probability Distributions

### 6.1 *The Geometric Random Variable*

Suppose that independent trials, each having a probability $p$, $0 < p < 1$, of being a success, are performed until a success occurs. If we let $X$ equal the number of trials required, then

$$P\{X = n\} = (1 - p)^{n-1}p \qquad n = 1, 2, \ldots \tag{6.1}$$

Equation (6.1) follows because, in order for $X$ to equal $n$, it is necessary and sufficient that the first $n - 1$ trials are failures and the $n$th trial is a success. Equation (6.1) then follows, since the outcomes of the successive trials are assumed to be independent.

Since

$$\sum_{n=1}^{\infty} P\{X = n\} = p \sum_{n=1}^{\infty} (1 - p)^{n-1} = \frac{p}{1 - (1 - p)} = 1$$

it follows that, with probability 1, a success will eventually occur. Any random variable $X$ whose probability mass function is given by Equation (6.1) is said to be a *geometric* (or a Pascal) random variable with parameter $p$.

**Example 6a.** An urn contains $N$ white and $M$ black balls. Balls are randomly selected, one at a time, until a black one is obtained. If we assume that each selected ball is replaced before the next one is drawn, what is the probability that (1) exactly $n$ draws are needed, and (2) at least $k$ draws are needed?

***Solution:*** If we let $X$ denote the number of draws needed to select a black ball, then $X$ satisfies Equation (6.1) with $p = M/(M + N)$. Hence

1. $$P\{X = n\} = \left(\frac{N}{M+N}\right)^{n-1} \frac{M}{M+N} = \frac{MN^{n-1}}{(M+N)^n}$$

2. $$\begin{aligned} P\{X \geq k\} &= \frac{M}{M+N} \sum_{n=k}^{\infty} \left(\frac{N}{M+N}\right)^{n-1} \\ &= \left(\frac{M}{M+N}\right)\left(\frac{N}{M+N}\right)^{k-1} \bigg/ \left[1 - \frac{N}{M+N}\right] \\ &= \left(\frac{N}{M+N}\right)^{k-1} \end{aligned}$$

Of course, (2) could have been obtained directly, since the probability that at least $k$ trials are necessary to obtain a success is equal to the probability that the first $k - 1$ trials are all failures. That is, for a geometric random variable

$$P\{X \geq k\} = (1 - p)^{k-1}$$ ∎

## 6.2 *The Negative Binomial Random Variable*

Suppose that independent trials, each having probability $p$, $0 < p < 1$, of being a success are performed until a total of $r$ successes is accumulated. If we let $X$ equal the number of trials required, then

$$P\{X = n\} = \binom{n-1}{r-1} p^r (1-p)^{n-r} \qquad n = r, r+1, \ldots \tag{6.2}$$

Equation (6.2) follows because, in order for the $r$th success to occur at the $n$th trial, there must be $r - 1$ successes in the first $n - 1$ trials, and the $n$th trial must be a success. The probability of the first event is

$$\binom{n-1}{r-1} p^{r-1} (1-p)^{n-r}$$

and the probability of the second is $p$; thus, by independence, Equation (6.2) is established. To verify that a total of $r$ successes must eventually be accumulated, we can either analytically prove that

$$\sum_{n=r}^{\infty} P\{X = n\} = \sum_{n=r}^{\infty} \binom{n-1}{r-1} p^r (1-p)^{n-r} = 1 \tag{6.3}$$

or we can give a probabilistic argument as follows: The number of trials required to obtain $r$ successes can be expressed as $Y_1 + Y_2 + \cdots + Y_r$, where $Y_1$ equals the number of trials required for the first success, $Y_2$ the number of additional trials after the first success until the second success occurs, $Y_3$ the number of additional trials until the third success, and so on. As the trials are independent and all have the same probability of a success, it follows that $Y_1, Y_2, \ldots, Y_r$ are all geometric random variables. Hence each is finite with probability 1, and so $\sum_{i=1}^{r} Y_i$ must also be finite, establishing Equation (6.3).

Any random variable $X$ whose probability mass function is given by Equation (6.2) is said to be a *negative binomial* random variable with parameters $(r, p)$. Note that a geometric random variable is just a negative binomial with parameter $(1, p)$.

In the next example we use the negative binomial to obtain another solution of the problem of the points.

**Example 6b.** If independent trials, each resulting in a success with probability $p$, are performed, what is the probability of $r$ successes occurring before $m$ failures?

***Solution:*** The solution will be arrived at by noting that $r$ successes will occur before $m$ failures if and only if the $r$th success occurs no later than the $r + m - 1$ trial. This follows because if the $r$th success occurs before or at the $r + m - 1$ trial, then it must have occurred before the $m$th failure, and conversely. Hence, from Equation (6.2), the desired probability is

$$\sum_{n=r}^{r+m-1} \binom{n-1}{r-1} p^r (1-p)^{n-r} \qquad \blacksquare$$

**Example 6c.** The Banach Match Problem. A pipe-smoking mathematician carries, at all times, 2 matchboxes, 1 in his left-hand pocket and 1 in his right-hand pocket. Each time he needs a match he is equally likely to take it from either pocket. Consider the moment when the mathematician first discovers that one of his matchboxes is empty. If it is assumed that both matchboxes initially contained $N$ matches, what is the probability that there are exactly $k$ matches in the other box, $k = 0, 1, \ldots, N$?

***Solution:*** Let $E$ denote the event that the mathematician first discovers that the right-hand matchbox is empty and there are $k$ matches in the left-hand box at the time. Now, this event will occur if and only if the $(N + 1)$th choice of the right-hand matchbox is made at the $N+1+N-k$ trial. Hence, from Equation (6.2) (with $p = \frac{1}{2}$, $r = N + 1$, $n = 2N - k + 1$), we see

$$P(E) = \binom{2N-k}{N}\left(\frac{1}{2}\right)^{2N-k+1}$$

As there is an equal probability that it is the left-hand box that is first discovered to be empty and there are $k$ matches in the right-hand box at that time, the desired result is

$$2P(E) = \binom{2N-k}{N}\left(\frac{1}{2}\right)^{2N-k}$$ ∎

## 6.3 *The Hypergeometric Random Variable*

Suppose that a sample of size $n$ is to be randomly chosen (without replacements) from an urn containing $N$ balls, of which $Np$ are white and $N - Np$ are black. If we let $X$ denote the number of white balls selected, then

$$P\{X = k\} = \frac{\binom{Np}{k}\binom{N-Np}{n-k}}{\binom{N}{n}} \qquad k = 0, 1, \ldots, \min(n, Np) \qquad (6.4)$$

Any random variable $X$, whose probability mass function is given by Equation (6.4) for some values on $n$, $N$, and $p$, is said to be a *hypergeometric* random variable.

**Example 6d.** An unknown number, say $N$, of animals inhabit a certain region. To obtain some information about the population size, ecologists often perform the following experiment: They first catch a number, say $r$, of these animals, mark them in some manner, and release them. After allowing the marked animals time to disperse throughout the region, a new catch of size, say $n$, is made. Let $X$ denote the number of marked animals in this second capture. If we assume that the population of animals in the region remained fixed between the time of the two catches and that each time an animal was caught it was equally likely to be any of the remaining uncaught animals, it follows that $X$ is a hypergeometric random variable such that

$$P\{X = i\} = \frac{\binom{r}{i}\binom{N-r}{n-i}}{\binom{N}{n}} \equiv P_i(N)$$

Suppose now that $X$ is observed to equal $i$. Then, as $P_i(N)$ represents the probability of the observed event when there are actually $N$ animals present in the region, it would appear that a reasonable estimate of $N$ would be the value of $N$ that maximizes $P_i(N)$. Such an estimate is called a *maximum-likelihood* estimate. (See Theoretical Exercises 8 and 12 for other examples of this type of estimation procedure.)

The maximization of $P_i(N)$ can most simply be done by first noting that

$$\frac{P_i(N)}{P_i(N-1)} = \frac{(N-r)(N-n)}{N(N-r-n+i)}$$

Now, the above ratio is greater than 1 if and only if

$$(N-r)(N-n) \geq N(N-r-n+i)$$

or, equivalently, if and only if

$$N \leq \frac{rn}{i}$$

Thus $P_i(N)$ is first increasing, and then decreasing, and reaches its maximum value at the largest integral value not exceeding $rn/i$. This value is thus the maximum likelihood estimate of $N$. For example, suppose that the initial catch consisted of $r = 50$ animals, which are marked and then released. If a subsequent catch consists of $n = 40$ animals of which $i = 4$ are marked, then we would estimate that there are some 500 animals in the region. (It should be noted that the above estimate could also have been obtained by assuming that the proportion of marked animals in the region, $r/N$, is approximately equal to the proportion of marked animals in our second catch, $i/n$.) ∎

**Example 6e.** A purchaser of electrical components buys them in lots of size 10. It is his policy to inspect 3 components randomly from a lot and to accept the lot only if all 3 are nondefective. If 30 percent of the lots have 4 defective components and 70 percent have only 1, what proportion of lots does the purchaser reject?

***Solution:*** Let $A$ denote the event that the purchaser accepts a lot. Now,

$$P(A) = P(A|\text{lot has 4 defectives})\frac{3}{10} + P(A|\text{lot has 1 defective})\frac{7}{10}$$

$$= \frac{\binom{4}{0}\binom{6}{3}}{\binom{10}{3}}\left(\frac{3}{10}\right) + \frac{\binom{1}{0}\binom{9}{3}}{\binom{10}{3}}\left(\frac{7}{10}\right)$$

$$= \frac{54}{100}$$

Hence 46 percent of the lots are rejected. ∎

### 6.4 *The Zeta (or Zipf) Distribution*

A random variable is said to have a zeta (sometimes called the Zipf) distribution if its probability mass function is given by

$$P\{X = k\} = \frac{C}{k^{\alpha+1}}, \qquad k = 1, 2, \ldots$$

for some value of $\alpha > 0$. Since the sum of the above probabilities must equal 1, it follows that

$$C = \left[\sum_{k=1}^{\infty}\left(\frac{1}{k}\right)^{\alpha+1}\right]^{-1}$$

The zeta distribution owes its name to the fact that the function

$$\zeta(s) = 1 + \left(\frac{1}{2}\right)^s + \left(\frac{1}{3}\right)^s + \cdots + \left(\frac{1}{k}\right)^s + \cdots$$

is known in mathematical disciplines as the Riemann zeta function (after the German mathematician G. F. B. Riemann).

The zeta distribution was used by the Italian economist Pareto to describe the distribution of family incomes in a given country. However, it was G. K. Zipf who applied these distributions in a wide variety of different areas and, in doing so, popularized their use.

## *Theoretical Exercises*

**1.** There are $N$ distinct types of coupons, and each time one is obtained it will, independently of past choices, be of type $i$ with probability $P_i$, $i = 1, \ldots, N$. Let $T$ denote the number one need select to obtain at least one of each type. Compute $P\{T = n\}$.

**2.** Prove Property 3 of a distribution function.

**3.** Express $P\{X \geq a\}$ in terms of the distribution function of $X$.

**4.** Prove or give a counterexample:

$$P\{X < b\} = \lim_{b_n \to b} P\{X < b_n\}$$

**5.** If $X$ has distribution function $F$, what is the distribution function of the random variable $\alpha X + \beta$, where $\alpha$ and $\beta$ are constants, $\alpha \neq 0$?

**6.** There are $n$ components lined up in a linear arrangement. Suppose that each component independently functions with probability $p$. What is the probability that no 2 neighboring components are both nonfunctional?

**7.** Consider $n$ independent sequential trials, each of which is successful with probability $p$. If there are a total of $k$ successes, show that each of the $n!/[k!(n-k)!]$ possible arrangements of the $k$ successes and $n-k$ failures is equally likely.

**8.** Let $X$ be a binomial random variable with parameters $(n, p)$. What value of $p$ maximizes $P\{X = k\}$, $k = 0, 1, \ldots, n$? This is an example of a statistical method used to estimate $p$ when a binomial $(n, p)$ random variable is observed to equal $k$. If we assume that $n$ is known, then we estimate $p$ by choosing that value of $p$ that maximizes $P\{X = k\}$. This is known as the method of maximum likelihood estimation.

**9.** A family has $n$ children with probability $\alpha p^n, n \geq 1$, where $\alpha \leq (1-p)/p$.
(a) What proportion of families has no children?
(b) If each child is equally likely to be a boy or a girl (independently of each other), what proportion of families consists of $k$ boys (and any number of girls)?

**10.** Suppose that $n$ independent tosses of a coin having probability $p$ of coming up heads are made. Show that the probability that an even number of heads results is $\frac{1}{2}[1 + (q-p)^n]$, where $q = 1 - p$. Do this by proving and then utilizing the identity

$$\sum_{i=0}^{[n/2]} \binom{n}{2i} p^{2i} q^{n-2i} = \frac{1}{2}[(p+q)^n + (q-p)^n]$$

when $[n/2]$ is the largest integer less than or equal to $n/2$. Compare this exercise with Theoretical Exercise 13 of Chapter 3.

**11.** Let $X$ be a Poisson random variable with parameter $\lambda$. Show that $P\{X = i\}$ increases monotonically and then decreases monotonically as

$i$ increases, reaching its maximum when $i$ is the largest integer not exceeding $\lambda$.

HINT: Consider $P\{X = i\}/P\{X = i - 1\}$.

**12.** Let $X$ be a Poisson random variable with parameter $\lambda$. Show that

$$P\{X \text{ is even}\} = \tfrac{1}{2}[1 + e^{-2\lambda}]$$

(a) by using the results of Problem 10 and the relationship between Poisson and Binomial random variables.
(b) Verify the above directly by making use of the expansion of $e^{-\lambda} + e^{\lambda}$.

**13.** Let $X$ be a Poisson random variable with parameter $\lambda$. What value of $\lambda$ maximizes $P\{X = k\}$, $k \geq 0$?

**14.** From a set of $n$ randomly chosen people let $E_{ij}$ denote the event that persons $i$ and $j$ have the same birthday. Assume that each person is equally likely to have any of the 365 days of the year as his birthday. Find
(i) $P(E_{3,4}|E_{1,2})$
(ii) $P(E_{1,3}|E_{1,2})$
(iii) $P(E_{2,3}|E_{1,2} \cap E_{1,3})$

What can you conclude from the above about the independence of the $\binom{n}{2}$ events $E_{ij}$?

**15.** An urn contains $2n$ balls, of which 2 are numbered 1, 2 are numbered 2, . . . , and 2 are numbered $n$. Balls are successively withdrawn 2 at a time without replacement. Let $T$ denote the first selection in which the balls withdrawn have the same number (and let it equal infinity if none of the pairs withdrawn have the same number). For $0 < \alpha < 1$, we want to show that

$$\lim_{n} P\{T > \alpha n\} = e^{-\alpha/2}$$

To verify the above, let $M_k$ denote the number of pairs withdrawn in the first $k$ selections, $k = 1, \ldots, n$.
(i) Argue that when $n$ is large, $M_k$ can be regarded as the number of successes in $k$ (approximately) independent trials.
(ii) When $n$ is large, approximate $P\{M_k = 0\}$.
(iii) Write the event $\{T > \alpha n\}$ in terms of the value of one of the variables $M_k$.
(iv) Verify the above limiting probability.

**16.** Suppose that the number of events that occur in a specified time is a Poisson random variable with parameter $\lambda$. If each event is counted with probability $p$, independently of every other event, show that the number of events that are counted is a Poisson random variable with parameter $\lambda p$. Also, give an intuitive argument as to why this should be so.

As an application of the preceding paragraph, suppose that the number of distinct uranium deposits in a given area is a Poisson random variable with parameter $\lambda = 10$. If, in a fixed period of time, each deposit is independently discovered with probability $\frac{1}{50}$, find the probability that (a) exactly 1, (b) at least 1, and (c) at most 1 deposit is discovered during that time.

**17.** Prove

$$\sum_{i=0}^{n} e^{-\lambda} \frac{\lambda^i}{i!} = \frac{1}{n!} \int_{\lambda}^{\infty} e^{-x} x^n \, dx$$

HINT: Use integration by parts.

**18.** If $X$ is a geometric random variable, show analytically that

$$P\{X = n + k \mid X > n\} = P\{X = k\}$$

Give a verbal argument using the interpretation of a geometric random variable as to why the above is true.

**19.** For a hypergeometric random variable with probability mass function given by (6.4) determine $P\{X = k + 1\}/P\{X = k\}$.

**20.** Balls numbered 1 through $N$ are in an urn. Suppose that $n$, $n \le N$, of them are randomly selected without replacement. Let $Y$ denote the largest number selected. Find the probability mass function of $Y$.

**21.** A jar contains $m + n$ chips, numbered $1, 2, \ldots, n + m$. A set of size $n$ is drawn. If we let $X$ denote the number of chips drawn having numbers that exceed all of the numbers of those remaining, compute the probability mass function of $X$.

**22.** A jar contains $n$ chips. Suppose that a boy successively draws a chip from the jar each time replacing the one drawn before drawing another. This continues until the boy draws a chip that he has previously drawn before. Let $X$ denote the number of draws, and compute its probability mass function.

**23.** Show that as $N \to \infty$

$$\frac{\binom{Np}{k}\binom{N-Np}{n-k}}{\binom{N}{n}} \to \binom{n}{k} p^k (1-p)^{n-k}$$

This illustrates the binomial approximation to the hypergeometric distribution. Further, present an intuitive explanation of why the equation is true.

## Problems

**1.** Two balls are randomly chosen from an urn containing 8 white, 4 black, and 2 orange balls. Suppose that we win \$2 for each black ball selected and we lost \$1 for each white ball selected. Let $X$ denote our winnings. What are the possible values of $X$, and what are the probabilities associated with each value?

**2.** Two fair dice are rolled. Let $X$ equal the product of the 2 dice. Compute $P\{X = i\}$ for $i = 1, 2, \ldots$.

**3.** Three dice are rolled. By assuming that each of the $6^3 = 216$ possible outcomes is equally likely, find the probabilities attached to the possible values that $X$ can take on, where $X$ is the sum of the 3 dice.

**4.** Five men and 5 women are ranked according to their scores on an examination. Assume that no two scores are alike and all 10! possible rankings are equally likely. Let $X$ denote the highest ranking achieved by a woman (for instance, $X = 2$ if the top ranked person was male and the next ranked person was female). Find $P\{X = i\}$, $i = 1, 2, 3, \ldots, 8, 9, 10$.

**5.** Let $X$ represent the difference between the number of heads and the number of tails obtained when a coin is tossed $n$ times. What are the possible values of $X$?

**6.** In Problem 5, if the coin is assumed fair, for $n = 3$, what are the probabilities associated with the values that $X$ can take on?

**7.** Suppose that a die is rolled twice. What are the possible values that the following random variables can take on:
(a) the maximum value to appear in the two rolls;
(b) the minimum value to appear on the two rolls;
(c) the sum of the two rolls;
(d) the value of the first roll minus the value of the second roll?

**8.** If the die in Problem 7 is assumed fair, calculate the probabilities associated with the random variables in parts (a) through (d).

**9.** Repeat Example 1b when the balls are selected with replacement.

**10.** In Example 1d compute the conditional probability that we win $i$ dollars, given that we win something; compute it for $i = 1, 2, 3$.

**11.** (a) An integer $N$ is to be selected at random from $\{1, 2, \ldots, (10)^3\}$ in the sense that each integer has the same probability of being selected. What is the probability that a selected number will be divisible by 3? by 5? by 7? by 15? by 105? How would your answer change if $(10)^3$ is replaced by $(10)^k$ as $k$ became larger and larger?

(b) An important function in number theory—one whose properties can be shown to be related to what is probably the most important unsolved problem of mathematics, the Riemann Hypothesis—is the Mobius function $\mu(n)$, defined for all positive integral values $n$ as follows: Factor $n$ into its prime factors. If there is a repeated prime factor, as in $12 = 2 \cdot 2 \cdot 3$ or $49 = 7 \cdot 7$, then $\mu(n)$ is defined to equal 0. On the other hand, if all the prime factors of $n$ are distinct, then set $\mu(n)$ equal to 1 if there is an odd number of prime factors and equal to $-1$ if there is an even number. For instance, $6 = 2 \cdot 3$, and so $\mu(6) = -1$; $30 = 2 \cdot 3 \cdot 5$, and so $\mu(30) = 1$. Now let $N$ be chosen at random from $\{1, 2, \ldots (10)^k\}$, where $k$ is large. Determine the distribution of $\mu(N)$ as $k \to \infty$.

HINT: To compute $P\{\mu(N) \neq 0\}$, use the identity

$$\prod_{i=1}^{\infty} \frac{P_i^2 - 1}{P_i^2} = \left(\frac{3}{4}\right)\left(\frac{8}{9}\right)\left(\frac{24}{25}\right)\left(\frac{48}{49}\right) \cdots = \frac{6}{\pi^2}$$

where $P_i$ is the $i$th smallest prime. (We do not include 1 as a prime.)

**12.** In the game of "two-finger Morra," 2 players show 1 or 2 fingers and simultaneously guess the number of fingers their opponent will show. If only one of the players guesses correctly, he wins an amount (in dollars) equal to the sum of the fingers shown by him and his opponent. If both players guess correctly or if neither guesses correctly, then no money is exchanged. Consider a specified player and denote by $X$ the amount of money he wins in a single game of two-finger Morra.

(a) If each player acts independently of the other, and if each player makes his choice of the number of fingers he will hold up and the number he will guess that his opponent will hold up in such a way that each of the 4 possibilities is equally likely, what are the possible values of $X$ and what are their associated probabilities?

(b) Suppose that each player acts independently of the other. If each player decides to hold up the same number of fingers that he guesses his opponent will hold up, and if each player is equally likely to hold up 1 or 2 fingers, what are the possible values of $X$ and their associated probabilities?

**13.** Suppose that the distribution function of $X$ is given by

$$F(b) = \begin{cases} 0 & b < 0 \\ \dfrac{b}{4} & 0 \le b < 1 \\ \dfrac{1}{2} + \dfrac{b-1}{4} & 1 \le b < 2 \\ \dfrac{11}{12} & 2 \le b < 3 \\ 1 & 3 \le b \end{cases}$$

Find $P\{X = i\}$, $i = 1, 2, 3$. Find $P\{\frac{1}{2} < X < \frac{3}{2}\}$.

**14.** Four independent flips of a fair coin are made. Let $X$ denote the number of heads obtained. Plot the probability mass function of the random variable $X - 2$.

**15.** If the distribution function of $X$ is given by

$$F(b) = \begin{cases} 0 & b < 0 \\ \frac{1}{2} & 0 \le b < 1 \\ \frac{3}{5} & 1 \le b < 2 \\ \frac{4}{5} & 2 \le b < 3 \\ \frac{9}{10} & 3 \le b < 3.5 \\ 1 & b \ge 3.5 \end{cases}$$

calculate the probability mass function of $X$.

**16.** A ball is drawn from an urn containing 3 white and 3 black balls. After the ball is drawn, it is then replaced and another ball is drawn. This goes on indefinitely. What is the probability that, of the first 4 balls drawn, exactly 2 are white?

**17.** On a multiple-choice exam with 3 possible answers for each of the 5 questions, what is the probability that a student would get 4 or more correct answers just by guessing?

**18.** A man claims to have extrasensory perception. As a test, a fair coin is flipped 10 times, and he is asked to predict the outcome in advance. Our man gets 7 out of 10 correct. What is the probability that he would have done at least this well if he had no ESP?

**19.** Suppose that airplane engines will fail, when in flight, with probability $1 - p$ independently from engine to engine. If an airplane needs a majority of its engines operative to make a successful flight, for what values of $p$ is a 5-engine plane preferable to a 3-engine one?

**20.** A communications channel transmits the digits 0 and 1. However, due to static, the digit transmitted is incorrectly received with probability .2. Suppose that we want to transmit an important message consisting of one binary digit. To reduce the chance of error, we transmit 00000 instead of 0 and 11111 instead of 1. If the receiver of the message uses "majority" decoding, what is the probability that the message will be wrong when decoded? What independence assumptions are you making?

**21.** A satellite system consists of $n$ components and functions on any given day if at least $k$ of the $n$ components function on that day. On a rainy day each of the components independently functions with probability $p_1$, whereas on a dry day they each independently function with probability $p_2$. If the probability of rain tomorrow is $\alpha$, what is the probability that the satellite system will function?

**22.** A student is getting ready to take an important oral examination and is concerned about her possibility of having an "on" day or an "off" day. She figures that if she has an "on" day, then each of her examiners will pass her, independently of each other, with probability .8, whereas, if she has an "off" day, this probability will be reduced to .4. Suppose that the student will pass the examination if a majority of the examiners pass her. If the student feels that she is twice as likely to have an "off" day as she is to have an "on" day, should she request an examination with 3 examiners or with 5 examiners?

**23.** Suppose that it takes at least 9 votes from a 12-member jury to convict a defendant. Suppose that the probability that a juror votes a guilty person innocent is .2, whereas the probability that the juror votes an innocent person guilty is .1. If each juror acts independently and if 65 percent of the defendants are guilty, find the probability that the jury renders a correct decision. What percentage of defendants is convicted?

**24.** In some military courts, 9 judges are appointed. However, both the prosecution and the defense attorneys are entitled to a peremptory challenge of any judge, in which case that judge is removed from the case and is not replaced. A defendant is declared guilty if the majority of judges cast votes of guilty, and he or she is declared innocent otherwise. Suppose that when the defendant is, in fact, guilty, each judge will (independently) vote guilty with probability .7, whereas, when the defendant is, in fact, innocent, this probability drops to .3.

(a) What is the probability that a guilty defendant is declared guilty when there are (i) 9, (ii) 8, and (iii) 7 judges?

(b) Repeat part (a) for an innocent defendant.

(c) If the prosecution attorney does not exercise the right to a peremptory challenge of a judge and if the defense is limited to at most two such challenges, how many challenges should the defense attorney make if he or she is 60 percent certain that the client is guilty?

**25.** It is known that diskettes produced by a certain company will be defective with probability .01, independently of each other. The company sells the diskettes in packages of size 10 and offers a money-back guarantee that at most 1 of the 10 diskettes in the package will be defective. If someone buys 3 packages, what is the probability that they will return exactly 1 of them?

**26.** Suppose that 10 percent of the chips produced by a computer hardware manufacturer are defective. If we order 100 such chips, will the number of defective ones we receive be a binomial random variable?

**27.** Suppose that the number of accidents occurring on a highway each day is a Poisson random variable with parameter $\lambda = 3$.
(a) Find the probability that 3 or more accidents occur today.
(b) Repeat part (a) under the assumption that at least 1 accident occurs today.

**28.** Compare the Poisson approximation with the correct binomial probability for the following cases:
(a) $P\{X = 2\}$ when $n = 8, p = .1$;
(b) $P\{X = 9\}$ when $n = 10, p = .95$;
(c) $P\{X = 0\}$ when $n = 10, p = .1$;
(d) $P\{X = 4\}$ when $n = 9, p = .2$.

**29.** If you buy a lottery ticket in 50 lotteries, in each of which your chance of winning a prize is $\frac{1}{100}$, what is the (approximate) probability that you will win a prize (a) at least once, (b) exactly once, and (c) at least twice?

**30.** The number of times that an individual contracts a cold in a given year is a Poisson random variable with parameter $\lambda = 5$. Suppose a new wonder drug (based on large quantities of vitamin C) has just been marketed that reduces the Poisson parameter to $\lambda = 3$ for 75 percent of the population. For the other 25 percent of the population the drug has no appreciable effect on colds. If an individual tries the drug for a year and has 2 colds in that time, how likely is it that the drug is beneficial for him or her?

**31.** The probability of being dealt a full house in a hand of poker is approximately .0014. Find an approximation for the probability that in 1000 hands of poker you will be dealt at least 2 full houses.

**32.** If $n$ married couples are seated at random at a round table, approximately what is the probability that no wife sits next to her husband?

When $n = 10$ compare your approximation with the exact answer given in Example 5j of Chapter 2.

**33.** People enter a gambling casino at a rate of 1 for every 2 minutes.
(a) What is the probability that no one enters between 12:00 and 12:05?
(b) What is the probability that at least 4 people enter the casino during that time?

**34.** The suicide rate in a certain state is 1 suicide per 100,000 inhabitants per month.
(a) Find the probability that in a city of 400,000 inhabitants within this state, there will be 8 or more suicides in a given month.
(b) What is the probability that there will be at least 2 months during the year that will have 8 or more suicides?
(c) Counting the present month as month number 1, what is the probability that the first month to have 8 or more suicides will be month number $i$, $i \geq 1$?
What assumptions are you making?

**35.** It is known that 10 percent of all patients who exhibit a specified set of symptoms have a certain disease, which is of epidemic proportions in a certain town. Final diagnosis of this disease depends on a blood test. However, as individual blood tests are quite expensive, the hematologist waits until $n$ patients exhibiting the symptoms visit him. He then combines the blood of the $n$ patients and runs a test on it. If none of the $n$ persons has the disease, then the composite blood test is negative. However, if at least one of the patients has the disease, then the blood test result will be positive, and the doctor will be forced to take individual blood tests to determine which of the $n$ patients has the disease.
(a) What is the probability that the composite blood test will be negative if (i) $n = 2$, (ii) $n = 4$, (iii) $n = 6$, and (iv) $n = 10$?
(b) Let $X$ denote the number of tests that the hematologist will have to run on the $n$ patients; determine the probability mass function of $X$ when (i) $n = 2$, (ii) $n = 4$, (iii) $n = 6$, and (iv) $n = 10$.

**36.** Each of 500 individuals in an army company independently has a certain disease with probability $1/10^3$. This disease will show up in a blood test, and to facilitate matters blood samples from all 500 are pooled and tested.
(a) What is the (approximate) probability that the blood test will be positive (and so at least one person has the disease)?
Suppose now that the blood test yields a positive result.
(b) What is the probability, under this circumstance, that more than one person has the disease?

One of the 500 people is Jones, who knows that he has the disease.

(c) What does Jones think is the probability that more than one person has the disease?

As the pooled test was positive, the authorities have decided to test each individual separately. The first $i - 1$ of these tests were negative, and the $i$th one—which was on Jones—was positive.

(d) Given the above, as a function of $i$, what is the probability that any of the remaining people have the disease?

**37.** Consider a roulette wheel consisting of 38 numbers—1 through 36, 0, and double 0. If Smith always bets that the outcome will be one of the numbers 1 through 12, what is the probability that Smith will lose his first 5 bets? What is the probability that his first win will occur on his fourth bet?

**38.** Two athletic teams play a series of games; the first team to win 4 games is declared the overall winner. Suppose that one of the teams is stronger than the other and wins each game with probability .6, independent of the outcomes of the other games. Find the probability that the stronger team wins the series in exactly $i$ games. Do it for $i = 4, 5, 6, 7$. Compare the probability that the stronger team wins with the probability that it would win a 2-out-of-3 series.

**39.** An interviewer is given a list of potential people she can interview. If the interviewer needs to interview 5 people and if each person (independently) agrees to be interviewed with probability $\frac{2}{3}$, what is the probability that her list of potential people will enable her to obtain her necessary number of interviews if the list consists of (a) 5 people, and (b) 8 people? For part (b) what is the probability that the interviewer will speak to exactly (c) 6 people, and (d) 7 people on the list?

**40.** A fair coin is continually flipped until heads appears for the tenth time. Let $X$ denote the number of tails that occur. Compute the probability mass function of $X$.

**41.** Solve the Banach match problem (Example 6c) when the left-hand matchbox originally contained $N_1$ matches and the right-hand box contained $N_2$ matches.

**42.** In Banach's matchbox problem find the probability that at the moment when the first box is emptied (as opposed to being found empty), the other box contains exactly $k$ matches.

**43.** A urn contains 4 white and 4 black balls. We randomly choose 4 balls. If 2 of them are white and 2 are black, we stop. If not, we replace the balls in the urn and again randomly select 4 balls. This continues until exactly 2 of the 4 chosen are white. What is the probability that we shall make exactly $n$ selections?

**44.** A game popular in Nevada gambling casinos is Keno, which is played as follows: Twenty numbers are selected at random by the casino from the set of numbers 1 through 80. A player can select from 1 to 15 numbers; a win occurs if some fraction of the player's chosen subset matches with any of the 20 numbers drawn by the house. The payoff is a function of the number of elements in the player's selection and the number of matches. For instance, if the player selects only 1 number, then he or she wins if this number is among the set of 20, and the payoff is \$2.2 won for every dollar bet. (As the player's probability of winning in this case is $\frac{1}{4}$, it is clear that the "fair" payoff should be \$3 won for every \$1 bet.) When the player selects 2 numbers, a payoff (of odds) of \$12 won for very \$1 bet is made when both numbers are among the 20,

(a) What would be the fair payoff in this case?
Let $P_{n,k}$ denote the probability that exactly $k$ of the $n$ numbers chosen by the player are among the 20 selected by the house.

(b) Compute $P_{n,k}$.

(c) The most typical wager at Keno consists of selecting 10 numbers. For such a bet the casino pays off as shown in the following table. Complete the last column of the table.

**Keno Payoffs in 10 Number Bets**

| *Number of matches* | *Dollars won for each \$1 bet* | *Fair amount to be paid* |
|---|---|---|
| 0–4 | −1 | |
| 5 | 1 | |
| 6 | 17 | |
| 7 | 179 | |
| 8 | 1,299 | |
| 9 | 2,599 | |
| 10 | 24,999 | |

**45.** In Example 6e, what percentage of $i$ defective lots does the purchaser reject? Do it for $i = 1, 4$. Given that a lot is rejected, what is the conditional probability that it contained 4 defective components?

**46.** A purchaser of transistor buys them in lots of 20. It is his policy to inspect 4 components from a lot randomly and to accept the lot only if all 4 are nondefective. If each component in a lot is, independently, defective with probability .1, what proportion of lots is rejected?

## Computing the Binomial Distribution Function (see p. 127)

```
10 PRINT"THE DISTRIBUTION FUNCTION OF A BINOMIAL(n,p) RANDOM VARIABLE"
20 PRINT "ENTER n"
30 INPUT N
40 PRINT "ENTER p"
50 INPUT P
60 PRINT "ENTER i"
70 INPUT I
80 S=(1-P)^N
90 IF S=0 GOTO 180
100 A=P/(1-P)
110 T=S
120 IF I=0 GOTO 390
130 FOR K=0 TO I-1
140 S=S*A*(N-K)/(K+1)
150 T=T+S
160 NEXT K
170 GOTO 390
180 J=I
190 IF J>N*P THEN J=INT(N*P)
200 FOR K=1 TO J
210 L=L+LOG(N+1-K)-LOG(J+1-K)
220 NEXT K
230 L=L+J*LOG(P)+(N-J)*LOG(1-P)
240 L=EXP(L)
250 B=(1-P)/P
260 F=1
270 FOR K=1 TO J
280 F=F*B*(J+1-K)/(N-J+K)
290 T=T+F
300 NEXT K
310 IF J=I GOTO 380
320 C=1/B
330 F=1
340 FOR K=1 TO I-J
350 F=F*C*(N+1-J-K)/(J+K)
360 T=T+F
370 NEXT K
380 T=(T+1)*L
390 PRINT "THE PROBABILITY IS";T
400 END
```

## Computing the Poisson Distribution Function (see p. 136)

```
10 PRINT "THE PROBABILITY THAT A POISSON VARIABLE  IS LESS THAN OR EQUAL TO i"
20 PRINT "ENTER THE MEAN OF THE RANDOM VARIABLE"
30 INPUT C
40 PRINT "ENTER THE DESIRED VALUE OF i"
50 INPUT I
60 S=EXP(-C)
70 IF S=0 GOTO 150
80 T=S
90 IF I=0 GOTO 340
100 FOR K=0 TO I-1
110 S=S*C/(K+1)
120 T=T+S
130 NEXT K
140 GOTO 340
150 J=I
160 IF J>C THEN J=INT(C)
170 FOR K=1 TO J
180 FAC=FAC+LOG(K)
190 NEXT K
200 L=-C-FAC+J*LOG(C)
210 L=EXP(L)
220 F=1
230 FOR K=1 TO J
240 F=F*(J+1-K)/C
250 T=T+F
260 NEXT K
270 IF J=I GOTO 330
280 F=1
290 FOR K=1 TO I-J
300 F=F*C/(K+J)
310 T=T+F
320 NEXT K
330 T=(T+1)*L
340 PRINT "THE PROBABILITY IS";T
350 END
```

# 5

# Continuous Random Variables

## 1 Introduction

In the previous chapter we considered discrete random variables, that is, random variables whose set of possible values is either finite or countably infinite. However, there also exist random variables whose set of possible values is uncountable. Two examples would be the time that a train arrives at a specified stop and the lifetime of a transistor. Let $X$ be such a random variable. We say that $X$ is a *continuous*[1] random variable if there exists a nonnegative function $f$, defined for all real $x \in (-\infty, \infty)$, having the property that for any set $B$ of real numbers

$$P\{X \in B\} = \int_B f(x)\,dx \tag{1.1}$$

The function $f$ is called the *probability density function* of the random variable $X$.

In words, Equation (1.1) states that the probability that $X$ will be in $B$ may be obtained by integrating the probability density function over the set $B$. Since $X$ must assume some value, $f$ must satisfy

$$1 = P\{X \in (-\infty, \infty)\} = \int_{-\infty}^{\infty} f(x)\,dx$$

All probability statements about $X$ can be answered in terms of $f$. For instance, letting $B = [a, b]$, we obtain from Equation (1.1) that

$$P\{a \leq X \leq b\} = \int_a^b f(x)\,dx \tag{1.2}$$

If we let $a = b$ in Equation (1.2), we obtain

$$P\{X = a\} = \int_a^a f(x)\,dx = 0$$

[1] Sometimes called absolutely continuous.

In words, this equation states that the probability that a continuous random variable will assume any fixed value is zero. Hence, for a continuous random variable,

$$P\{X < a\} = P\{X \le a\} = F(a) = \int_{-\infty}^{a} f(x)\, dx$$

**Example 1a.** Suppose that $X$ is a continuous random variable whose probability density function is given by

$$f(x) = \begin{cases} C(4x - 2x^2) & 0 < x < 2 \\ 0 & \text{otherwise} \end{cases}$$

1. What is the value of $C$?
2. Find $P\{X > 1\}$.

***Solution:*** Since $f$ is a probability density function, we must have that $\int_{-\infty}^{\infty} f(x)\, dx = 1$, implying that

$$C \int_0^2 (4x - 2x^2)\, dx = 1$$

or

$$C\left[2x^2 - \frac{2x^3}{3}\right]\Bigg|_{x=0}^{x=2} = 1$$

or

$$C = \frac{3}{8}$$

Hence

$$P\{X > 1\} = \int_1^{\infty} f(x)\, dx = \tfrac{3}{8}\int_1^2 (4x - 2x^2)\, dx = \tfrac{1}{2}$$ ∎

**Example 1b.** The amount of time, in hours, that a computer functions before breaking down is a continuous random variable with probability density function given by

$$f(x) = \begin{cases} \lambda e^{-x/100} & x \ge 0 \\ 0 & x < 0 \end{cases}$$

What is the probability that a computer will function between 50 and 150 hours before breaking down? What is the probability that it will function less than 100 hours?

***Solution:*** Since

$$1 = \int_{-\infty}^{\infty} f(x)\,dx = \lambda \int_{0}^{\infty} e^{-x/100}\,dx$$

we obtain

$$1 = -\lambda(100)e^{-x/100}\Big|_0^{\infty} = 100\lambda \qquad \text{or} \qquad \lambda = \frac{1}{100}$$

Hence the probability that a computer will function between 50 and 150 hours before breaking down is given by

$$P\{50 < X < 150\} = \int_{50}^{150} \frac{1}{100} e^{-x/100}\,dx = -e^{-x/100}\Big|_{50}^{150}$$
$$= e^{-1/2} - e^{-3/2} \approx .384$$

Similarly,

$$P\{X < 100\} = \int_0^{100} \frac{1}{100} e^{-x/100}\,dx = -e^{-x/100}\Big|_0^{100} = 1 - e^{-1} \approx .633$$

In other words, approximately 63.3 percent of the time a computer will fail before registering 100 hours of use. ▮

**Example 1c.** The lifetime in hours of a certain kind of radio tube is a random variable having a probability density function given by

$$f(x) = \begin{cases} 0 & x \le 100 \\ \dfrac{100}{x^2} & x > 100 \end{cases}$$

What is the probability that exactly 2 of 5 such tubes in a radio set will have to be replaced within the first 150 hours of operation? Assume that the events $E_i$, $i = 1, 2, 3, 4, 5$, that the $i$th such tube will have to be replaced within this time, are independent.

***Solution:*** Now,

$$P(E_i) = \int_0^{150} f(x)\,dx$$
$$= 100 \int_{100}^{150} x^{-2}\,dx$$
$$= \frac{1}{3}$$

Hence, from the independence of the events $E_i$, it follows that the desired probability is

$$\binom{5}{2}\left(\frac{1}{3}\right)^2\left(\frac{2}{3}\right)^3 = \frac{80}{243}$$ ∎

The relationship between the cumulative distribution $F$ and the probability density $f$ is expressed by

$$F(a) = P\{X \in (-\infty, a]\} = \int_{-\infty}^{a} f(x)\, dx$$

Differentiating both sides of the above yields

$$\frac{d}{da} F(a) = f(a)$$

That is, the density is the derivative of the cumulative distribution function. A somewhat more intuitive interpretation of the density function may be obtained from Equation (1.2) as follows:

$$P\left\{a - \frac{\varepsilon}{2} \le X \le a + \frac{\varepsilon}{2}\right\} = \int_{a-\varepsilon/2}^{a+\varepsilon/2} f(x)\, dx \approx \varepsilon f(a)$$

when $\varepsilon$ is small and when $f(\cdot)$ is continuous at $x = a$. In other words, the probability that $X$ will be contained in an interval of length $\varepsilon$ around the point $a$ is approximately $\varepsilon f(a)$. From this, we see that $f(a)$ is a measure of how likely it is that the random variable will be near $a$.

There are several important classes of continuous random variables that appear frequently in probability theory; the next few sections will be devoted to a study of some of them.

## 2 The Uniform Random Variable

A random variable is said to be *uniformly* distributed over the interval (0, 1) if its probability density function is given by

$$f(x) = \begin{cases} 1 & 0 < x < 1 \\ 0 & \text{otherwise} \end{cases} \tag{2.1}$$

Note that Equation (2.1) is a density function, since $f(x) \ge 0$ and $\int_{-\infty}^{\infty} f(x)\, dx = \int_0^1 dx = 1$. Because $f(x) > 0$ only when $x \in (0, 1)$, it follows that $X$ must assume a value in $(0, 1)$. Also, since $f(x)$ is constant for $x \in (0, 1)$, $X$ is just as likely to be near any value in $(0, 1)$ as any other value. To check this, note that for any $0 < a < b < 1$,

$$P\{a \leq X \leq b\} = \int_a^b f(x)\,dx = b - a$$

In other words, the probability that $X$ is in any particular subinterval of $(0, 1)$ equals the length of that subinterval.

In general, we say that $X$ is a uniform random variable on the interval $(\alpha, \beta)$ if its probability density function is given by

$$f(x) = \begin{cases} \dfrac{1}{\beta - \alpha} & \text{if } \alpha < x < \beta \\ 0 & \text{otherwise} \end{cases} \tag{2.2}$$

Since $F(a) = \int_{-\infty}^{a} f(x)\,dx$, we obtain from Equation (2.2) that the distribution function of a uniform random variable on the interval $(\alpha, \beta)$ is given by

$$F(a) = \begin{cases} 0 & a \leq \alpha \\ \dfrac{a - \alpha}{\beta - \alpha} & \alpha < a < \beta \\ 1 & a \geq \beta \end{cases}$$

Figure 5.1 presents a graph of $f(a)$ and $F(a)$.

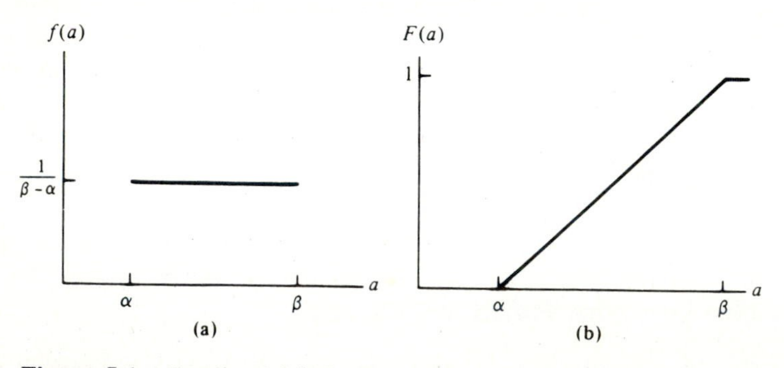


**Figure 5.1** Graph of (a) $f(a)$ and (b) $F(a)$ for a uniform $(\alpha, \beta)$ random variable

**Example 2a.** If $X$ is uniformly distributed over $(0, 10)$, calculate the probability that (1) $X < 3$, (2) $X > 6$, and (3) $3 < X < 8$.

***Solution***

1.
$$P\{X < 3\} = \int_0^3 \tfrac{1}{10}\,dx = \tfrac{3}{10}$$

2. $$P\{X > 6\} = \int_6^{10} \tfrac{1}{10}\, dx = \tfrac{4}{10}$$

3. $$P\{3 < X < 8\} = \int_3^{8} \tfrac{1}{10}\, dx = \tfrac{1}{2}$$ ∎

**Example 2b.** Buses arrive at a specified stop at 15-minute intervals starting at 7 A.M. That is, they arrive at 7, 7:15, 7:30, 7:45, and so on. If a passenger arrives at the stop at a time that is uniformly distributed between 7 and 7:30, find the probability that he waits (1) less than 5 minutes for a bus. (2) More than 10 minutes for a bus.

***Solution:*** Let $X$ denote the number of minutes past 7 that the passenger arrives at the stop. Since $X$ is a uniform random variable over the interval $(0, 30)$, it follows that the passenger will have to wait less than 5 minutes if (and only if) he arrives between 7:10 and 7:15 or between 7:25 and 7:30. Hence the desired probability for (1) is

$$P\{10 < X < 15\} + P\{25 < X < 30\} = \int_{10}^{15} \tfrac{1}{30}\, dx + \int_{25}^{30} \tfrac{1}{30}\, dx = \tfrac{1}{3}$$

Similarly, he would have to wait more than 10 minutes if he arrives between 7 and 7:05 or between 7:15 and 7:20, and so the probability for (2) is

$$P\{0 < X < 5\} + P\{15 < X < 20\} = \tfrac{1}{3}$$ ∎

The next example was first considered by the French mathematician L. F. Bertrand in 1889 and is often referred to as "Bertrand's paradox." It represents our initial introduction to a subject commonly referred to as geometrical probability.

**Example 2c.** Consider a "random chord" of a circle. What is the probability that the length of the chord will be greater than the side of the equilateral triangle inscribed in that circle?

***Solution:*** The problem as stated is incapable of solution because it is not clear what is meant by a random chord. To give meaning to this phrase, we shall reformulate the problem in two distinct ways.

The first formulation is as follows: The position of the chord can be determined by its distance from the center of the circle. This distance can vary between 0 and $r$, the radius of the circle. Now, the length of the chord will be greater than the side of the equilateral triangle inscribed in the circle if its distance from the center is less than $r/2$. Hence, by assuming that a random chord is one whose distance $D$ from the center is uniformly distributed between 0 and $r$, the probability that it is greater than the side of an inscribed equilateral triangle is

$$P\left\{D < \frac{r}{2}\right\} = \frac{r/2}{r} = \frac{1}{2}$$

For our second formulation of the problem consider an arbitrary chord of the circle; through one end of the chord draw a tangent. The angle $\theta$ between the chord and the tangent, which can vary from 0 to 180°, determines the position of the chord. (See Figure 5.2.) Furthermore, the length of the chord will be greater than the side of the inscribed equilateral triangle if the angle $\theta$ is between 60 and 120°. Hence, assuming that a random chord is one whose angle $\theta$ is uniformly distributed between 0 and 180°, we see that the desired answer in this formulation is

$$P\{60 < \theta < 120\} = \frac{120 - 60}{180} = \frac{1}{3}$$

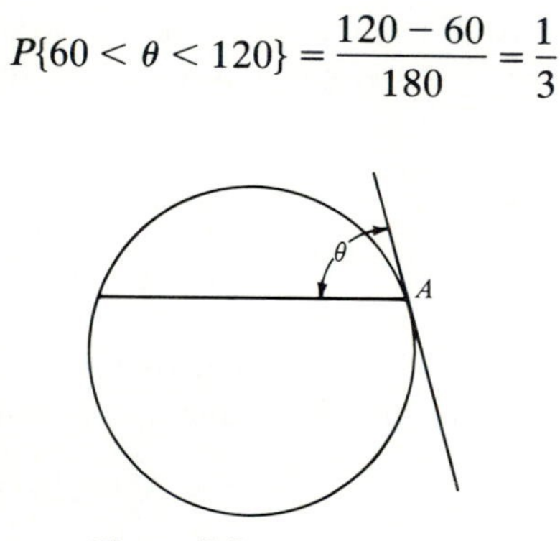


**Figure 5.2**

It should be noted that random experiments could be performed in such a way that $\frac{1}{2}$ or $\frac{1}{3}$ would be the correct probability. For instance, if a circular disk of radius $r$ is thrown on a table ruled with parallel lines a distance $2r$ apart, then one and only one of these lines would cross the disk and form a chord. All distances from this chord to the center of the disk would be equally likely so that the desired probability that the chord's length will be greater than the side of an inscribed equilateral triangle is $\frac{1}{2}$. On the other hand, if the experiment consisted of rotating a needle freely about a point $A$ on the edge (see Figure 5.2) of the circle, the desired answer would be $\frac{1}{3}$. ∎

## 3 Normal Random Variables

We say that $X$ is a normal random variable, or simply that $X$ is normally distributed, with parameters $\mu$ and $\sigma^2$ if the density of $X$ is given by

$$f(x) = \frac{1}{\sqrt{2\pi}\,\sigma} e^{-(x-\mu)^2/2\sigma^2} \qquad -\infty < x < \infty$$

This density function is a bell-shaped curve that is symmetric about $\mu$. (See Figure 5.3.) The values $\mu$ and $\sigma^2$ represent, in some sense, the average value and the possible variation of $X$. (These concepts will be made precise in Chapter 7.)

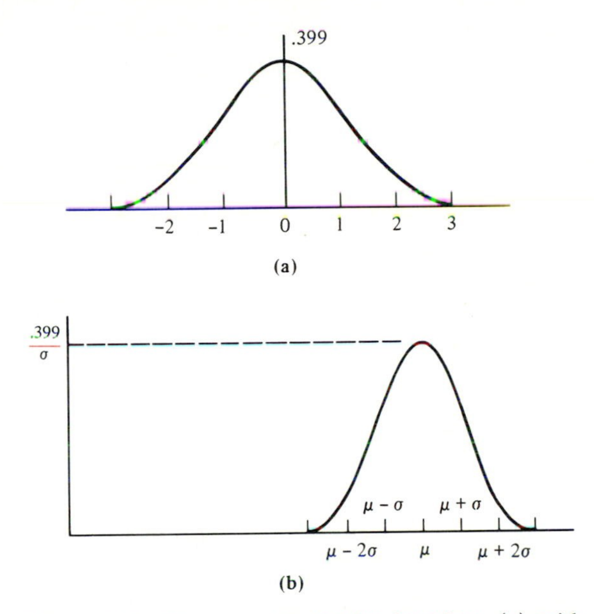


**Figure 5.3** The normal density function: (a) with $\mu = 0$, $\sigma = 1$; and (b) arbitrary $\mu$, $\sigma^2$

The normal distribution was introduced by the French mathematician Abraham de Moivre in 1733 and was used by him to approximate probabilities associated with binomial random variables when the binomial parameter $n$ is large. This result was later extended by Laplace and others and is now encompassed in a probability theorem known as the central limit theorem, which will be discussed in Chapter 8. The central limit, one of the two most important results in probability theory,[2] gives a theoretical base to the often noted empirical observation that, in practice, many random phenomena obey, at least approximately, a normal probability distribution. Some examples of this behavior are the height of a man, the velocity in any direction of a molecule in gas, and the error made in measuring a physical quantity.

To prove that $f(x)$ is indeed a probability density function, we need to show that

$$\frac{1}{\sqrt{2\pi}\,\sigma} \int_{-\infty}^{\infty} e^{-(x-\mu)^2/2\sigma^2}\, dx = 1$$

[2] The other being the strong law of large numbers.

By making the substitution $y = (x - \mu)/\sigma$, we see that

$$\frac{1}{\sqrt{2\pi}\,\sigma}\int_{-\infty}^{\infty} e^{-(x-\mu)^2/2\sigma^2}\,dx = \frac{1}{\sqrt{2\pi}}\int_{-\infty}^{\infty} e^{-y^2/2}\,dy$$

and hence we must show that

$$\int_{-\infty}^{\infty} e^{-y^2/2}\,dy = \sqrt{2\pi}$$

Toward this end, let $I = \int_{-\infty}^{\infty} e^{-y^2/2}\,dy$. Then

$$\begin{aligned} I^2 &= \int_{-\infty}^{\infty} e^{-y^2/2}\,dy \int_{-\infty}^{\infty} e^{-x^2/2}\,dx \\ &= \int_{-\infty}^{\infty}\int_{-\infty}^{\infty} e^{-(y^2+x^2)/2}\,dy\,dx \end{aligned}$$

We now evaluate the double integral by means of a change of variables to polar coordinates. (That is, let $x = r\cos\theta$, $y = r\sin\theta$, $dy\,dx = r\,d\theta\,dr$.) Thus

$$\begin{aligned} I^2 &= \int_0^{\infty}\int_0^{2\pi} e^{-r^2/2} r\,d\theta\,dr \\ &= 2\pi\int_0^{\infty} re^{-r^2/2}\,dr \\ &= -2\pi e^{-r^2/2}\Big|_0^{\infty} \\ &= 2\pi \end{aligned}$$

Hence $I = \sqrt{2\pi}$, and the result is proved.

An important fact about normal random variables is that if $X$ is normally distributed with parameters $\mu$ and $\sigma^2$, then $Y = \alpha X + \beta$ is normally distributed with parameters $\alpha\mu + \beta$ and $\alpha^2\sigma^2$. This follows because $F_Y$,[3] the cumulative distribution function of the random variable $Y$, is given, when $\alpha > 0$, by

$$\begin{aligned} F_Y(a) &= P\{Y \le a\} \\ &= P\{\alpha X + \beta \le a\} \\ &= P\left\{X \le \frac{a-\beta}{\alpha}\right\} \end{aligned}$$

[3] When there is more than one random variable under consideration, we shall denote the cumulative distribution function of a random variable $Z$ by $F_Z$. Similarly, we shall denote the density of $Z$ by $f_Z$.

$$= F_X\left(\frac{a - \beta}{\alpha}\right)$$

$$= \int_{-\infty}^{[(a-\beta)/\alpha]} \frac{1}{\sqrt{2\pi}\,\sigma} e^{-(x-\mu)^2/2\sigma^2}\,dx$$

$$= \int_{-\infty}^{a} \frac{1}{\sqrt{2\pi}\,\alpha\sigma} \exp\left\{-\frac{[y-(\alpha\mu+\beta)]^2}{2\alpha^2\sigma^2}\right\} dy \qquad (3.1)$$

where Equation (3.1) is obtained by the change of variables $y = \alpha x + \beta$. However, since $F_Y(a) = \int_{-\infty}^{a} f_Y(y)\,dy$, it follows from Equation (3.1) that the probability density function of $Y, f_Y(y)$, is given by

$$f_Y(y) = \frac{1}{\sqrt{2\pi}\,\alpha\sigma} \exp\left\{-\frac{[y-(\alpha\mu+\beta)]^2}{2(\alpha\sigma)^2}\right\}$$

Hence $Y$ is normally distributed with parameters $\alpha\mu + \beta$ and $(\alpha\sigma)^2$.

An important implication of the preceding result is that if $X$ is normally distributed with parameters $\mu$ and $\sigma^2$, then $Z = (X - \mu)/\sigma$ is normally distributed with parameters 0 and 1. Such a random variable $Z$ is said to have the *standard*, or *unit*, normal distribution.

It is traditional to denote the cumulative distribution function of a standard normal random variable by $\Phi(x)$. That is,

$$\Phi(x) = \frac{1}{\sqrt{2\pi}} \int_{-\infty}^{x} e^{-y^2/2}\,dy$$

The values of $\Phi(x)$ for nonnegative $x$ are given in Table 5.1. For negative values of $x$, $\Phi(x)$ can be obtained from Equation (3.2):

$$\Phi(-x) = 1 - \Phi(x) \qquad -\infty < x < \infty \qquad (3.2)$$

The proof of Equation (3.2), which follows from the symmetry of the standard normal density, is left as an exercise. This equation states that if $Z$ is a standard normal random variable, then

$$P\{Z \le -x\} = P\{Z > x\} \qquad -\infty < x < \infty$$

Since $Z = (X - \mu)/\sigma$ is a standard normal random variable whenever $X$ is normally distributed with parameters $\mu$ and $\sigma^2$, it follows that the distribution function of $X$ can be expressed as

$$F_X(a) = P\{X \le a\}$$

$$= P\left(\frac{X-\mu}{\sigma} \le \frac{a-\mu}{\sigma}\right)$$

$$= \Phi\left(\frac{a-\mu}{\sigma}\right)$$

**Table 5.1 Area $\Phi(x)$ Under the Standard Normal Curve to the Left of $x$**

| $x$ | .00 | .01 | .02 | .03 | .04 | .05 | .06 | .07 | .08 | .09 |
|---|---|---|---|---|---|---|---|---|---|---|
| .0 | .5000 | .5040 | .5080 | .5120 | .5160 | .5199 | .5239 | .5279 | .5319 | .5359 |
| .1 | .5398 | .5438 | .5478 | .5517 | .5557 | .5596 | .5636 | .5675 | .5714 | .5753 |
| .2 | .5793 | .5832 | .5871 | .5910 | .5948 | .5987 | .6026 | .6064 | .6103 | .6141 |
| .3 | .6179 | .6217 | .6255 | .6293 | .6331 | .6368 | .6406 | .6443 | .6480 | .6517 |
| .4 | .6554 | .6591 | .6628 | .6664 | .6700 | .6736 | .6772 | .6808 | .6844 | .6879 |
| .5 | .6915 | .6950 | .6985 | .7019 | .7054 | .7088 | .7123 | .7157 | .7190 | .7224 |
| .6 | .7257 | .7291 | .7324 | .7357 | .7389 | .7422 | .7454 | .7486 | .7517 | .7549 |
| .7 | .7580 | .7611 | .7642 | .7673 | .7704 | .7734 | .7764 | .7794 | .7823 | .7852 |
| .8 | .7881 | .7910 | .7939 | .7967 | .7995 | .8023 | .8051 | .8078 | .8106 | .8133 |
| .9 | .8159 | .8186 | .8212 | .8238 | .8264 | .8289 | .8315 | .8340 | .8365 | .8389 |
| 1.0 | .8413 | .8438 | .8461 | .8485 | .8508 | .8531 | .8554 | .8577 | .8599 | .8621 |
| 1.1 | .8643 | .8665 | .8686 | .8708 | .8729 | .8749 | .8770 | .8790 | .8810 | .8830 |
| 1.2 | .8849 | .8869 | .8888 | .8907 | .8925 | .8944 | .8962 | .8980 | .8997 | .9015 |
| 1.3 | .9032 | .9049 | .9066 | .9082 | .9099 | .9115 | .9131 | .9147 | .9162 | .9177 |
| 1.4 | .9192 | .9207 | .9222 | .9236 | .9251 | .9265 | .9279 | .9292 | .9306 | .9319 |
| 1.5 | .9332 | .9345 | .9357 | .9370 | .9382 | .9394 | .9406 | .9418 | .9429 | .9441 |
| 1.6 | .9452 | .9463 | .9474 | .9484 | .9495 | .9505 | .9515 | .9525 | .9535 | .9545 |
| 1.7 | .9554 | .9564 | .9573 | .9582 | .9591 | .9599 | .9608 | .9616 | .9625 | .9633 |
| 1.8 | .9641 | .9649 | .9656 | .9664 | .9671 | .9678 | .9686 | .9693 | .9699 | .9706 |
| 1.9 | .9713 | .9719 | .9726 | .9732 | .9738 | .9744 | .9750 | .9756 | .9761 | .9767 |
| 2.0 | .9772 | .9778 | .9783 | .9788 | .9793 | .9798 | .9803 | .9808 | .9812 | .9817 |
| 2.1 | .9821 | .9826 | .9830 | .9834 | .9838 | .9842 | .9846 | .9850 | .9854 | .9857 |
| 2.2 | .9861 | .9864 | .9868 | .9871 | .9875 | .9878 | .9881 | .9884 | .9887 | .9890 |
| 2.3 | .9893 | .9896 | .9898 | .9901 | .9904 | .9906 | .9909 | .9911 | .9913 | .9916 |
| 2.4 | .9918 | .9920 | .9922 | .9925 | .9927 | .9929 | .9931 | .9932 | .9934 | .9936 |
| 2.5 | .9938 | .9940 | .9941 | .9943 | .9945 | .9946 | .9948 | .9949 | .9951 | .9952 |
| 2.6 | .9953 | .9955 | .9956 | .9957 | .9959 | .9960 | .9961 | .9962 | .9963 | .9964 |
| 2.7 | .9965 | .9966 | .9967 | .9968 | .9969 | .9970 | .9971 | .9972 | .9973 | .9974 |
| 2.8 | .9974 | .9975 | .9976 | .9977 | .9977 | .9978 | .9979 | .9979 | .9980 | .9981 |
| 2.9 | .9981 | .9982 | .9982 | .9983 | .9984 | .9984 | .9985 | .9985 | .9986 | .9986 |
| 3.0 | .9987 | .9987 | .9987 | .9988 | .9988 | .9989 | .9989 | .9989 | .9990 | .9990 |
| 3.1 | .9990 | .9991 | .9991 | .9991 | .9992 | .9992 | .9992 | .9992 | .9993 | .9993 |
| 3.2 | .9993 | .9993 | .9994 | .9994 | .9994 | .9994 | .9994 | .9995 | .9995 | .9995 |
| 3.3 | .9995 | .9995 | .9995 | .9996 | .9996 | .9996 | .9996 | .9996 | .9996 | .9997 |
| 3.4 | .9997 | .9997 | .9997 | .9997 | .9997 | .9997 | .9997 | .9997 | .9997 | .9998 |

**Example 3a.** If $X$ is a normal random variable with parameters $\mu = 3$ and $\sigma^2 = 9$, find (1) $P\{2 < X < 5\}$, (2) $P\{X > 0\}$, and (3) $P\{|X - 3| > 6\}$.

***Solution***

1.
$$P\{2 < X < 5\} = P\left\{\frac{2-3}{3} < \frac{X-3}{3} < \frac{5-3}{3}\right\} = P\left\{-\frac{1}{3} < Z < \frac{2}{3}\right\}$$
$$= \Phi\left(\frac{2}{3}\right) - \Phi\left(-\frac{1}{3}\right)$$
$$= \Phi\left(\frac{2}{3}\right) - \left[1 - \Phi\left(\frac{1}{3}\right)\right] = .3779$$

2.
$$P\{X > 0\} = P\left\{\frac{X-3}{3} > \frac{0-3}{3}\right\} = P\{Z > -1\}$$
$$= 1 - \Phi(-1)$$
$$= \Phi(1)$$
$$= .8413$$

3.
$$P\{|X - 3| > 6\} = P\{X > 9\} + P\{X < -3\}$$
$$= P\left\{\frac{X-3}{3} > \frac{9-3}{3}\right\} + P\left\{\frac{X-3}{3} < \frac{-3-3}{3}\right\}$$
$$= P\{Z > 2\} + P\{Z < -2\}$$
$$= 1 - \Phi(2) + \Phi(-2)$$
$$= 2[1 - \Phi(2)]$$
$$= .0456$$

∎

**Example 3b.** An examination is often regarded as being good (in the sense of determining a valid grade spread for those taking it) if the test scores of those taking the examination can be approximated by a normal density function. (In other words, a graph of the frequency of grade scores should have approximately the bell-shaped form of the normal density.) The instructor often uses the test scores to estimate the normal parameters $\mu$ and $\sigma^2$ and then assigns the letter grade A to those whose test score is greater than $\mu + \sigma$, B to those whose score is between $\mu$ and $\mu + \sigma$, C to those whose score is between $\mu - \sigma$ and $\mu$, D to those whose score is between $\mu - 2\sigma$ and $\mu - \sigma$, and F to those getting a score below

$\mu - 2\sigma$. (This is sometimes referred to as grading "on the curve.") Since

$$P\{X > \mu + \sigma\} = P\left\{\frac{X - \mu}{\sigma} > 1\right\} = 1 - \Phi(1) = .1587$$

$$P\{\mu < X < \mu + \sigma\} = P\left\{0 < \frac{X - \mu}{\sigma} < 1\right\} = \Phi(1) - \Phi(0) = .3413$$

$$P\{\mu - \sigma < X < \mu\} = P\left\{-1 < \frac{X - \mu}{\sigma} < 0\right\}$$

$$= \Phi(0) - \Phi(-1) = .3413$$

$$P\{\mu - 2\sigma < X < \mu - \sigma\} = P\left\{-2 < \frac{X - \mu}{\sigma} < -1\right\}$$

$$= \Phi(2) - \Phi(1) = .1359$$

$$P\{X < \mu - 2\sigma\} = P\left\{\frac{X - \mu}{\sigma} < -2\right\} = \Phi(-2) = .0228$$

it follows that approximately 16 percent of the class will receive an A grade on the examination, 34 percent a B grade, 34 percent a C grade, and 14 percent a D grade; 2 percent will fail. ▮

**Example 3c.** An expert witness in a paternity suit testifies that the length (in days) of pregnancy (that is, the time from impregnation to the delivery of the child) is approximately normally distributed with parameters $\mu = 270$ and $\sigma^2 = 100$. The defendant in the suit is able to prove that he was out of the country during a period that began 290 days before the birth of the child and ended 240 days before the birth. If the defendant was, in fact, the father of the child, what is the probability that the mother could have had the very long or very short pregnancy indicated by the testimony?

***Solution:*** Let $X$ denote the length of the pregnancy and assume that the defendant is the father. Then the probability that the birth could occur within the indicated period is

$$P\{X > 290 \text{ or } X < 240\} = P\{X > 290\} + P\{X < 240\}$$

$$= P\left\{\frac{X - 270}{10} > 2\right\} + P\left\{\frac{X - 270}{10} < -3\right\}$$

$$= 1 - \Phi(2) + 1 - \Phi(3)$$

$$= .0241$$

▮

**Example 3d.** Suppose that a binary message—either 0 or 1—must be transmitted by wire from location $A$ to location $B$. However, the data sent over the wire is subject to a channel noise disturbance, and so to reduce the possibility of error, the value 2 is sent over the wire when the message is 1 and the value $-2$ is sent when the message is 0. If $x$, $x = \pm 2$, is the value sent at location $A$, then $R$, the value received at location $B$, is given by $R = x + N$, where $N$ is the channel noise disturbance. When the message is received at location $B$ the receiver decodes it according to the following rule:

$$\text{if } R \geq .5, \text{ then 1 is concluded}$$
$$\text{if } R < .5, \text{ then 0 is concluded}$$

As the channel noise is often normally distributed, we will determine the error probabilities when $N$ is a unit normal random variable.

There are two types of errors that can occur: One is that the message 1 can be incorrectly concluded to be 0, and the other that 0 is concluded to be 1. The first type of error will occur if the message is 1 and $2 + N < .5$, whereas the second will occur if the message is 0 and $-2 + N \geq .5$.

Hence

$$\begin{aligned} P\{\text{error}\,|\,\text{message is } 1\} &= P\{N < -1.5\} \\ &= 1 - \Phi(1.5) = .0668 \end{aligned}$$

and

$$\begin{aligned} P\{\text{error}\,|\,\text{message is } 0\} &= P\{N \geq 2.5\} \\ &= 1 - \Phi(2.5) = .0062 \end{aligned}$$ ∎

The following inequality for $\Phi(x)$ is of theoretical importance:

$$\frac{1}{\sqrt{2\pi}}\left(\frac{1}{x} - \frac{1}{x^3}\right)e^{-x^2/2} < 1 - \Phi(x) < \frac{1}{\sqrt{2\pi}}\frac{1}{x}e^{-x^2/2} \qquad \text{for all } x > 0 \tag{3.3}$$

To prove inequality (3.3), we first note the obvious inequality

$$(1 - 3y^{-4})e^{-y^2/2} < e^{-y^2/2} < (1 + y^{-2})e^{-y^2/2}$$

implying that

$$\int_x^\infty (1 - 3y^{-4})e^{-y^2/2}\,dy < \int_x^\infty e^{-y^2/2}\,dy < \int_x^\infty (1 + y^{-2})e^{-y^2/2}\,dy \tag{3.4}$$

However,

$$\frac{d}{dy}[(y^{-1} - y^{-3})e^{-y^2/2}] = -(1 - 3y^{-4})e^{-y^2/2}$$

$$\frac{d}{dy}[y^{-1}e^{-y^2/2}] = -(1 + y^{-2})e^{-y^2/2}$$

and so, for $x > 0$,

$$-(y^{-1} - y^{-3})e^{-y^2/2}\Big|_x^{\infty} < \int_x^{\infty} e^{-y^2/2}\,dy < -y^{-1}e^{-y^2/2}\Big|_x^{\infty}$$

or

$$(x^{-1} - x^{-3})e^{-x^2/2} < \int_x^{\infty} e^{-y^2/2}\,dy < x^{-1}e^{-x^2/2}$$

establishing Equation (3.3).

It also follows from the inequality (3.3) that

$$1 - \Phi(x) \sim \frac{1}{x\sqrt{2\pi}}\,e^{-x^2/2}$$

for large $x$. [The notation $a(x) \sim b(x)$ for large $x$ means that $\lim_{x\to\infty} a(x)/b(x) = 1$.]

## 3.1 *The Normal Approximation to the Binomial Distribution*

The following theorem is known as the DeMoivre–Laplace limit theorem. It was originally proved for the special case $p = \frac{1}{2}$ by DeMoivre in 1733 and then extended to general $p$ by Laplace in 1812.

---

### *DeMoivre–Laplace Limit Theorem*

*If $S_n$ denotes the number of successes that occur when $n$ independent trials, each resulting in a success with probability $p$, are performed then, for any $a < b$,*

$$P\left\{a \le \frac{S_n - np}{\sqrt{np(1-p)}} \le b\right\} \to \Phi(b) - \Phi(a)$$

*as $n \to \infty$.*

---

As the above theorem is only a special case of the central limit theorem, which will be presented in Chapter 8, we shall not present a proof.

It should be noted that we now have two possible approximations to binomial probabilities: the Poisson approximation, which yields a good approximation when $n$ is large and $np$ moderate, and the normal approximation, which can be shown to be quite good when $np(1-p)$ is large. [The normal approximation will, in general, be quite good for values of $n$ satisfying $np(1-p) \geq 10$.]

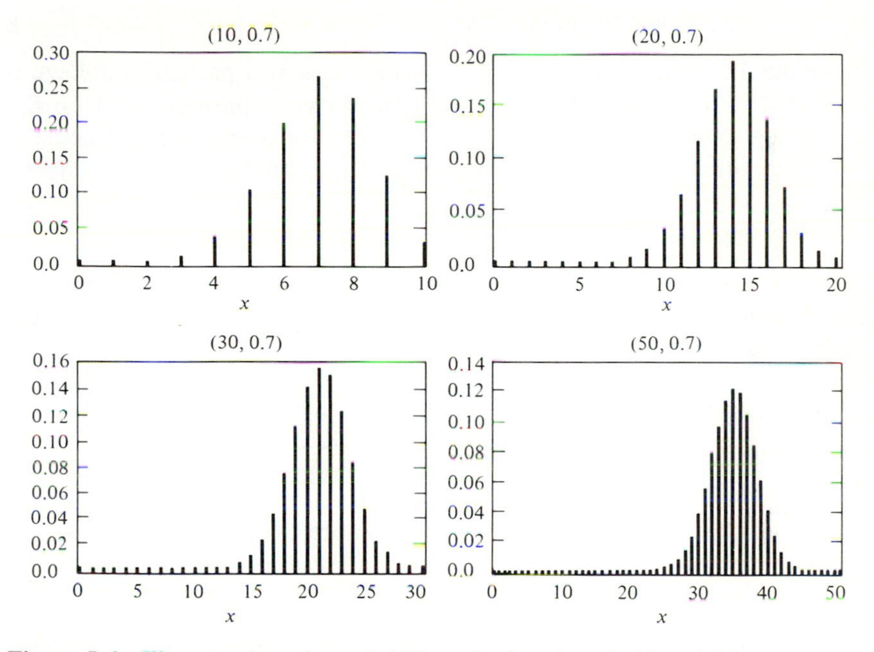


**Figure 5.4** Illustrates how the probability mass function of a binomial $(n, p)$ random variable becomes more and more "normal" as $n$ becomes larger and larger.

**Example 3e.** Let $X$ be the number of times that a fair coin, flipped 40 times, lands heads. Find the probability that $X = 20$. Use the normal approximation and then compare it to the exact solution.

***Solution:*** Since the binomial is a discrete random variable and the normal a continuous random variable, the best approximation is obtained by writing the desired probability as

$$\begin{aligned} P\{X = 20\} &= P\{19.5 < X < 20.5\} \\ &= P\left\{\frac{19.5 - 20}{\sqrt{10}} < \frac{X - 20}{\sqrt{10}} < \frac{20.5 - 20}{\sqrt{10}}\right\} \\ &= P\left\{-.16 < \frac{X - 20}{\sqrt{10}} < .16\right\} \\ &\approx \Phi(.16) - \Phi(-.16) = .1272 \end{aligned}$$

The exact result is

$$P\{X = 20\} = \binom{40}{20}\left(\frac{1}{2}\right)^{40}$$

which can be shown to equal .1254. ▮

**Example 3f.** The ideal size of a first-year class at a particular college is 150 students. The college, knowing from past experience that on the average only 30 percent of those accepted for admission will actually attend, uses a policy of approving the applications of 450 students. Compute the probability that more than 150 first-year students attend this college.

***Solution:*** Let $X$ denote the number of students that attend; then $X$ is a binomial random variable with parameters $n = 450$ and $p = .3$. The normal approximation yields that

$$\begin{aligned} P\{X \geq 150.5\} &= P\left\{\frac{X - (450)(.3)}{\sqrt{450(.3)(.7)}} \geq \frac{150.5 - (450)(.3)}{\sqrt{450(.3)(.7)}}\right\} \approx 1 - \Phi(1.59) \\ &= .0559 \end{aligned}$$

Hence less than 6 percent of the time do more than 150 of the first 450 accepted actually attend. (What independence assumptions have we made?) ▮

**Example 3g.** To determine the effectiveness of a certain diet in reducing the amount of cholesterol in the blood stream, 100 people are put on the diet. After they have been on the diet for a sufficient length of time, their cholesterol count will be taken. The nutritionist running this experiment has decided to endorse the diet if at least 65 percent of the people have a lower cholesterol count after going on the diet. What is the probability that the nutritionist endorses the new diet if, in fact, it has no effect on the cholesterol level?

***Solution:*** Let us assume that if the diet has no effect on the cholesterol count, then, strictly by chance, each person's count will be lower than it was before the diet with probability $\frac{1}{2}$. Hence, if $X$ is the number of people whose count is lowered, then the probability that the nutritionist will endorse the diet when it actually has no effect on the cholesterol count is

$$\begin{aligned} \sum_{i=65}^{100} \binom{100}{i}\left(\frac{1}{2}\right)^{100} &= P\{X \geq 64.5\} \\ &= P\left\{\frac{X - (100)(\frac{1}{2})}{\sqrt{100(\frac{1}{2})(\frac{1}{2})}} \geq 2.9\right\} \\ &\approx 1 - \Phi(2.9) \\ &= .0019 \end{aligned}$$

▮

## 4 Exponential Random Variables

A continuous random variable whose probability density function is given, for some $\lambda > 0$, by

$$f(x) = \begin{cases} \lambda e^{-\lambda x} & \text{if } x \geq 0 \\ 0 & \text{if } x < 0 \end{cases}$$

is said to be an *exponential* random variable (or, more simply, is said to be exponentially distributed) with parameter $\lambda$. The cumulative distribution function $F(a)$ of an exponential random variable is given by

$$\begin{aligned} F(a) &= P\{X \leq a\} \\ &= \int_0^a \lambda e^{-\lambda x}\, dx \\ &= -e^{-\lambda x}\Big|_0^a \\ &= 1 - e^{-\lambda a} \qquad a \geq 0 \end{aligned}$$

Note that $F(\infty) = \int_0^\infty \lambda e^{-\lambda x}\, dx = 1$, as, of course, it must. The parameter $\lambda$ will be shown in Chapter 7 to equal the reciprocal of the "average value" of the random variable.

The exponential distribution often arises, in practice, as being the distribution of the amount of time until some specific event occurs. For instance, the amount of time (starting from now) until an earthquake occurs, or until a new war breaks out, or until a telephone call you receive turns out to be a wrong number are all random variables that tend in practice to have exponential distributions. (For a theoretical explanation for this the reader should consult Section 5 of Chapter 4, and, in particular, Example 5d of that section.)

**Example 4a.** Suppose that the length of a phone call in minutes is an exponential random variable with parameter $\lambda = \frac{1}{10}$. If someone arrives immediately ahead of you at a public telephone booth, find the probability that you will have to wait (1) more than 10 minutes, and (2) between 10 and 20 minutes.

***Solution:*** Letting $X$ denote the length of the call made by the person in the booth, we have that the desired probabilities are

1. $$P\{X > 10\} = \int_{10}^{\infty} \tfrac{1}{10} e^{-x/10}\, dx = -e^{-x/10}\Big|_{10}^{\infty} = e^{-1} \approx .368$$

2. $$\begin{aligned} P\{10 < X < 20\} &= \int_{10}^{20} \tfrac{1}{10} e^{-x/10}\, dx = -e^{-x/10}\Big|_{10}^{20} \\ &= e^{-1} - e^{-2} \approx .233 \end{aligned}$$

■

We say that a nonnegative random variable $X$ is *memoryless* if

$$P\{X > s + t | X > t\} = P\{X > s\} \qquad \text{for all } s, t \geq 0 \tag{4.1}$$

If we think of $X$ as being the lifetime of some instrument, Equation (4.1) states that the probability that the instrument survives for at least $s + t$ hours, given that it has survived $t$ hours, is the same as the initial probability that it survives for at least $s$ hours. In other words, if the instrument is alive at age $t$, the distribution of the remaining amount of time that it survives is the same as the original lifetime distribution (that is, it is as if the instrument does not remember that it has already been in use for a time $t$).

The condition (4.1) is equivalent to

$$\frac{P\{X > s + t, X > t\}}{P\{X > t\}} = P\{X > s\}$$

or

$$P\{X > s + t\} = P\{X > s\}P\{X > t\} \tag{4.2}$$

Since Equation (4.2) is satisfied when $X$ is exponentially distributed (for $e^{-\lambda(s+t)} = e^{-\lambda s}e^{-\lambda t}$), it follows that exponentially distributed random variables are memoryless.

**Example 4b.** Consider a post office that is staffed by two clerks. Suppose that when Mr. Smith enters the system, he discovers that Ms. Jones is being served by one of the clerks and Mr. Brown by the other. Suppose also that Mr. Smith is told that his service will begin as soon as either Jones or Brown leaves. If the amount of time that a clerk spends with a customer is exponentially distributed with parameter $\lambda$, what is the probability that, of the three customers, Mr. Smith is the last to leave the post office?

***Solution:*** The answer is obtained by reasoning as follows: Consider the time at which Mr. Smith first finds a free clerk. At this point either Ms. Jones or Mr. Brown would have just left and the other one would still be in service. However, by the lack of memory of the exponential, it follows that the additional amount of time that this other person (either Jones or Brown) would still have to spend in the post office is exponentially distributed with parameter $\lambda$. That is, it is the same as if service for this person were just starting at this point. Hence, by symmetry, the probability that the remaining person finishes before Smith must equal $\frac{1}{2}$. ∎

It turns out that not only is the exponential distribution memoryless, but it is also the unique distribution possessing this property. To see this, suppose that $X$ is memoryless and let $\bar{F}(x) = P\{X > x\}$. Then, by Equation (4.2), it follows that

$$\bar{F}(s+t) = \bar{F}(s)\bar{F}(t)$$

That is, $\bar{F}(\cdot)$ satisfies the functional equation

$$g(s+t) = g(s)g(t)$$

However, it turns out that the only right continuous solution of this functional equation is[4]

$$g(x) = e^{-\lambda x} \tag{4.3}$$

and, since a distribution function is always right continuous, we must have

$$\bar{F}(x) = e^{-\lambda x} \qquad \text{or} \qquad F(x) = P\{X \le x\} = 1 - e^{-\lambda x}$$

which shows that $X$ is exponentially distributed.

**Example 4c.** Suppose that the number of miles that a car can run before its battery wears out is exponentially distributed with an average value of 10,000 miles. If a person desires to take a 5000-mile trip, what is the probability that he or she will be able to complete the trip without having to replace the car battery? What can be said when the distribution is not exponential?

***Solution:*** It follows, by the memoryless property of the exponential distribution, that the remaining lifetime (in thousands of miles) of the battery is exponential with parameter $\lambda = \frac{1}{10}$. Hence the desired probability is

$$P(\text{remaining lifetime} > 5\} = 1 - F(5) = e^{-5\lambda} = e^{-1/2} \approx .604$$

However, if the lifetime distribution $F$ is not exponential, then the relevant probability is

$$P\{\text{lifetime} > t + 5 \mid \text{lifetime} > t\} = \frac{1 - F(t+5)}{1 - F(t)}$$

[4] Equation (4.3) as follows: If $g(s+t) = g(s)g(t)$, then

$$g\left(\frac{2}{n}\right) = g\left(\frac{1}{n} + \frac{1}{n}\right) = g^2\left(\frac{1}{n}\right)$$

and repeating this yields $g(m/n) = g^m(1/n)$. Also

$$g(1) = g\left(\frac{1}{n} + \frac{1}{n} + \cdots + \frac{1}{n}\right) = g^n\left(\frac{1}{n}\right) \qquad \text{or} \qquad g\left(\frac{1}{n}\right) = (g(1))^{1/n}$$

Hence $g(m/n) = (g(1))^{m/n}$, which implies, since $g$ is right continuous, that $g(x) = (g(1))^x$. Since $g(1) = (g(\frac{1}{2}))^2 \ge 0$, we obtain $g(x) = e^{-\lambda x}$, where $\lambda = -\log(g(1))$.

where $t$ is the number of miles that the battery had been in use prior to the start of the trip. Therefore, if the distribution is not exponential, additional information is needed (namely, $t$) before the desired probability can be calculated. ∎

A variation of the exponential distribution is the distribution of a random variable that is equally likely to be either positive or negative and whose absolute value is exponentially distributed with parameter $\lambda$, $\lambda \geq 0$. Such a random variable is said to have a *Laplacian* distribution[5] and its density is given by

$$\begin{aligned} f(x) &= \tfrac{1}{2}\lambda e^{-\lambda x}, & x > 0 \\ &\phantom{=}\ \tfrac{1}{2}\lambda e^{\lambda x}, & x < 0 \\ &= \tfrac{1}{2}\lambda e^{-\lambda |x|}, & -\infty < x < \infty \end{aligned}$$

Its distribution function is given by

$$F(x) = \begin{cases} \frac{1}{2}\displaystyle\int_{-\infty}^{x} \lambda e^{\lambda x}\,dx, & x < 0 \\ \frac{1}{2}\displaystyle\int_{-\infty}^{0} \lambda e^{\lambda x}\,dx + \frac{1}{2}\int_{0}^{x} \lambda e^{-\lambda x}\,dx, & x > 0 \end{cases}$$

$$= \begin{cases} \frac{1}{2} e^{\lambda x}, & x < 0 \\ 1 - \frac{1}{2} e^{-\lambda x}, & x > 0 \end{cases}$$

**Example 4d.** Let us reconsider Example 3d, which supposes that a binary message is to be transmitted from $A$ to $B$, with the value 2 being sent when the message is 1 and $-2$ when it is 0. However, suppose now that rather than be a standard normal random variable, the channel noise $N$ is a Laplacian random variable with parameter $\lambda = 1$. Again suppose that if $R$ is the value received at location $B$, then the message is decoded as follows:

if $R \geq .5$, then 1 is concluded
if $R < .5$, then 0 is concluded

In this case, where the noise is Laplacian with parameter $\lambda = 1$, the 2 types of errors will have probabilities given by:

$$\begin{aligned} P\{\text{error} \mid \text{message 1 is sent}\} &= P\{N < -1.5\} \\ &= \tfrac{1}{2} e^{-1.5} \\ &= .1116 \end{aligned}$$

[5] It also is sometimes called the double exponential random variable.

$$\begin{aligned} P\{\text{error} \mid \text{message 0 is sent}\} &= P\{N \geq 2.5\} \\ &= \tfrac{1}{2}e^{-2.5} \\ &= .041 \end{aligned}$$

Hence, on comparing this with the results of Example 3d, we see that the error probabilities are higher when the noise is Laplacian with $\lambda = 1$ than when it is a unit normal random variable. (As will be shown in Chapter 7, both of these random variables have an average value of 0 and an average squared deviation from 0 of 1.) ∎

## 4.1 *Hazard Rate Functions*

Consider a positive continuous random variable $X$ that we interpret as being the lifetime of some item, having distribution function $F$ and density $f$. The *hazard rate* (sometimes called the *failure rate*) function $\lambda(t)$ of $F$ is defined by

$$\lambda(t) = \frac{f(t)}{\bar{F}(t)}, \qquad \bar{F} = 1 - F$$

To interpret $\lambda(t)$, suppose that the item has survived for $t$ hours and we desire the probability that it will not survive for an additional time $dt$. That is, consider $P\{X \in (t, t + dt) \mid X > t\}$. Now

$$\begin{aligned} P\{X \in (t, t + dt) \mid X > t\} &= \frac{P\{X \in (t, t + dt), X > t\}}{P\{X > t\}} \\ &= \frac{P\{X \in (t, t + dt)\}}{P\{X > t\}} \\ &\approx \frac{f(t)}{\bar{F}(t)}\, dt \end{aligned}$$

That is, $\lambda(t)$ represents the conditional probability intensity that a $t$-unit-old item will fail.

Suppose now that the lifetime distribution is exponential. Then, by the memoryless property, it follows that the distribution of remaining life for a $t$-year-old item is the same as for a new item. Hence $\lambda(t)$ should be constant. This checks out, since

$$\begin{aligned} \lambda(t) &= \frac{f(t)}{\bar{F}(t)} \\ &= \frac{\lambda e^{-\lambda t}}{e^{-\lambda t}} \\ &= \lambda \end{aligned}$$

Thus the failure rate function for the exponential distribution is constant. The parameter $\lambda$ is often referred to as the *rate* of the distribution.

It turns out that the failure rate function $\lambda(t)$ uniquely determines the distribution $F$. To prove this, note that by definition

$$\lambda(t) = \frac{\frac{d}{dt}F(t)}{1 - F(t)}$$

Integrating both sides yields

$$\log(1 - F(t)) = -\int_0^t \lambda(t)\,dt + k$$

or

$$1 - F(t) = e^k \exp\left\{-\int_0^t \lambda(t)\,dt\right\}$$

Letting $t = 0$ shows that $k = 0$ and thus

$$F(t) = 1 - \exp\left\{-\int_0^t \lambda(t)\,dt\right\} \tag{4.4}$$

Hence a distribution function of a positive continuous random variable can be specified by giving its hazard rate function. For instance, if a random variable has a linear hazard rate function—that is, if

$$\lambda(t) = a + bt$$

then its distribution function is given by

$$F(t) = 1 - e^{-at - bt^2/2}$$

and differentiation yields that its density is

$$f(t) = (a + bt)e^{-(at + bt^2/2)}, \qquad t \geq 0$$

When $a = 0$, the above is known as the Rayleigh density function.

**Example 4e.** One often hears that the death rate of a person that smokes is, at each age, twice that of a nonsmoker. What does this mean? Does it mean that a nonsmoker has twice the probability of surviving a given number of years as does a smoker of the same age?

***Solution:*** If $\lambda_s(t)$ denotes the hazard rate of a smoker of age $t$ and $\lambda_n(t)$ that of a nonsmoker of age $t$, then the above is equivalent to the statement that

$$\lambda_s(t) = 2\lambda_n(t).$$

The probability that an $A$ year old nonsmoker will survive until age $B$, $A < B$ is

$$
\begin{aligned}
&P\{A \text{ year old nonsmoker reaches age } B\} \\
&= P\{\text{nonsmokers lifetime} > B \mid \text{nonsmokers lifetime} > A\} \\
&= \frac{1 - F_{\text{non}}(B)}{1 - F_{\text{non}}(A)} \\
&= \frac{\exp\left\{-\int_0^B \lambda_n(t)\,dt\right\}}{\exp\left\{-\int_0^A \lambda_n(t)\,dt\right\}} \qquad \text{from (4.4)} \\
&= \exp\left\{-\int_A^B \lambda_n(t)\,dt\right\}
\end{aligned}
$$

whereas the corresponding probability for a smoker is, by the same reasoning,

$$
\begin{aligned}
P\{A \text{ year old smoker reaches age } B\} &= \exp\left\{-\int_A^B \lambda_s(t)\,dt\right\} \\
&= \exp\left\{-2\int_A^B \lambda_n(t)\,dt\right\} \\
&= \left[\exp\left\{-\int_A^B \lambda_n(t)\,dt\right\}\right]^2
\end{aligned}
$$

In other words, of two individuals of the same age, one of whom is a smoker and other a nonsmoker, the probability that the smoker survives to any given age is the *square* (not one-half) of the corresponding probability for a nonsmoker. For instance, if $\lambda_n(t) = \frac{1}{30}$, $t \leq 50 \leq 60$, then the probability that a 50 year old nonsmoker reaches age 60 is $e^{-1/3} = .7165$ whereas the corresponding probability for a smoker is $e^{-2/3} = .5134$. ∎

## 5 Other Continuous Distributions

### 5.1 *The Gamma Distribution*

A random variable is said to have a gamma distribution with parameters $(t, \lambda)$, $\lambda > 0$, and $t > 0$ if its density function is given by

$$
f(x) = \begin{cases} \dfrac{\lambda e^{-\lambda x}(\lambda x)^{t-1}}{\Gamma(t)} & x \geq 0 \\ 0 & x < 0 \end{cases}
$$

where

$$\Gamma(t) = \int_0^\infty e^{-y} y^{t-1}\, dy$$

The integration by parts of $\Gamma(t)$ yields that

$$\begin{aligned}\Gamma(t) &= -e^{-y} y^{t-1} \Big|_0^\infty + \int_0^\infty e^{-y}(t-1) y^{t-2}\, dy \\ &= (t-1)\int_0^\infty e^{-y} y^{t-2}\, dy \qquad (5.1) \\ &= (t-1)\Gamma(t-1)\end{aligned}$$

For integral values of $t$, say $t = n$, we obtain by applying Equation (5.1) repeatedly that

$$\begin{aligned}\Gamma(n) &= (n-1)\Gamma(n-1) \\ &= (n-1)(n-2)\Gamma(n-2) \\ &= \cdots \\ &= (n-1)(n-2)\cdots 3\cdot 2\Gamma(1)\end{aligned}$$

Since $\Gamma(1) = \int_0^\infty e^{-x}\, dx = 1$, it follows that for integral values of $n$,

$$\Gamma(n) = (n-1)!$$

When $t$ is a positive integer, say $t = n$, the gamma distribution with parameters $(t, \lambda)$ often arises, in practice, as the distribution of the amount of time one has to wait until a total of $n$ events has occurred. More specifically, if events are occurring randomly in time and in accordance with the three axioms of Section 5, Chapter 4, then it turns out that the amount of time one has to wait until a total of $n$ events has occurred will be a gamma random variable with parameters $(n, \lambda)$. To prove this, let $T_n$ denote the time at which the $n$th event occurs, and note that $T_n$ is less than or equal to $t$ if and only if the number of events that have occurred by time $t$ is at least $n$. That is, with $N(t)$ equal to the number of events in $[0, t]$,

$$\begin{aligned}P\{T_n \le t\} &= P\{N(t) \ge n\} \\ &= \sum_{j=n}^{\infty} P\{N(t) = j\} \\ &= \sum_{j=n}^{\infty} \frac{e^{-\lambda t}(\lambda t)^j}{j!}\end{aligned}$$

where the final identity follows, since the number of events in $[0, t]$ has a Poisson distribution with parameter $\lambda t$. Differentiation of the above yields that the density function of $T_n$ is as follows:

$$f(t) = \sum_{j=n}^{\infty} \frac{e^{-\lambda t} j(\lambda t)^{j-1}\lambda}{j!} - \sum_{j=n}^{\infty} \frac{\lambda e^{-\lambda t}(\lambda t)^j}{j!}$$

$$= \sum_{j=n}^{\infty} \frac{\lambda e^{-\lambda t}(\lambda t)^{j-1}}{(j-1)!} - \sum_{j=n}^{\infty} \frac{\lambda e^{-\lambda t}(\lambda t)^j}{j!}$$

$$= \frac{\lambda e^{-\lambda t}(\lambda t)^{n-1}}{(n-1)!}$$

n=t=1

Hence $T_n$ is gamma distribution with parameters $(n, \lambda)$. (This distribution is often referred to in the literature as the $n$-Erlang distribution.) Note that when $n = 1$, this distribution reduces to the exponential.

The gamma distribution with $\lambda = \frac{1}{2}$ and $t = n/2$ ($n$ being a positive integer) is called the $\chi_n^2$ (read "chi-square") distribution with $n$ degrees of freedom. The chi-squared distribution often arises in practice as being the distribution of the error involved in attempting to hit a target in $n$ dimensional space when each coordinate error is normally distributed. This distribution will be studied in Chapter 6 where its relation to the normal distribution will be detailed.

## 5.2 *The Weibull Distribution*

The Weibull distribution is widely used in engineering practice due to its versatility. It was originally proposed for the interpretation of fatigue data, but now its use has extended to many other engineering problems. In particular, it is widely used, in the field of life phenomena, as the distribution of the lifetime of some object, particularly when the "weakest link" model is appropriate for the object. That is, consider an object consisting of many parts and suppose that the object experiences death (failure) when any of its parts fail. Under these conditions, it has been shown (both theoretically and empirically) that a Weibull distribution provides a close approximation to the distribution of the lifetime of the item.

The Weibull distribution function has the form

$$F(x) = \begin{cases} 0 & x \le v \\ 1 - \exp\left\{ -\left(\dfrac{x - v}{\alpha}\right)^{\beta} \right\} & x > v \end{cases} \tag{5.2}$$

A random variable whose cumulative distribution function is given by Equation (5.2) is said to be a Weibull random variable with parameters $v$,

$\alpha$, and $\beta$. Differentiation yields that the density is

$$f(x) = \begin{cases} 0 & x \leq v \\ \left(\dfrac{\beta}{\alpha}\right)\left(\dfrac{x-v}{\alpha}\right)^{\beta-1} \exp\left\{-\left(\dfrac{x-v}{\alpha}\right)^{\beta}\right\} & x > v \end{cases}$$

## 5.3 *The Cauchy Distribution*

A random variable is said to have a Cauchy distribution with parameter $\theta$, $-\infty < \theta < \infty$, if its density is given by

$$f(x) = \frac{1}{\pi} \frac{1}{[1 + (x - \theta)^2]} \qquad -\infty < x < \infty$$

**Example 5a.** Suppose that a narrow beam flashlight is spun around its center, which is located a unit distance from the $x$-axis. (See Figure 5.4.) When the flashlight has stopped spinning, consider the point $X$ at which the beam intersects the $x$-axis. (If the beam is not pointing toward the $x$-axis, repeat the experiment.)

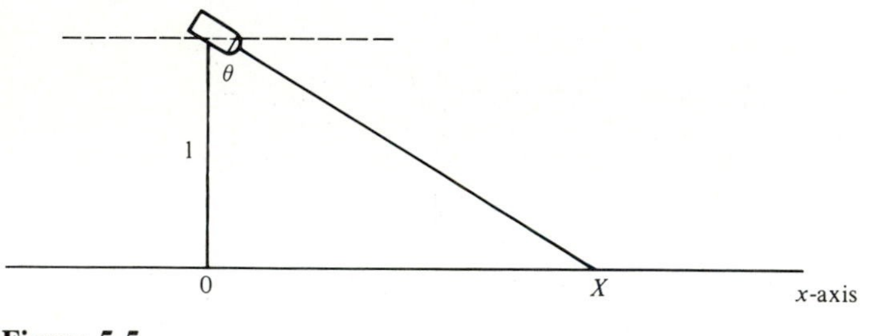

**Figure 5.5**

As indicated in Figure 5.5, the point $X$ is determined by the angle $\theta$ between the flashlight and the $y$-axis, which from the physical situation appears to be uniformly distributed between $-\pi/2$ and $\pi/2$. The distribution function of $X$ is thus given by

$$\begin{aligned} F(x) &= P\{X \leq x\} \\ &= P\{\tan \theta \leq x\} \\ &= P\{\theta \leq \tan^{-1} x\} \\ &= \frac{1}{2} + \frac{1}{\pi} \tan^{-1} x \end{aligned}$$

where the last equality follows since $\theta$, being uniform over $(-\pi/2, \pi/2)$, yields that

$$P\{\theta \le a\} = \frac{a - (-\pi/2)}{\pi} = \frac{1}{2} + \frac{a}{\pi}, \qquad -\frac{\pi}{2} < a < \frac{\pi}{2}$$

Hence the density function of $X$ is given by

$$f(x) = \frac{d}{dx} F(x) = \frac{1}{\pi(1 + x^2)} \qquad -\infty < x < \infty$$

and we see that $X$ has the Cauchy distribution.[6] ∎

## 5.4 *The Beta Distribution*

A random variable is said to have a beta distribution if its density is given by

$$f(x) = \begin{cases} \dfrac{1}{B(a, b)} x^{a-1}(1 - x)^{b-1} & 0 < x < 1 \\ 0 & \text{otherwise} \end{cases}$$

where

$$B(a, b) = \int_0^1 x^{a-1}(1 - x)^{b-1}\, dx$$

The beta distribution can be used to model a random phenomenon whose set of possible values is some finite interval $[c, d]$—which by letting $c$ denote the origin and taking $d - c$ as a unit measurement can be transformed into the interval $[0, 1]$.

[6] That $d/dx \tan^{-1} x = 1/1 + x^2$ can be seen as follows: If $y = \tan^{-1} x$, then $\tan y = x$, and so

$$1 = \frac{d}{dx}(\tan y) = \frac{d}{dy}(\tan y)\frac{dy}{dx} = \frac{d}{dy}\left(\frac{\sin y}{\cos y}\right)\frac{dy}{dx}$$
$$= \left(\frac{\cos^2 y + \sin^2 y}{\cos^2 y}\right)\frac{dy}{dx}$$

or

$$\frac{dy}{dx} = \cos^2 y = \frac{\cos^2 y}{\sin^2 y + \cos^2 y} = \frac{1}{\tan^2 y + 1} = \frac{1}{x^2 + 1}$$

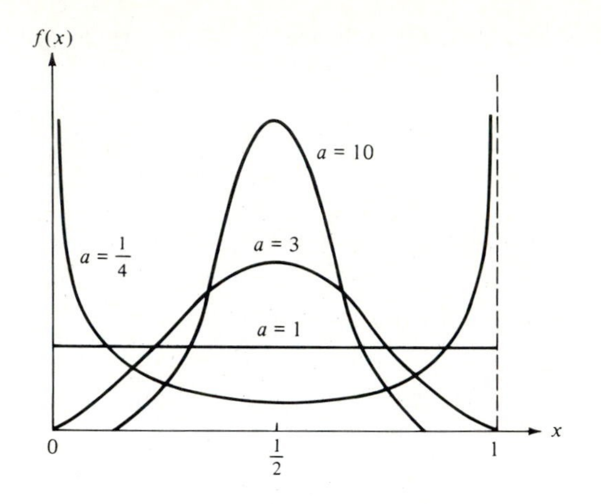


**Figure 5.6** Beta densities with parameters $(a, b)$ when $a = b$

When $a = b$, the beta density is symmetric about $\frac{1}{2}$, giving more and more weight to regions about $\frac{1}{2}$ as the common value $a$ increases. (See Figure 5.6.) When $b > a$, the density is skewed to the left (in the sense that smaller values become more likely); and it is skewed to the right when $a > b$. (See Figure 5.7.)

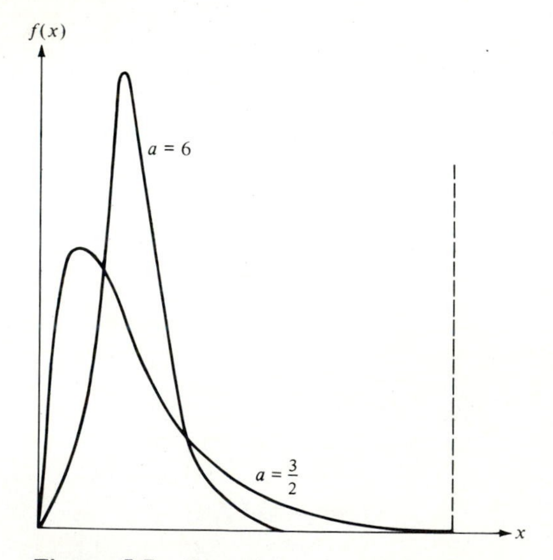


**Figure 5.7** Beta densities with parameters $(a, b)$ when

$$\frac{a}{a+b} = \frac{1}{20}$$

## 6 The Distribution of a Function of a Random Variable

It is often the case that we know the probability distribution of a random variable and are interested in determining the distribution of some function of it. For instance, suppose that we know the distribution of $X$ and want to find the distribution of $g(X)$. To do so, it is necessary to express the event that $g(X) \leq y$ in terms of $X$ being in some set. We illustrate by the following examples.

**Example 6a.** Let $X$ be uniformly distributed over $(0, 1)$. We obtain the distribution of the random variable $Y$, defined by $Y = X^n$ as follows: For $0 \leq y \leq 1$,

$$\begin{aligned} F_Y(y) &= P\{Y \leq y\} \\ &= P\{X^n \leq y\} \\ &= P\{X \leq y^{1/n}\} \\ &= F_X(y^{1/n}) \\ &= y^{1/n} \end{aligned}$$

Thus, for instance, the density function of $Y$ is given by

$$f_Y(y) = \begin{cases} \dfrac{1}{n}\, y^{-[(n-1)/n]} & 0 \leq y \leq 1 \\ 0 & \text{otherwise} \end{cases}$$ ∎

**Example 6b.** If $X$ is a continuous random variable with probability density $f_X$, then the distribution of $Y = X^2$ is obtained as follows: For $y \geq 0$,

$$\begin{aligned} F_Y(y) &= P\{Y \leq y\} \\ &= P\{X^2 \leq y\} \\ &= P\{-\sqrt{y} \leq X \leq \sqrt{y}\} \\ &= F_X(\sqrt{y}) - F_X(-\sqrt{y}) \end{aligned}$$

Differentiation yields

$$f_Y(y) = \frac{1}{2\sqrt{y}}[f_X(\sqrt{y}) + f_X(-\sqrt{y})]$$ ∎

**Example 6c.** If $X$ has a probability density $f_X$, then $Y = |X|$ has a density function that is obtained as follows: For $y \geq 0$

$$\begin{aligned} F_Y(y) &= P\{Y \leq y\} \\ &= P\{|X| \leq y\} \\ &= P\{-y \leq X \leq y\} \\ &= F_X(y) - F_X(-y) \end{aligned}$$

Hence, on differentiation, we obtain

$$f_Y(y) = f_X(y) + f_X(-y) \qquad y \geq 0$$ ∎

The method employed in Examples 6a through 6c can be used to prove Theorem 6.1.

---

## Theorem 6.1

*Let $X$ be a continuous random variable having probability density function $f_X$. Suppose that $g(x)$ is a strictly monotone (increasing or decreasing) differentiable (and thus continuous) function of $x$. Then the random variable $Y$ defined by $Y = g(X)$ has a probability density function given by*

$$f_Y(y) = \begin{cases} f_X[g^{-1}(y)] \left| \dfrac{d}{dy} g^{-1}(y) \right| & \text{if } y = g(x) \text{ for some } x \\ 0 & \text{if } y \neq g(x) \text{ for all } x \end{cases}$$

*where $g^{-1}(y)$ is defined to equal that value of $x$ such that $g(x) = y$.*

---

The proof of Theorem 6.1 is left as an exercise.

## Theoretical Exercises

**1.** Give a counterexample to the following proposition: If $P(E_a) = 1$ for all $0 < a < 1$, then $P\left(\bigcap_a E_a\right) = 1$, where the intersection is over all $a \in (0, 1)$.

HINT: Let $X$ be uniformly distributed over $(0, 1)$. Define $E_a$ in terms of $X$.

**2.** The speed of a molecule in a uniform gas at equilibrium is a random variable whose probability density function is given by

$$f(x) = \begin{cases} ax^2 e^{-bx^2} & x \geq 0 \\ 0 & x < 0 \end{cases}$$

where $b = m/2kT$ and $k$, $T$, $m$ denote, respectively, Boltzmann's constant, the absolute temperature, and the mass of the molecule. Evaluate $a$ in terms of $b$.

**3.** Prove Equation (3.2).

**4.** If $Z$ is a unit normal random variable, show that, for every $a > 0$,

$$\lim_{x \to \infty} \frac{P\{Z \geq x + a/x\}}{P\{Z \geq x\}} = e^{-a}$$

**5.** The median of a continuous random variable having distribution function $F$ is that value $m$ such that $F(m) = \frac{1}{2}$. That is, a random variable is just as likely to be larger than its median as it is to be smaller. Find the median of $X$ if $X$ is
(a) uniformly distributed over $(a, b)$;
(b) normal with parameters $\mu, \sigma^2$;
(c) exponential with rate $\lambda$.

**6.** The mode of a continuous random variable having density $f$ is the value of $x$ for which $f(x)$ attains its maximum. Compute the mode of $X$ in cases (a), (b), and (c) of Theoretical Exercise 5.

**7.** If $X$ is an exponential random variable with parameter $\lambda$, and $c > 0$, show that $cX$ is exponential with parameter $\lambda/c$.

**8.** Compute the hazard rate function of $X$ when $X$ is uniformly distributed over $(0, a)$.

**9.** If $X$ has hazard rate function $\lambda_X(t)$, compute the hazard rate function of $aX$ where $a$ is a positive constant.

**10.** Verify that the gamma density function integrates to 1.

**11.** Compute $\Gamma(n + \frac{1}{2})$ for $n = 1, 2, \ldots$.

**12.** Compute the hazard rate function of a gamma random variable with parameters $(t, \lambda)$ and show it is increasing when $t \geq 1$ and decreasing when $t \leq 1$.

**13.** Compute the hazard rate function of a Weibull random variable and show it is increasing when $\beta \geq 1$ and decreasing when $\beta \leq 1$.

**14.** Show that a plot of $\log(\log(1 - F(x))^{-1})$ against $\log x$ will be a straight line with slope $\beta$ when $F(\cdot)$ is a Weibull distribution function. Show also that approximately 63.2 percent of all observations from such a distribution will be less than $\alpha$. Assume that $v = 0$.

**15.** Let

$$Y = \left(\frac{X - v}{\alpha}\right)^{\beta}$$

Show that if $X$ is a Weibull random variable with parameters $v, \alpha, \beta$, then $Y$ is an exponential random variable with parameter $\lambda = 1$, and vice versa.

**16.** Show that

$$B(a, b) = \frac{\Gamma(a)\Gamma(b)}{\Gamma(a+b)}$$

**17.** If $X$ is uniformly distributed over $(a, b)$, what random variable, having a linear relation with $X$, is uniformly distributed over $(0, 1)$?

**18.** Consider the Beta distribution with parameters $(a, b)$. Show that:
(a) when $a > 1$, $b > 1$, the density is unimodal (that is, it has a unique mode) with mode equal to $(a-1)/(a+b-2)$;
(b) when $a \leq 1$, $b \leq 1$, $a + b < 2$, the density is either unimodal with mode at 0 or 1 or U-shaped with modes at both 0 and 1;
(c) when $a = 1 = b$, all points in $[0, 1]$ are modes.

**19.** Let $X$ be a continuous random variable having cumulative distribution function $F$. Define the random variable $Y$ by $Y = F(X)$. Show that $Y$ is uniformly distributed over $(0, 1)$.

**20.** Let $X$ have probability density $f_X$. Find the probability density function of the random variable $Y$, defined by $Y = aX + b$.

**21.** Find the probability density function of $Y = e^X$ when $X$ is normally distributed with parameters $\mu$ and $\sigma^2$. The random variable $Y$ is said to have a lognormal distribution (since $\log Y$ has a normal distribution) with parameters $\mu$ and $\sigma^2$.

**22.** Prove Theorem 6.1.

**23.** Let $X$ and $Y$ be independent random variables that are both equally likely to be either $1, 2, \ldots, (10)^N$, where $N$ is very large. Let $D$ denote the greatest common divisor of $X$ and $Y$, and let $Q_k = P\{D = k\}$

(a) Give a heuristic argument that $Q_k = \dfrac{1}{k^2} Q_1$.

HINT: Note that in order for $D$ to equal $k$, $k$ must divide both $X$ and $Y$ and also $X/k$ and $Y/k$ must be relatively prime. (That is, they must have a greatest common divisor equal to 1.)

(b) Use (a) to show that

$$Q_1 = P\{X \text{ and } Y \text{ are relatively prime}\} = \frac{1}{\sum_{k=1}^{\infty} 1/k^2}$$

It is a well-known identity that $\sum_1^\infty 1/k^2 = \pi^2/6$ and so $Q_1 = 6/\pi^2$ (In number theory this is known as the Legendre theorem.)

(c) Now argue that

$$Q_1 = \prod_{i=1}^{\infty} \left( \frac{P_i^2 - 1}{P_i^2} \right),$$

where $P_i$ is the $i$th smallest prime greater than 1.

HINT: $X$ and $Y$ will be relatively prime if they have no common prime factors.

Hence, from (b), we see that $\prod_{i=1}^{\infty} \left( \frac{P_i^2 - 1}{P_i^2} \right) = \frac{6}{\pi^2}$, which was noted without explanation in Problem 11 of Chapter 4. (The relationship between this problem and Problem 11 of Chapter 4 is that $X$ and $Y$ are relatively prime if $XY$ has no multiple prime factors.)

## Problems

**1.** Let $X$ be a random variable with probability density function

$$f(x) = \begin{cases} c(1 - x^2) & -1 < x < 1 \\ 0 & \text{otherwise} \end{cases}$$

(a) What is the value of $c$?
(b) What is the cumulative distribution function of $X$?

**2.** A system consisting of one original unit plus a spare can function for a random amount of time $X$. If the density of $X$ is given (in units of months) by

$$f(x) = \begin{cases} Cxe^{-x/2} & x > 0 \\ 0 & x \le 0 \end{cases}$$

what is the probability that the system functions for at least 5 months?

**3.** Consider the function

$$f(x) = \begin{cases} C(2x - x^3) & 0 < x < \frac{5}{2} \\ 0 & \text{otherwise} \end{cases}$$

Could $f$ be a probability density function? If so, determine $C$. Repeat if $f(x)$ were given by

$$f(x) = \begin{cases} C(2x - x^2) & 0 < x < \frac{5}{2} \\ 0 & \text{otherwise} \end{cases}$$

**4.** The probability density function of $X$, the lifetime of a certain type of electronic device (measured in hours), is given by

$$f(x) = \begin{cases} \dfrac{10}{x^2} & x > 10 \\ 0 & x \le 10 \end{cases}$$

(a) Find $P\{X > 20\}$.
(b) What is the cumulative distribution function of $X$?
(c) What is the probability that of 6 such types of devices at least 3 will function for at least 15 hours? What assumptions are you making?

**5.** A filling station is supplied with gasoline once a week. If its weekly volume of sales in thousands of gallons is a random variable with probability density function

$$f(x) = \begin{cases} 5(1-x)^4 & 0 < x < 1 \\ 0 & \text{otherwise} \end{cases}$$

what need the capacity of the tank be so that the probability of the supply's being exhausted in a given week is .01?

**6.** Trains headed for destination $A$ arrive at the train station at 15-minute intervals starting at 7 A.M., whereas trains headed for destination $B$ arrive at 15-minute intervals starting at 7:05 A.M. If a certain passenger arrives at the station at a time uniformly distributed between 7 and 8 A.M. and then gets on the first train that arrives, what proportion of time does he or she go to destination $A$? What if the passenger arrives at a time uniformly distributed between 7:10 and 8:10 A.M.?

**7.** A point is chosen at random on a line segment of length $L$. Interpret this statement and find the probability that the ratio of the shorter to the longer segment is less than $\frac{1}{4}$.

**8.** A bus travels between the two cities $A$ and $B$, which are 100 miles apart. If the bus has a breakdown, the distance from the breakdown to city $A$ has a uniform distribution over $(0, 100)$. There is a bus service station in city $A$, in $B$, and in the center of the route between $A$ and $B$. It is suggested that it would be more efficient to have the three stations located 25, 50, and 75 miles, respectively, from $A$. Do you agree? Why?

**9.** You arrive at a bus stop at 10 o'clock, knowing that the bus will arrive at some time uniformly distributed between 10 and 10:30. What is the probability that you will have to wait longer than 10 minutes? If at 10:15 the bus has not yet arrived, what is the probability that you will have to wait at least an additional 10 minutes?

**10.** If $X$ is a normal random variable with parameters $\mu = 10$, $\sigma^2 = 36$, compute
(a) $P\{X > 5\}$; (b) $P\{4 < X < 16\}$; (c) $P\{X < 8\}$;
(d) $P\{X < 20\}$; (e) $P\{X > 16\}$.

**11.** The annual rainfall (in inches) in a certain region is normally distributed with $\mu = 40$, $\sigma = 4$. What is the probability that starting with this year, it will take over 10 years before a year occurs having a rainfall of over 50 inches? What assumptions are you making?

**12.** Suppose that the height, in inches, of a 25-year-old man is a normal random variable with parameters $\mu = 71$, $\sigma^2 = 6.25$. What percentage of 25-year-old men are over 6 feet 2 inches tall? What percentage of men in the 6-footer club are over 6 foot 5 inches?

**13.** The width of a slot of a duralumin forging is (in inches) normally distributed with $\mu = .9000$ and $\sigma = .0030$. The specification limits were given as $.9000 \pm .0050$. What percentage of forgings will be defective? What is the maximum allowable value of $\sigma$ that will permit no more than 1 in 100 defectives when the widths are normally distributed with $\mu = .9000$ and $\sigma$?

**14.** One thousand independent rolls of a fair die will be made. Compute an approximation to the probability that number 6 will appear between 150 and 200 times. If number 6 appears exactly 200 times, find the probability that number 5 will appear less than 150 times.

**15.** The lifetimes of interactive computer chips produced by a certain semiconductor manufacturer are normally distributed with parameters $\mu = 1.4 \times 10^6$ hours and $\sigma = 3 \times 10^5$ hours. What is the approximate probability that a batch of 100 chips will contain at least 20 whose lifetimes are less than $1.8 \times 10^6$?

**16.** Use the computer program provided at the end of Chapter 4 to compute $P\{X \leq 25\}$ when $X$ is a binomial random variable with parameters $n = 300$, $p = .1$. Now compare this with its (a) Poisson and (b) normal approximation. In using the normal approximation, write the desired probability as $P\{X < 25.5\}$ so as to utilize the continuity correction. (You will need the program at the end of Chapter 4 to compute the Poisson approximation).

**17.** Two types of coins are produced at a factory: a fair coin and a biased one that comes up heads 55 percent of the time. We have one of these coins but do not know whether it is a fair coin or a biased one. In order to ascertain which type of coin we have, we shall perform the following statistical test: We shall toss the coin 1000 times. If the coin lands on heads 525 or more times, then we shall conclude that it is a biased coin, whereas, if it lands heads less than 525 times, then we shall conclude that it is the fair coin. If the coin is actually fair, what is the probability that we shall reach a false conclusion? What would it be if the coin were biased?

**18.** In 10,000 independent tosses of a coin, the coin landed heads 5800 times. Is is reasonable to assume that the coin is not fair? Explain.

**19.** An image is partitioned into 2 regions—one white and the other black. A reading taken from a randomly chosen point in the white section will give a reading that is normally distributed with parameters (4, 4), whereas one taken from a randomly chosen point in the black region will have a normally distributed reading with parameters (6, 9). A point is randomly chosen on the image and has a reading of 5. If the fraction of the image that is black is $\alpha$, for what value of $\alpha$ would the probability of making an error be the same whether one concluded the point was in the black region or in the white region?

**20.** The time (in hours) required to repair a machine is an exponentially distributed random variable with parameter $\lambda = \frac{1}{2}$.
(a) What is the probability that a repair time exceeds 2 hours?
(b) What is the conditional probability that a repair takes at least 10 hours, given that its duration exceeds 9 hours?

**21.** The number of years a radio functions is exponentially distributed with parameter $\lambda = \frac{1}{8}$. If Jones buys a used radio, what is the probability that it will be working after an additional 8 years?

**22.** Jones figures that the total number of thousands of miles that an auto can be driven before it would need to be junked is an exponential random variable with parameter $\frac{1}{20}$. Smith has a used car that he claims has been driven only 10,000 miles. If Jones purchases the car, what is the probability that she would get at least 20,000 additional miles out of it? Repeat under the assumption that the lifetime mileage of the car is not exponentially distributed but rather is (in thousands of miles) uniformly distributed over (0, 40).

**23.** The lung cancer hazard rate of a $t$-year old male smoker, $\lambda(t)$, is such that

$$\lambda(t) = .027 + .00025(t - 40)^2, \quad t \geq 40$$

Assuming that a 40 year old male smoker survives all other hazards, what is the probability that he survives to (a) age 50, (b) age 60, without contracting lung cancer?

**24.** Suppose the life distribution of an item has hazard rate function $\lambda(t) = t^3, t > 0$.
(a) What is the probability the item survives to age 2?
(b) What is the probability that the item's lifetime is between .4 and 1.4?
(c) What is the probability a 1-year-old item will survive to age 2?

**25.** If $X$ is uniformly distributed over $(-1, 1)$, find
(a) $P\{|X| > \frac{1}{2}\}$;

(b) $P\{\sin(\pi X/2) > \frac{1}{3}\}$;
(c) the density function of the random variable $|X|$.

**26.** If $Y$ is uniformly distributed over $(0, 5)$, what is the probability that the roots of the equation $4x^2 + 4xY + Y + 2 = 0$ are both real?

**27.** If $X$ is an exponential random variable with parameter $\lambda = 1$, compute the probability density function of the random variable $Y$ defined by $Y = \log X$.

**28.** If $X$ is uniformly distributed over $(0, 1)$, find the density function of $Y = e^X$.

**29.** Find the distribution of $R = A \sin \theta$, where $A$ is a fixed constant and $\theta$ is uniformly distributed on $(-\pi/2, \pi/2)$. Such a random variable $R$ arises in the theory of ballistics. If a projectile is fired from the origin at an angle $\alpha$ from the earth with a velocity $V$, then the point $R$ at which it returns to the earth can be expressed as $R = (v^2/g)(\sin 2\alpha)$, where $g$ is the gravitational constant, equal to 980 centimeters per second squared.

# 6

# Jointly Distributed Random Variables

## 1 Joint Distribution Functions

Thus far, we have only concerned ourselves with probability distributions for single random variables. However, we are often interested in probability statements concerning two or more random variables. In order to deal with such probabilities, we define, for any two random variables $X$ and $Y$, the *joint cumulative probability distribution function* of $X$ and $Y$ by

$$F(a, b) = P\{X \leq a, Y \leq b\} \qquad -\infty < a, b < \infty$$

The distribution of $X$ can be obtained from the joint distribution of $X$ and $Y$ as follows:

$$\begin{aligned} F_X(a) &= P\{X \leq a\} \\ &= P\{X \leq a, Y < \infty\} \\ &= P\left(\lim_{b\to\infty} \{X \leq a, Y \leq b\}\right) \\ &= \lim_{b\to\infty} P\{X \leq a, Y \leq b\} \\ &= \lim_{b\to\infty} F(a, b) \\ &\equiv F(a, \infty) \end{aligned}$$

The reader should note that we have, in the preceding set of equalities, once again made use of the fact that probability is a continuous set (that is, event) function. Similarly, the cumulative distribution function of $Y$ is given by

$$\begin{aligned} F_Y(b) &= P\{Y \leq b\} \\ &= \lim_{a\to\infty} F(a, b) \\ &\equiv F(\infty, b) \end{aligned}$$

The distribution functions $F_X$ and $F_Y$ are sometimes referred to as the *marginal* distributions of $X$ and $Y$.

All joint probability statements about $X$ and $Y$ can, in theory, be answered in terms of their joint distribution function. For instance, suppose we wanted to compute the joint probability that $X$ is greater than $a$ and $Y$ greater than $b$. This could be done as follows.

$$\begin{aligned} P\{X > a, Y > b\} &= 1 - P(\{X > a, Y > b\}^c) \\ &= 1 - P(\{X > a\}^c \cup \{Y > b\}^c) \\ &= 1 - P(\{X \le a\} \cup \{Y \le b\}) \\ &= 1 - [P\{X \le a\} + P\{Y \le b\} - P\{X \le a, Y \le b\}] \\ &= 1 - F_X(a) - F_Y(b) + F(a, b) \end{aligned} \tag{1.1}$$

Equation (1.1) is a special case of Equation (1.2), whose verification is left as an exercise.

$$\begin{aligned} &P\{a_1 < X \le a_2, b_1 < Y \le b_2\} \\ &\quad = F(a_2, b_2) + F(a_1, b_1) - F(a_1, b_2) - F(a_2, b_1) \end{aligned} \tag{1.2}$$

whenever $a_1 < a_2$, $b_1 < b_2$.

In the case when $X$ and $Y$ are both discrete random variables, it is convenient to define the *joint probability mass function* of $X$ and $Y$ by

$$p(x, y) = P\{X = x, Y = y\}$$

The probability mass function of $X$ can be obtained from $p(x, y)$ by

$$\begin{aligned} p_X(x) &= P\{X = x\} \\ &= \sum_{y:p(x,y)>0} p(x, y) \end{aligned}$$

Similarly,

$$p_Y(y) = \sum_{x:p(x,y)>0} p(x, y)$$

**Example 1a.** Suppose that 3 balls are randomly selected from an urn containing 3 red, 4 white, and 5 blue balls. If we let $X$ and $Y$ denote, respectively, the number of red and white balls chosen, then the joint probability mass function of $X$ and $Y$, $p(i, j) = P\{X = i, Y = j\}$, is given by

$$p(0, 0) = \binom{5}{3} \Big/ \binom{12}{3} = \frac{10}{220}$$

$$p(0, 1) = \binom{4}{1}\binom{5}{2} \Big/ \binom{12}{3} = \frac{40}{220}$$

$$p(0, 2) = \binom{4}{2}\binom{5}{1} \Big/ \binom{12}{3} = \frac{30}{220}$$

$$p(0, 3) = \binom{4}{3} \Big/ \binom{12}{3} = \frac{4}{220}$$

$$p(1, 0) = \binom{3}{1}\binom{5}{2} \Big/ \binom{12}{3} = \frac{30}{220}$$

$$p(1, 1) = \binom{3}{1}\binom{4}{1}\binom{5}{1} \Big/ \binom{12}{3} = \frac{60}{220}$$

$$p(1, 2) = \binom{3}{1}\binom{4}{2} \Big/ \binom{12}{3} = \frac{18}{220}$$

$$p(2, 0) = \binom{3}{2}\binom{5}{1} \Big/ \binom{12}{3} = \frac{15}{220}$$

$$p(2, 1) = \binom{3}{2}\binom{4}{1} \Big/ \binom{12}{3} = \frac{12}{220}$$

$$p(3, 0) = \binom{3}{3} \Big/ \binom{12}{3} = \frac{1}{220}$$

These probabilities can most easily be expressed in tabular form as in Table 6.1.

**Table 6.1** $P\{X = i, Y = j\}$

| $i$ \ $j$ | 0 | 1 | 2 | 3 | *Row sum* $= P\{X = i\}$ |
|---|---|---|---|---|---|
| 0 | $\frac{10}{220}$ | $\frac{40}{220}$ | $\frac{30}{220}$ | $\frac{4}{220}$ | $\frac{84}{220}$ |
| 1 | $\frac{30}{220}$ | $\frac{60}{220}$ | $\frac{18}{220}$ | 0 | $\frac{108}{220}$ |
| 2 | $\frac{15}{220}$ | $\frac{12}{220}$ | 0 | 0 | $\frac{27}{220}$ |
| 3 | $\frac{1}{220}$ | 0 | 0 | 0 | $\frac{1}{220}$ |
| *Column sum* $= P\{Y = j\}$ | $\frac{56}{220}$ | $\frac{112}{220}$ | $\frac{48}{220}$ | $\frac{4}{220}$ | |

The reader should note that the probability mass function of $X$ is obtained by computing the row sums, whereas the probability mass function of $Y$ is obtained by computing the column sums. As the individual probability mass functions of $X$ and $Y$ thus appear in the margin of such a table, they are often referred to as being the marginal probability mass functions of $X$ and $Y$, respectively. ∎

**Example 1b.** Suppose that 15 percent of the families in a certain community have no children, 20 percent have 1, 35 percent have 2, and 30 percent have 3; and suppose, further, that in each family, each child is equally likely (independently) to be a boy or a girl. If a family is chosen at random from this community, then $B$, the number of boys, and $G$, the number of girls, in this family will have the joint probability mass function shown in Table 6.2.

**Table 6.2** $P\{B = i, G = j\}$

| $i$ \ $j$ | 0 | 1 | 2 | 3 | *Row sum* = $P\{B = i\}$ |
|---|---|---|---|---|---|
| 0 | .15 | .10 | .0875 | .0375 | .3750 |
| 1 | .10 | .175 | .1125 | 0 | .3875 |
| 2 | .0875 | .1125 | 0 | 0 | .2000 |
| 3 | .0375 | 0 | 0 | 0 | .0375 |
| *Column sum* = $P\{G = j\}$ | .375 | .3875 | .2000 | .0375 | |

***Solution:*** These probabilities are obtained as follows:

$$P\{B = 0, G = 0\} = P\{\text{no children}\} = .15$$

$$\begin{aligned} P\{B = 0, G = 1\} &= P\{1 \text{ girl and total of } 1 \text{ child}\} \\ &= P\{1 \text{ child}\}\, P\{1 \text{ girl} \mid 1 \text{ child}\} = (.20)(\tfrac{1}{2}) \end{aligned}$$

$$\begin{aligned} P\{B = 0, G = 2\} &= P\{2 \text{ girls and total of } 2 \text{ children}\} \\ &= P\{2 \text{ children}\} P\{2 \text{ girls} \mid 2 \text{ children}\} = (.35)(\tfrac{1}{2})^2 \end{aligned}$$

We leave the verification of the remaining probabilities in Table 6.2 to the reader. ∎

We say that $X$ and $Y$ are *jointly continuous* if there exists a function $f(x, y)$ defined for all real $x$ and $y$, having the property that for every set

$C$ of pairs of real numbers (that is, $C$ is a set in the two-dimensional plane)

$$P\{(X, Y) \in C\} = \iint_{(x,y)\in C} f(x, y)\,dx\,dy \tag{1.3}$$

The function $f(x, y)$ is called the *joint probability density function* of $X$ and $Y$. If $A$ and $B$ are any sets of real numbers, then by defining $C = \{(x, y)\colon x \in A,\ y \in B\}$, we see from Equation (1.3) that

$$P\{X \in A, Y \in B\} = \int_B \int_A f(x, y)\,dx\,dy \tag{1.4}$$

Because

$$\begin{aligned} F(a, b) &= P\{X \in (-\infty, a], Y \in (-\infty, b]\} \\ &= \int_{-\infty}^{b} \int_{-\infty}^{a} f(x, y)\,dx\,dy \end{aligned}$$

it follows, upon differentiation, that

$$f(a, b) = \frac{\partial^2}{\partial a\,\partial b} F(a, b)$$

wherever the partial derivatives are defined. Another interpretation of the joint density function is obtained from Equation (1.4) as follows:

$$\begin{aligned} P\{a < X < a + da, b < Y < b + db\} &= \int_{b}^{a+db} \int_{a}^{a+da} f(x, y)\,dx\,dy \\ &\approx f(a, b)\,da\,db \end{aligned}$$

when $da$ and $db$ are small and $f(x, y)$ is continuous at $a, b$. Hence $f(a, b)$ is a measure of how likely it is that the random vector $(X, Y)$ will be near $(a, b)$.

If $X$ and $Y$ are jointly continuous, they are individually continuous, and their probability density functions can be obtained as follows:

$$\begin{aligned} P\{X \in A\} &= P\{X \in A, Y \in (-\infty, \infty)\} \\ &= \int_A \int_{-\infty}^{\infty} f(x, y)\,dy\,dx \\ &= \int_A f_X(x)\,dx \end{aligned}$$

where

$$f_X(x) = \int_{-\infty}^{\infty} f(x, y)\,dy$$

is thus the probability density function of $X$. Similarly, the probability density function of $Y$ is given by

$$f_Y(y) = \int_{-\infty}^{\infty} f(x, y)\, dx$$

**Example 1c.** The joint density function of $X$ and $Y$ is given by

$$f(x, y) = \begin{cases} 2e^{-x}e^{-2y} & 0 < x < \infty, 0 < y < \infty \\ 0 & \text{otherwise} \end{cases}$$

Compute (1) $P\{X > 1,\ Y < 1\}$; (2) $P\{X < Y\}$; and (3) $P\{X < a\}$.

***Solution***

1.
$$\begin{aligned} P\{X > 1, Y < 1\} &= \int_0^1 \int_1^{\infty} 2e^{-x}e^{-2y}\, dx\, dy \\ &= \int_0^1 2e^{-2y}\left(-e^{-x}\Big|_1^{\infty}\right) dy \\ &= e^{-1}\int_0^1 2e^{-2y}\, dy \\ &= e^{-1}(1 - e^{-2}) \end{aligned}$$

2.
$$\begin{aligned} P\{X < Y\} & \iint\limits_{(x,y):x<y} 2e^{-x}e^{-2y}\, dx\, dy \\ &= \int_0^{\infty}\int_0^{y} 2e^{-x}e^{-2y}\, dx\, dy \\ &= \int_0^{\infty} 2e^{-2y}(1 - e^{-y})\, dy \\ &= \int_0^{\infty} 2e^{-2y}\, dy - \int_0^{\infty} 2e^{-3y}\, dy \\ &= 1 - \tfrac{2}{3} \\ &= \tfrac{1}{3} \end{aligned}$$

3.
$$\begin{aligned} P\{X < a\} &= \int_0^{a}\int_0^{\infty} 2e^{-2y}e^{-x}\, dy\, dx \\ &= \int_0^{a} e^{-x}\, dx \\ &= 1 - e^{-a} \end{aligned}$$

∎

**Example 1d.** Consider a circle of radius $R$ and suppose that a point within the circle is randomly chosen in such a manner that all regions within the circle of equal area are equally likely to contain the point. (In other words, the point is uniformly distributed within the circle.) If we let the center of the circle denote the origin and define $X$ and $Y$ to be the coordinates of the point chosen (Figure 6.1), it follows, since $(X, Y)$ is equally likely to be near each point in the circle, that the joint density function of $X$ and $Y$ is given by

$$f(x, y) = \begin{cases} c & \text{if } x^2 + y^2 \leq R^2 \\ 0 & \text{if } x^2 + y^2 > R^2 \end{cases}$$

for some value of $c$.

1. Determine $c$.
2. Find the marginal density functions of $X$ and $Y$.
3. Compute the probability that the distance from the origin of the point selected is not greater than $a$.

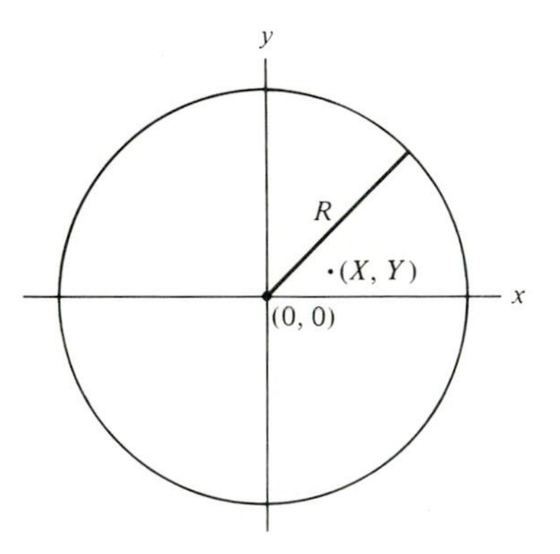


**Figure 6.1** Joint probability distribution

***Solution***

1. Because

$$\int_{-\infty}^{\infty} \int_{-\infty}^{\infty} f(x, y)\, dy\, dx = 1$$

it follows that

$$c \iint_{x^2+y^2 \leq R^2} dy\, dx = 1$$

We can evaluate $\iint_{x^2+y^2\le R^2} dy\,dx$ either by using polar coordinates, or more simply, by noting that it represents the area of the circle and is thus equal to $\pi R^2$. Hence

$$c = \frac{1}{\pi R^2}$$

2.

$$\begin{aligned} f_X(x) &= \int_{-\infty}^{\infty} f(x, y)\, dy \\ &= \frac{1}{\pi R^2} \int_{x^2+y^2\le R^2} dy \\ &= \frac{1}{\pi R^2} \int_{-\sqrt{R^2-x^2}}^{+\sqrt{R^2-x^2}} dy \qquad x^2 \le R^2 \\ &= \frac{2}{\pi R^2} \sqrt{R^2 - x^2} \qquad x^2 \le R^2 \end{aligned}$$

and it equals 0 when $x^2 > R^2$. By symmetry the marginal density of $Y$ is given by

$$\begin{aligned} f_Y(y) &= \frac{2}{\pi R^2} \sqrt{R^2 - y^2} \qquad y^2 \le R^2 \\ &= 0 \qquad y^2 > R^2 \end{aligned}$$

3. The distribution function of $Z = \sqrt{X^2 + Y^2}$, the distance from the origin, is obtained as follows: for $0 \le a \le R$,

$$\begin{aligned} F_Z(a) &= P\{\sqrt{X^2 + Y^2} \le a\} \\ &= P\{X^2 + Y^2 \le a^2\} \\ &= \iint_{x^2+y^2\le a^2} f(x, y)\, dy\, dx \\ &= \frac{1}{\pi R^2} \iint_{x^2+y^2\le a^2} dy\, dx \\ &= \frac{\pi a^2}{\pi R^2} \\ &= \frac{a^2}{R^2} \end{aligned}$$

where we have used the fact that $\iint_{x^2+y^2\le a^2} dy\,dx$ is the area of a circle of radius $a$ and thus is equal to $\pi a^2$. ∎

**Example 1e.** The joint density of $X$ and $Y$ is given by

$$f(x, y) = \begin{cases} e^{-(x+y)} & 0 < x < \infty, 0 < y < \infty \\ 0 & \text{otherwise} \end{cases}$$

Find the density function of the random variable $X/Y$.

***Solution:*** We start by computing the distribution function of $X/Y$. For $a < 0$,

$$\begin{aligned} F_{X/Y}(a) &= P\left\{\frac{X}{Y} \le a\right\} \\ &= \iint_{x/y \le a} e^{-(x+y)}\, dx\, dy \\ &= \int_0^\infty \int_0^{ay} e^{-(x+y)}\, dx\, dy \\ &= \int_0^\infty (1 - e^{-ay})e^{-y}\, dy \\ &= \left[-e^{-y} + \frac{e^{-(a+1)y}}{a+1}\right]\Bigg|_0^\infty \\ &= 1 - \frac{1}{a+1} \end{aligned}$$

Differentiation yields that the density function of $X/Y$ is given by $f_{X/Y}(a) = 1/(a+1)^2$, $0 < a < \infty$. ∎

We can also define joint probability distributions for $n$ random variables in exactly the same manner as we did for $n = 2$. For instance, the joint cumulative probability distribution function $F(a_1, a_2, \ldots, a_n)$ of the $n$ random variables $X_1, X_2, \ldots, X_n$ is defined by

$$F(a_1, a_2, \ldots, a_n) = P\{X_1 \le a_1, X_2 \le a_2, \ldots, X_n \le a_n\}$$

Further, the $n$ random variables are said to be jointly continuous if there exists a function $f(x_1, x_2, \ldots, x_n)$, called the joint probability density function, such that for any set $C$ in $n$-space

$$P\{(X_1, X_2, \ldots, X) \in C\} = \iint \cdots \int_{(x_1,\ldots,x_n) \in C} f(x_1, \ldots, x_n)\, dx_1\, dx_2 \cdots dx_n$$

In particular, for any $n$ sets of real numbers $A_1, A_2, \ldots, A_n$

$$\begin{aligned} &P\{X_1 \in A_1, X_2 \in A_2, \ldots, X_n \in A_n\} \\ &\quad = \int_{A_n} \int_{A_{n-1}} \cdots \int_{A_1} f(x_1, \ldots, x_n)\, dx_1\, dx_2 \cdots dx_n \end{aligned}$$

**Example 1f.** Multinomial Distribution. One of the most important joint distributions is the multinomial, which arises when a sequence of $n$ independent and identical experiments is performed. Suppose that each experiment can result in any one of $r$ possible outcomes, with respective probabilities $p_1, p_2, \ldots, p_r$, $\sum_{i=1}^{r} p_i = 1$. If we denote by $X_i$, the number of the $n$ experiments that result in outcome number $i$, then

$$P\{X_1 = n_1, X_2 = n_2, \ldots, X_r = n_r\} = \frac{n!}{n_1!n_2!\ldots n_r!} p_1^{n_1} p_2^{n_2} \cdots p_r^{n_r} \tag{1.5}$$

whenever $\sum_{i=1}^{r} n_i = n$.

Equation (1.5) is verified by noting that any sequence of outcomes for the $n$ experiments that leads to outcome $i$ occurring $n_i$ times for $i = 1, 2, \ldots, r$, will, by the assumed independence of experiments, have probability $p_1^{n_1} p_2^{n_2} \cdots p_r^{n_r}$ of occurring. As there are $n!/n_1!\,n_2! \ldots n_r!$ such sequences of outcomes (there are $n!/n_1! \ldots n_r!$ different permutations of $n$ things of which $n_1$ are alike, $n_2$ are alike, $\ldots$, $n_r$ are alike), Equation (1.5) is established. The joint distribution whose joint probability mass function is specified by Equation (1.5) is called the multinomial distribution. The reader should note that when $r = 2$, the multinomial reduces to the binomial distribution.

As an application of the multinomial, suppose that a fair die is rolled 9 times. The probability that 1 appears three times, 2 and 3 twice each, 4 and 5 once each, and 6 not at all is

$$\frac{9!}{3!\,2!\,2!\,1!\,1!\,0!}\left(\frac{1}{6}\right)^3\left(\frac{1}{6}\right)^2\left(\frac{1}{6}\right)^2\left(\frac{1}{6}\right)^1\left(\frac{1}{6}\right)^1\left(\frac{1}{6}\right)^0 = \frac{9!}{3!\,2!\,2!}\left(\frac{1}{6}\right)^9 \qquad ∎$$

## 2 Independent Random Variables

The random variables $X$ and $Y$ are said to be *independent* if for any two sets of real numbers $A$ and $B$

$$P\{X \in A, Y \in B\} = P\{X \in A\}P\{Y \in B\} \tag{2.1}$$

In other words, $X$ and $Y$ are independent if, for all $A$ and $B$, the events $E_A = \{X \in A\}$ and $F_B = \{Y \in B\}$ are independent.

It can be shown by using the three axioms of probability that Equation (2.1) will follow if and only if for all $a$, $b$

$$P\{X \le a, Y \le b\} = P\{X \le a\}P\{Y \le b\}$$

Hence, in terms of the joint distribution function $F$ of $X$ and $Y$, we have that $X$ and $Y$ are independent if

$$F(a, b) = F_X(a)F_Y(b) \qquad \text{for all } a, b$$

When $X$ and $Y$ are discrete random variables, the condition of independence (2.1) is equivalent to

$$p(x, y) = p_X(x)p_Y(y) \qquad \text{for all } x, y \tag{2.2}$$

The equivalence follows because, if (2.1) is satisfied, then we obtain Equation (2.2) by letting $A$ and $B$ be, respectively, the one point sets $A = \{x\}$, $B = \{y\}$. Furthermore, if Equation (2.2) is valid, then for any sets $A$, $B$

$$\begin{aligned} P\{X \in A, Y \in B\} &= \sum_{y \in B} \sum_{x \in A} p(x, y) \\ &= \sum_{y \in B} \sum_{x \in A} p_X(x)p_Y(y) \\ &= \sum_{y \in B} p_Y(y) \sum_{x \in A} p_X(x) \\ &= P\{Y \in B\}P\{X \in A\} \end{aligned}$$

and thus Equation (2.1) is established.

In the jointly continuous case the condition of independence is equivalent to

$$f(x, y) = f_X(x)f_Y(y) \qquad \text{for all } x, y$$

Thus, loosely speaking, $X$ and $Y$ are independent if knowing the value of one does not change the distribution of the other. Random variables that are not independent are said to be *dependent.*

**Example 2a.** Suppose that $n + m$ independent trials, having a common success probability $p$, are performed. If $X$ is the number of successes in the first $n$ trials, and $Y$ the number of successes in the final $m$ trials, then $X$ and $Y$ are independent, since knowing the number of successes in the first $n$ trials does not affect the distribution of the number of successes in the final $m$ trials (by the assumption of independent trials). In fact, for integral $x$ and $y$,

$$\begin{aligned} P\{X = x, Y = y\} &= \binom{n}{x} p^x(1-p)^{n-x} \binom{m}{y} p^y(1-p)^{m-y}, \qquad \begin{array}{l} 0 \le x \le n \\ 0 \le y \le m \end{array} \\ &= P\{X = x\}P\{Y = y\} \end{aligned}$$

On the other hand, $X$ and $Z$ will be dependent where $Z$ is the total number of successes in the $n + m$ trials. (Why is this?) ∎

**Example 2b.** Suppose that the number of people that enter a post office on a given day is a Poisson random variable with parameter $\lambda$. Show that if each person that enters the post office is a male with probability

$p$ and a female with probability $1 - p$, then the number of males and females entering the post office are independent Poisson random variables with respective parameters $\lambda p$ and $\lambda(1 - p)$.

***Solution:*** Let $X$ and $Y$ denote, respectively, the number of males and females that enter the post office. We shall show the independence of $X$ and $Y$ by establishing Equation (2.2). To obtain an expression for $P\{X = i, Y = j\}$, we condition on $X + Y$ as follows:

$$P\{X = i, Y = j\} = P\{X = i, Y = j \mid X + Y = i + j\}P\{X + Y = i + j\} + P\{X = i, Y = j \mid X + Y \neq i + j\}P\{X + Y \neq i + j\}$$

[The reader should note that this equation is merely a special case of the formula $P(E) = P(E \mid F)P(F) + P(E \mid F^c)P(F^c)$.] As $P\{X = i, Y = j \mid X + Y \neq i + j\}$ is clearly 0, we obtain

$$P\{X = i, Y = j\} = P\{X = i, Y = j \mid X + Y = i + j\}P\{X + Y = i + j\} \tag{2.3}$$

Now, as $X + Y$ is the total number that enter the post office, it follows, by assumption, that

$$P\{X + Y = i + j\} = e^{-\lambda}\frac{\lambda^{i+j}}{(i + j)!} \tag{2.4}$$

Furthermore, given that $i + j$ people do enter the post office, since each person entering will be male with probability $p$, it follows that the probability that exactly $i$ of them will be male(and thus $j$ of them female) is just the binomial probability $\binom{i+j}{i}p^i(1 - p)^j$. That is,

$$P\{X = i, Y = j \mid X + Y = i + j\} = \binom{i+j}{i}p^i(1 - p)^j \tag{2.5}$$

Substituting Equations (2.4) and (2.5) into Equation (2.3) yields

$$\begin{aligned} P\{X = i, Y = j\} &= \binom{i+j}{i}p^i(1 - p)^j e^{-\lambda}\frac{\lambda^{i+j}}{(i + j)!} \\ &= e^{-\lambda}\frac{(\lambda p)^i}{i!\,j!}[\lambda(1 - p)]^j \\ &= \frac{e^{-\lambda p}(\lambda p)^i}{i!}e^{-\lambda(1-p)}\frac{[\lambda(1 - p)]^j}{j!} \end{aligned} \tag{2.6}$$

Hence

$$P\{X = i\} = e^{-\lambda p}\frac{(\lambda p)^i}{i!}\sum_j e^{-\lambda(1-p)}\frac{[\lambda(1 - p)]^j}{j!} = e^{-\lambda p}\frac{(\lambda p)^i}{i!} \tag{2.7}$$

and similarly,

$$P\{Y = j\} = e^{-\lambda(1-p)} \frac{[\lambda(1-p)]^j}{j!} \tag{2.8}$$

Equations (2.6), (2.7), and (2.8) establish the desired result. ∎

**Example 2c.** A man and a woman decide to meet at a certain location. If each person independently arrives at a time uniformly distributed between 12 noon and 1 P.M., find the probability that the first to arrive has to wait longer than 10 minutes.

***Solution:*** If we let $X$ and $Y$ denote, respectively, the time past 12 that the man and the woman arrive, then $X$ and $Y$ are independent random variables, each of which is uniformly distributed over $(0, 60)$. The desired probability, $P\{X + 10 < Y\} + P\{Y + 10 < X\}$, which by symmetry equals $2P\{X + 10 < Y\}$, is obtained as follows:

$$\begin{aligned}
2P\{X + 10 < Y\} &= 2 \cdot \iint\limits_{x+10<y} f(x, y)\, dx\, dy \\
&= 2 \iint\limits_{x+10<y} f_X(x) f_Y(y)\, dx\, dy \\
&= 2 \int_{10}^{60} \int_0^{y-10} \left(\frac{1}{60}\right)^2 dx\, dy \\
&= \frac{2}{(60)^2} \int_{10}^{60} (y - 10)\, dy \\
&= \frac{25}{36}
\end{aligned}$$
∎

Our next example presents the oldest problem dealing with geometrical probabilities. It was first considered and solved by Buffon, a French naturalist of the eighteenth century, and is usually referred to as Buffon's needle problem.

**Example 2d.** Buffon's Needle Problem. A table is ruled with equidistant parallel lines a distance $D$ apart. A needle of length $L$, where $L \leq D$, is randomly thrown on the table. What is the probability that the needle will intersect one of the lines (the other possibility being that the needle will be completely contained in the strip between two lines)?

***Solution:*** Let us determine the position of the needle by specifying the distance $X$ from the middle point of the needle to the nearest parallel line, and the angle $\theta$ between the needle and the projected line of length $X$. (See Figure 6.2.) The needle will intersect a line if the hypotenuse of the right triangle in Figure 6.2 is less than $L/2$, that is, if

$$\frac{X}{\cos\theta} < \frac{L}{2} \quad \text{or} \quad X < \frac{L}{2}\cos\theta$$

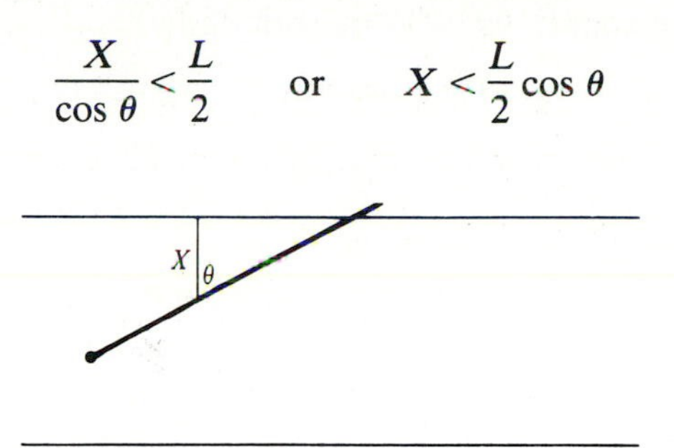


**Figure 6.2**

As $X$ varies between 0 and $D/2$ and $\theta$ between 0 and $\pi/2$, it is reasonable to assume that they are independent, uniformly distributed random variables over these respective ranges. Hence

$$\begin{aligned} P\left\{X < \frac{L}{2}\cos\theta\right\} &= \iint\limits_{x<L/2\cos y} f_X(x)f_\theta(y)\,dx\,dy \\ &= \frac{4}{\pi D}\int_0^{\pi/2}\int_0^{L/2\cos y} dx\,dy \\ &= \frac{4}{\pi D}\int_0^{\pi/2}\frac{L}{2}\cos y\,dy \\ &= \frac{2L}{\pi D} \end{aligned}$$

∎

**Example 2e.** Characterization of the Normal Distribution. Let $X$ and $Y$ denote the horizontal and vertical miss distance when a bullet is fired at a target, and assume that

1. $X$ and $Y$ are independent continuous random variables having differentiable density functions.
2. The joint density $f(x, y) = f_X(x)f_Y(y)$ of $X$ and $Y$ depends on $(x, y)$ only through $x^2 + y^2$.

Loosely put, Assumption (2) states that the probability of the bullet landing on any point of the $x\ y$ plane depends only on the distance of the point from the target and not on its angle of orientation. An equivalent way of phrasing Assumption (2) is to say that the joint density function is rotation invariant.

It is a rather interesting fact that Assumptions (1) and (2) imply that $X$ and $Y$ are normally distributed random variables. To prove this, note first that the assumptions yield the relation

$$f(x, y) = f_X(x)f_Y(y) = g(x^2 + y^2) \tag{2.9}$$

for some function $g$. Differentiating Equation (2.9) with respect to $x$ yields

$$f'_X(x)f_Y(y) = 2xg'(x^2 + y^2) \tag{2.10}$$

Dividing Equation (2.10) by Equation (2.9) gives

$$\frac{f'_X(x)}{f_X(x)} = \frac{2xg'(x^2 + y^2)}{g(x^2 + y^2)}$$

or

$$\frac{f'_X(x)}{2xf_X(x)} = \frac{g'(x^2 + y^2)}{g(x^2 + y^2)} \tag{2.11}$$

As the value of the left-hand side of Equation (2.11) depends only on $x$, whereas the value of the right-hand side depends on $x^2 + y^2$, it follows that the left-hand side must be the same for all $x$. To see this, consider any $x_1, x_2$ and let $y_1, y_2$ be such that $x_1^2 + y_1^2 = x_2^2 + y_2^2$. Then, from Equation (2.11), we obtain

$$\frac{f'_X(x_1)}{2x_1 f_X(x_1)} = \frac{g'(x_1^2 + y_1^2)}{g(x_1^2 + y_1^2)} = \frac{g'(x_2^2 + y_2^2)}{g(x_2^2 + y_2^2)} = \frac{f'_X(x_2)}{2x_2 f_X(x_2)}$$

Hence

$$\frac{f'_X(x)}{xf_X(x)} = c \qquad \text{or} \qquad \frac{d}{dx}(\log f_X(x)) = cx$$

which implies, upon integration of both sides, that

$$\log f_X(x) = a + \frac{cx^2}{2} \qquad \text{or} \qquad f_X(x) = ke^{cx^2/2}$$

Since $\int_{-\infty}^{\infty} f_X(x)\, dx = 1$, it follows that $c$ is necessarily negative, and we may write $c = -1/\sigma^2$. Hence

$$f_X(x) = ke^{-x^2/2\sigma^2}$$

That is, $X$ is a normal random variable with parameters $\mu = 0$ and $\sigma^2$. A similar argument can be applied to $f_Y(y)$ to show that

$$f_Y(y) = \frac{1}{\sqrt{2\pi}\bar{\sigma}} e^{-y^2/2\bar{\sigma}^2}$$

Furthermore, it follows from Assumption (2) that $\sigma^2 = \bar{\sigma}^2$, and thus $X$ and $Y$ are independent, identically distributed normal random variables with parameters $\mu = 0$ and $\sigma^2$. ∎

The concept of independence may, of course, be defined for more than two random variables. In general, the $n$ random variables $X_1, X_2, \ldots, X_n$ are said to be independent if, for all sets of real numbers $A_1, A_2, \ldots, A_n$,

$$P\{X_1 \in A_1, X_2 \in A_2, \ldots, X_n \in A_n\} = \prod_{i=1}^{n} P\{X_i \in A_i\}$$

As before, it can be shown that this condition is equivalent to

$$P\{X_1 \le a_1, X_2 \le a_2, \ldots, X_n \le a_n\}$$
$$= \prod_{i=1}^{n} P\{X_i \le a_i\} \qquad \text{for all } a_1, a_2, \ldots, a_n$$

Finally, we say that an infinite collection of random variables is independent if every finite subcollection of them is independent.

**Example 2f.** How Can a Computer Choose a Random Subset? Most computers are able to generate the value of, or *simulate*, a uniform (0, 1) random variable by means of a built-in subroutine that (to a high degree of approximation) produces such "random numbers." As a result, it is quite easy for the computer to simulate an indicator (that is, a Bernoulli) random variable. Suppose $I$ is an indicator variable such that

$$P\{I = 1\} = p = 1 - P\{I = 0\}$$

The computer can simulate $I$ by choosing a uniform (0, 1) random number $U$ and then letting

$$I = \begin{cases} 1 & \text{if } U < p \\ 0 & \text{if } U > p \end{cases}$$

Suppose that we are interested in having the computer select $k$, $k \le n$, of the numbers $1, 2, \ldots, n$ in such a way that each of the $\binom{n}{k}$ subsets of size $k$ is equally likely to be chosen. We now present a method that will enable the computer to solve this task. To generate such a subset, we will first simulate, in sequence, $n$ indicator variables $I_1, I_2, \ldots, I_n$, of which exactly $k$ will equal 1. Those $i$ for which $I_i = 1$ will then constitute the desired subset.

To generate the random variables $I_1, \ldots, I_n$, start by simulating $n$ independent uniform (0, 1) random variables $U_1, U_2, \ldots, U_n$. Now define

$$I_1 = \begin{cases} 1 & \text{if } U_1 < \dfrac{k}{n} \\ 0 & \text{otherwise} \end{cases}$$

and then recursively, once $I_1, \ldots, I_i$ is determined, set

$$I_{i+1} = \begin{cases} 1 & \text{if } U_{i+1} < \dfrac{k - (I_1 + \cdots + I_i)}{n - i} \\ 0 & \text{otherwise} \end{cases}$$

In words, at the $i + 1$ stage we set $I_{i+1}$ equal to 1 (and thus put $i + 1$ into the desired subset) with a probability equal to the remaining number of places in the subset $\left(\text{namely, } k - \sum_{j=1}^{i} I_j\right)$ divided by the remaining number of possibilities (namely, $n - i$). Hence the joint distribution of $I_1, I_2, \ldots, I_n$ is determined from

$$P\{I_1 = 1\} = \frac{k}{n}$$

$$P\{I_{i+1} = 1 \mid I_1, \ldots, I_i\} = \frac{k - \sum_{j=1}^{i} I_j}{n - i} \qquad 1 < i < n$$

The proof that the above results in all subsets of size $k$ being equally likely to be chosen is by induction on $k + n$. It is immediate when $k + n = 2$ (that is, when $k = 1$, $n = 1$), and so assume it to be true whenever $k + n \leq l$. Now suppose that $k + n = l + 1$ and consider any subset of size $k$—say $i_1 \leq i_2 \leq \cdots \leq i_k$—and consider the following two cases.

**Case 1.** $i_1 = 1$

$$P\{I_1 = I_{i_2} = \cdots = I_{i_k} = 1, I_j = 0 \text{ otherwise}\}$$

$$= P\{I_1 = 1\}P\{I_{i_2} = \cdots = I_{i_k} = 1, I_j = 0 \text{ otherwise} \mid I_1 = 1\}$$

Now given that $I_1 = 1$, the remaining elements of the subset are chosen as if a subset of size $k - 1$ were to be chosen from the $n - 1$ elements $2, 3, \ldots, n$. Hence, by the induction hypothesis, the conditional probability that this will result in a given subset of size $k - 1$ being selected is $1 \Big/ \binom{n-1}{k-1}$. Hence

$$P\{I_1 = I_{i_2} = \cdots = I_{i_k} = 1, I_j = 0 \text{ otherwise}\}$$
$$= \frac{k}{n} \frac{1}{\binom{n-1}{k-1}} = \frac{1}{\binom{n}{k}}$$

**Case 2.** $i_1 \neq 1$

$$P\{I_{i_1} = I_{i_2} = \cdots = I_{i_k} = 1, I_j = 0 \text{ otherwise}\}$$
$$= P\{I_{i_1} = \cdots = I_{i_k} = 1, I_j = 0 \text{ otherwise} \mid I_1 = 0\}P\{I_1 = 0\}$$
$$= \frac{1}{\binom{n-1}{k}}\left(1 - \frac{k}{n}\right) = \frac{1}{\binom{n}{k}}$$

where the induction hypothesis was used to evaluate the above conditional probability.

Hence in all cases the probability that a given subset of size $k$ will be the subset chosen is $1\Big/\binom{n}{k}$. ∎

REMARK. The foregoing method for generating a random subset has a very low memory requirement. A faster algorithm that requires somewhat more memory is presented in Section 1 of Chapter 10 (This latter algorithm uses the last $k$ elements of a random permutation of $1, 2, \ldots, n$). ∎

**Example 2g.** Let $X$, $Y$, $Z$ be independent and uniformly distributed over $(0, 1)$. Compute $P\{X \geq YZ\}$.

***Solution:*** Since

$$f_{X,Y,Z}(x, y, z) = f_X(x)f_Y(y)f_Z(z) = 1 \qquad 0 \leq x \leq 1, 0 \leq y \leq 1, 0 \leq z \leq 1$$

we have

$$\begin{aligned}
P\{X \geq YZ\} &= \iiint\limits_{x \geq yz} f_{X,Y,Z}(x, y, z)\, dx\, dy\, dz \\
&= \int_0^1 \int_0^1 \int_{yz}^1 dx\, dy\, dz \\
&= \int_0^1 \int_0^1 (1 - yz)\, dy\, dz \\
&= \int_0^1 \left(1 - \frac{z}{2}\right) dz \\
&= \frac{3}{4}
\end{aligned}$$ ∎

**Example 2h.** A Probabilistic Interpretation of Half-Life. Let $N(t)$ denote the number of nuclei contained in a radioactive mass of material at time $t$. The concept of half-life is often defined in a deterministic fashion by stating that it is an empirical fact that for some value $h$, called the half-life,

$$N(t) = 2^{-t/h} N(0), \quad t > 0.$$

(Note that $N(h) = N(0)/2$). Since the above implies that for any nonnegative $s$ and $t$

$$N(t + s) = 2^{-(s+t)/h} N(0) = 2^{-t/h} N(s)$$

it follows that no matter how much time $s$ has already elapsed, in an additional time $t$ the number of existing nuclei will decrease by the factor $2^{-t/h}$.

Since the above deterministic relationship results from observations of radioactive masses containing huge numbers of nuclei it would seem that it might be consistent with a probabilistic interpretation. The clue to deriving the appropriate probability model for half-life resides in the empirical observation that the proportion of decay in any time interval depends neither on the total number of nuclei at the beginning at the interval nor on the location of this interval (since $N(t + s)/N(s)$ depends neither on $N(s)$ nor on $s$). Thus, it appears that the individual nuclei act independently and with a memoryless life distribution. Thus, since the unique life distribution which is memoryless is the exponential distribution, and since exactly one-half of a given amount of mass decays every $h$ time units, we propose the following probabilistic model for radioactive extinction:

***Probabilistic Interpretation of the Half-Life h:*** The lifetimes of the individual nuclei are independent random variables having a life distribution that is exponential with median equal to $h$. That is, if $L$ represents the lifetime of a given nucleus then

$$P\{L < t\} = 1 - 2^{-t/h}$$

(As $P\{L < h\} = \frac{1}{2}$ and the above can be written as $P\{L < t\} = 1 - \exp\left\{-t\,\frac{\log 2}{h}\right\}$ we see that $L$ indeed has an exponential distribution with median $h$).

It should be noted that, under the above probabilistic interpretation of half-life, if one starts with $N(0)$ nuclei at time 0 then $N(t)$, the number that remain at time $t$, will have a binomial distribution with parameters $n = N(0)$ and $p = 2^{-t/h}$. Results of Chapter 8 will show that this interpretation of half-life is consistent with the deterministic model when considering the proportion of a large number of nuclei that decay over a given time frame. However, the difference between the deterministic and probabilistic interpretation becomes apparent when one considers the actual number of decayed nuclei. We will now indicate this with regard to the question of whether protons decay.

There appears to be some controversy over whether or not protons decay. Indeed, one theory appears to predict that protons should decay with a half-life of about $h = 10^{30}$ years. To empirically check this, it has been suggested that one follow a large number of protons for, say, one or two years and determine whether any of them decay within this time period. (Clearly, it would not be feasible to follow a mass of protons for $10^{30}$ years to see whether one-half of it decays). Let us suppose that we are able to keep track of $N(0) = 10^{30}$ protons for $c$ years. The number of decays predicted by the deterministic model would then be given by

$$\begin{aligned} N(0) - N(c) &= h(1 - 2^{-c/h}) \\ &= \frac{1 - 2^{-c/h}}{1/h} \\ &\approx \lim_{x\to 0} (1 - 2^{-cx})/x \qquad \text{since } 1/h = 10^{-30} \approx 0 \\ &= \lim_{x\to 0} (c2^{-cx} \log 2) \qquad \text{by L'hopital's rule} \\ &= c\log 2 \approx .6931c \end{aligned}$$

For instance, in 2 years the deterministic model predicts that there should be 1.3863 decays, and it would thus appear to be a serious blow to the hypothesis that protons decay with a half-life of $10^{30}$ years if no decays are observed over these 2 years.

Let us now contrast the conclusions above with those obtained from the probabilistic model. Again let us consider the hypothesis that the half-life of protons is $h = 10^{30}$ years, and suppose that we follow $h$ protons for $c$ years. Since there are a huge number of independent protons, each of which will have a very small probability of decaying within this time period, it follows that the number of protons that decay will have (to a very strong approximation) a Poisson distribution with parameter equal to $h(1 - 2^{-c/h}) \approx c \log 2$. Thus,

$$\begin{aligned} P\{0 \text{ decays}\} &= e^{-c\log 2} \\ &= e^{-\log(2^c)} = 1/2^c \end{aligned}$$

and, in general,

$$P\{n \text{ decays}\} = [c \log 2]^n/(2^c n!), \qquad n \geq 0$$

Thus, we see that even though the average number of decays over 2 years is (as predicted by the deterministic model) 1.3863 there is one chance in 4 that there will not be any decays, thereby indicating that such a result in no way invalidates the original hypothesis of proton decay. ∎

## 3 Sums of Independent Random Variables

It is often important to be able to calculate the distribution of $X + Y$ from the distributions of $X$ and $Y$ when $X$ and $Y$ are independent. Suppose that $X$ and $Y$ are independent, continuous random variables having probability density functions $f_X$ and $f_Y$. The cumulative distribution function of $X + Y$ is obtained as follows:

$$
\begin{aligned}
F_{X+Y}(a) &= P\{X + Y \leq a\} \\
&= \iint_{x+y\leq a} f_X(x) f_Y(y)\, dx\, dy \\
&= \int_{-\infty}^{\infty} \int_{-\infty}^{a-y} f_X(x) f_Y(y)\, dx\, dy \\
&= \int_{-\infty}^{\infty} \int_{-\infty}^{a-y} f_X(x)\, dx\, f_Y(y)\, dy \\
&= \int_{-\infty}^{\infty} F_X(a-y) f_Y(y)\, dy \qquad (3.1)
\end{aligned}
$$

The cumulative distribution function $F_{X+Y}$ is called the *convolution* of the distributions $F_X$ and $F_Y$ (the cumulative distribution functions of $X$ and $Y$, respectively).

By differentiating Equation (3.1), we obtain that the probability density function $f_{X+Y}$ of $X + Y$ is given by

$$
\begin{aligned}
f_{X+Y}(a) &= \frac{d}{da} \int_{-\infty}^{\infty} F_X(a-y) f_Y(y)\, dy \\
&= \int_{-\infty}^{\infty} \frac{d}{da} F_X(a-y) f_Y(y)\, dy \\
&= \int_{-\infty}^{\infty} f_X(a-y) f_Y(y)\, dy \qquad (3.2)
\end{aligned}
$$

**Example 3a.** Sum of Two Independent Uniform Random Variables. If $X$ and $Y$ are independent random variables, both uniformly distributed on $(0, 1)$, calculate the probability density of $X + Y$.

***Solution:*** From Equation (3.2), since

$$
f_X(a) = f_Y(a) = \begin{cases} 1 & 0 < a < 1 \\ 0 & \text{otherwise} \end{cases}
$$

we obtain

$$f_{X+Y}(a) = \int_0^1 f_X(a-y)\,dy$$

For $0 \le a \le 1$, this yields

$$f_{X+Y}(a) = \int_0^a dy = a$$

For $1 < a < 2$, we get

$$f_{X+Y}(a) = \int_{a-1}^1 dy = 2-a$$

Hence

$$f_{X+Y}(a) = \begin{cases} a & 0 \le a \le 1 \\ 2-a & 1 < a < 2 \\ 0 & \text{otherwise} \end{cases}$$

∎

Recall that a gamma random variable has a density of the form

$$f(y) = \frac{\lambda e^{-\lambda y}(\lambda y)^{t-1}}{\Gamma(t)} \qquad 0 < y < \infty$$

An important property of this family of distributions is that for a fixed value of $\lambda$, it is closed under convolutions.

---

### Proposition 3.1

*If $X$ and $Y$ are independent gamma random variables with respective parameters $(s, \lambda)$ and $(t, \lambda)$, then $X + Y$ is a gamma random variable with parameters $(s + t, \lambda)$.*

---

**Proof:** Using Equation (3.2), we obtain

$$\begin{aligned} f_{X+Y}(a) &= \frac{1}{\Gamma(s)\Gamma(t)} \int_0^a \lambda e^{-\lambda(a-y)}[\lambda(a-y)]^{s-1}\lambda e^{-\lambda y}(\lambda y)^{t-1}\,dy \\ &= Ke^{-\lambda a}\int_0^a (a-y)^{s-1}y^{t-1}\,dy \\ &= Ke^{-\lambda a}a^{s+t-1}\int_0^1 (1-x)^{s-1}x^{t-1}\,dx \qquad \text{by letting } x = \frac{y}{a} \\ &= Ce^{-\lambda a}a^{s+t-1} \end{aligned}$$

where $C$ is a constant that does not depend on $a$. But as the above is a density function and thus must integrate to 1, the value of $C$ is determined, and we have

$$f_{X+Y}(a) = \frac{\lambda e^{-\lambda a}(\lambda a)^{s+t-1}}{\Gamma(s+t)}$$

Hence the result is proved. ∎

It is now a simple matter to establish, by using Proposition 3.1 and induction, that if $X_i$, $i = 1, \ldots, n$ are independent gamma random variables with respective parameters $(t_i, \lambda)$, $i = 1, \ldots, n$, then $\sum_{i=1}^{n} X_i$ is gamma with parameters $\left(\sum_{i=1}^{n} t_i, \lambda\right)$. We leave the proof of this as an exercise.

**Example 3b.** Let $X_1, X_2, \ldots, X_n$ be $n$ independent exponential random variables each having parameter $\lambda$. Then, as an exponential random variable with parameter $\lambda$ is the same as a gamma random variable with parameters $(1, \lambda)$, we see from Proposition 3.1 that $X_1 + X_2 + \cdots + X_n$ is a gamma random variable with parameters $(n, \lambda)$. ∎

If $Z_1, Z_2, \ldots, Z_n$ are independent unit normal random variables, then $Y \equiv \sum_{i=1}^{n} Z_i^2$ is said to have the *chi-square* (sometimes seen as $\chi^2$) distribution with $n$ degrees of freedom. Let us compute its density function. When $n = 1$, $Y = Z_1^2$, and from Example 6b of Chapter 5 we see that its probability density function is given by

$$\begin{aligned} f_{Z^2}(y) &= \frac{1}{2\sqrt{y}}\,[f_Z(\sqrt{y}) + f_Z(-\sqrt{y})] \\ &= \frac{1}{2\sqrt{y}}\,2\left(\frac{1}{\sqrt{2\pi}}\,e^{-y/2}\right) \\ &= \frac{\frac{1}{2}e^{-(1/2)y}(\frac{1}{2}y)^{1/2-1}}{\sqrt{\pi}} \end{aligned}$$

But we recognize the above as the gamma distribution with parameters $(\frac{1}{2}, \frac{1}{2})$. [A by-product of this analysis is that $\Gamma(\frac{1}{2}) = \sqrt{\pi}$.] But as each $Z_i^2$ is gamma $(\frac{1}{2}, \frac{1}{2})$, we obtain from Proposition 3.1 that the $\chi^2$ distribution with $n$ degrees of freedom is just the gamma distribution with parameters $[n/2, \frac{1}{2}]$

and hence has a probability density function given by

$$f_{\chi^2}(y) = \frac{\frac{1}{2}e^{-y/2}\left(\frac{y}{2}\right)^{n/2-1}}{\Gamma\left(\frac{n}{2}\right)} \qquad y > 0$$

$$= \frac{e^{-y/2}y^{n/2-1}}{2^{n/2}\Gamma\left(\frac{n}{2}\right)} \qquad y > 0$$

When $n$ is an even integer, $\Gamma(n/2) = [(n/2) - 1]!$, whereas when $n$ is odd, $\Gamma(n/2)$ can be obtained from iterating the relationship $\Gamma(t) = (t-1)\Gamma(t-1)$ and then using the previously obtained result that $\Gamma(\frac{1}{2}) = \sqrt{\pi}$. [For instance, $\Gamma(\frac{5}{2}) = \frac{3}{2}\Gamma(\frac{3}{2}) = \frac{3}{2}\frac{1}{2}\Gamma(\frac{1}{2}) = \frac{3}{4}\sqrt{\pi}$.]

The chi-squared distribution often arises in practice as being the distribution of the square of the error involved when one attempts to hit a target in $n$-dimensional space when the coordinate errors are taken to be independent unit normal random variables. It is also important in statistical analysis.

We can also use Equation (3.2) to prove the following important result about normal random variables.

---

*Proposition* 3.2

*If $X_i$, $i = 1, \ldots, n$, are independent random variables that are normally distributed with respective parameters $\mu_i$, $\sigma_i^2$; $i = 1, \ldots, n$, then $\sum_{i=1}^n X_i$ is normally distributed with parameters $\sum_{i=1}^n \mu_i$* and *$\sum_{i=1}^n \sigma_i^2$.*

---

We leave the proof of Proposition 3.2 as an exercise.

Rather than attempt to derive a general expression for the distribution of $X + Y$ in the discrete case, we shall consider some examples.

**Example 3c.** Sums of Independent Poisson Random Variables. If $X$ and $Y$ are independent Poisson random variables with respective parameters $\lambda_1$ and $\lambda_2$, compute the distribution of $X + Y$.

***Solution:*** Because the event $\{X + Y = n\}$ may be written as the union of the disjoint events $\{X = k, Y = n - k\}$, $0 \le k \le n$, we have

$$
\begin{aligned}
P\{X + Y = n\} &= \sum_{k=0}^{n} P\{X = k, Y = n - k\} \\
&= \sum_{k=0}^{n} P\{X = k\}P\{Y = n - k\} \\
&= \sum_{k=0}^{n} e^{-\lambda_1} \frac{\lambda_1^k}{k!} e^{-\lambda_2} \frac{\lambda_2^{n-k}}{(n-k)!} \\
&= e^{-(\lambda_1+\lambda_2)} \sum_{k=0}^{n} \frac{\lambda_1^k \lambda_2^{n-k}}{k!(n-k)!} \\
&= \frac{e^{-(\lambda_1+\lambda_2)}}{n!} \sum_{k=0}^{n} \frac{n!}{k!(n-k)!} \lambda_1^k \lambda_2^{n-k} \\
&= \frac{e^{-(\lambda_1+\lambda_2)}}{n!} (\lambda_1 + \lambda_2)^n
\end{aligned}
$$

In words, $X_1 + X_2$ has a Poisson distribution with parameter $\lambda_1 + \lambda_2$. ∎

**Example 3d.** Sums of Independent Binomial Random Variables. Let $X$ and $Y$ be independent binomial random variables with respective parameters $(n, p)$ and $(m, p)$. Calculate the distribution of $X + Y$.

***Solution:*** Without any computation at all we can immediately conclude, by recalling the interpretation of a binomial random variable, that $X + Y$ is binomial with parameters $(n + m, p)$. This follows because $X$ represents the number of successes in $n$ independent trials, each of which results in a success with probability $p$; similarly, $Y$ represents the number of successes in $m$ independent trials, each trial being a success with probability $p$. Hence, as $X$ and $Y$ are assumed independent, it follows that $X + Y$ represents the number of successes in $n + m$ independent trials when each trial has a probability $p$ of being a success. Therefore, $X + Y$ is a binomial random variable with parameters $(n + m, p)$. To check this result analytically, note that

$$
\begin{aligned}
P\{X + Y = k\} &= \sum_{i=0}^{n} P\{X = i, Y = k - i\} \\
&= \sum_{i=0}^{n} P\{X = i\}P\{Y = k - i\} \\
&= \sum_{i=0}^{n} \binom{n}{i} p^i q^{n-i} \binom{m}{k-i} p^{k-i} q^{m-k+i}
\end{aligned}
$$

where $q = 1 - p$ and where $\binom{r}{j} = 0$ when $j > r$. Hence

$$P\{X + Y = k\} = p^k q^{n+m-k} \sum_{i=0}^{n} \binom{n}{i} \binom{m}{k-i}$$

and the result follows upon application of the combinatorial identity

$$\binom{n+m}{k} = \sum_{i=0}^{n} \binom{n}{i} \binom{m}{k-i}$$ ∎

## 4 Conditional Distributions—Discrete Case

Recall that for any two events $E$ and $F$, the conditional probability of $E$ given $F$ is defined, provided that $P(F) > 0$, by

$$P(E|F) = \frac{P(EF)}{P(F)}$$

Hence, if $X$ and $Y$ are discrete random variables, it is natural to define the conditional probability mass function of $X$ given that $Y = y$, by

$$\begin{aligned} p_{X|Y}(x|y) &= P\{X = x | Y = y\} \\ &= \frac{P\{X = x, Y = y\}}{P\{Y = y\}} \\ &= \frac{p(x, y)}{p_Y(y)} \end{aligned}$$

for all values of $y$ such that $p_Y(y) > 0$. Similarly, the conditional probability distribution function of $X$ given that $Y = y$ is defined, for all $y$ such that $p_Y(y) > 0$, by

$$\begin{aligned} F_{X|Y}(x|y) &= P\{X \leq x | Y = y\} \\ &= \sum_{a \leq x} p_{X|Y}(a|y) \end{aligned}$$

In other words, the definitions are exactly the same as in the unconditional case except that everything is now conditional on the event that $Y = y$. If $X$ is independent of $Y$, then the conditional mass function and distribution function are the same as the unconditional ones. This follows because if $X$

is independent of $Y$, then

$$\begin{aligned}
p_{X|Y}(x|y) &= P\{X = x | Y = y\} \\
&= \frac{P\{X = x, Y = y\}}{P\{Y = y\}} \\
&= \frac{P\{X = x\}P\{Y = y\}}{P\{Y = y\}} \\
&= P\{X = x\}
\end{aligned}$$

**Example 4a.** Suppose that $p(x, y)$, the joint probability mass function of $X$ and $Y$, is given by

$$p(0,0) = .4 \qquad p(0,1) = .2 \qquad p(1,0) = .1 \qquad p(1,1) = .3$$

Calculate the conditional probability mass function of $X$, given that $Y = 1$.

***Solution:*** We first note that

$$p_Y(1) = \sum_x p(x, 1) = p(0, 1) + p(1, 1) = .5$$

Hence

$$p_{X|Y}(0|1) = \frac{p(0, 1)}{p_Y(1)} = \frac{2}{5}$$

and

$$p_{X|Y}(1|1) = \frac{p(1, 1)}{p_Y(1)} = \frac{3}{5}$$

∎

**Example 4b.** If $X$ and $Y$ are independent Poisson random variables with respective parameters $\lambda_1$ and $\lambda_2$, calculate the conditional distribution of $X$, given that $X + Y = n$.

***Solution:*** We calculate the conditional probability mass function of $X$ given that $X + Y = n$ as follows:

$$\begin{aligned}
P\{X = k | X + Y = n\} &= \frac{P\{X = k, X + Y = n\}}{P\{X + Y = n\}} \\
&= \frac{P\{X = k, Y = n - k\}}{P\{X + Y = n\}} \\
&= \frac{P\{X = k\}P\{Y = n - k\}}{P\{X + Y = n\}}
\end{aligned}$$

where the last equality follows from the assumed independence of $X$ and $Y$. Recalling (Example 3c) that $X + Y$ has a Poisson distribution with parameter $\lambda_1 + \lambda_2$, we see that the above equals

$$P\{X = k | X + Y = n\} = \frac{e^{-\lambda_1}\lambda_1^k}{k!} \frac{e^{-\lambda_2}\lambda_2^{n-k}}{(n-k)!}\left[\frac{e^{-(\lambda_1+\lambda_2)}(\lambda_1 + \lambda_2)^n}{n!}\right]^{-1}$$

$$= \frac{n!}{(n-k)!\,k!} \frac{\lambda_1^k \lambda_2^{n-k}}{(\lambda_1 + \lambda_2)^n}$$

$$= \binom{n}{k}\left(\frac{\lambda_1}{\lambda_1 + \lambda_2}\right)^k \left(\frac{\lambda_2}{\lambda_1 + \lambda_2}\right)^{n-k}$$

In other words, the conditional distribution of $X$, given that $X + Y = n$, is the binomial distribution with parameters $n$ and $\lambda_1/(\lambda_1 + \lambda_2)$. ∎

## 5 Conditional Distributions—Continuous Case

If $X$ and $Y$ have a joint probability density function $f(x, y)$, then the conditional probability density function of $X$, given that $Y = y$, is defined for all values of $y$ such that $f_Y(y) > 0$, by

$$f_{X|Y}(x|y) = \frac{f(x, y)}{f_Y(y)}$$

To motivate this definition, multiply the left-hand side by $dx$ and the right-hand side by $(dx\,dy)/dy$ to obtain

$$f_{X|Y}(x|y)\,dx = \frac{f(x, y)\,dx\,dy}{f_Y(y)\,dy}$$

$$\approx \frac{P\{x \le X \le x + dx, y \le Y \le y + dy\}}{P\{y \le Y \le y + dy\}}$$

$$= P\{x \le X \le x + dx | y \le Y \le y + dy\}$$

In other words, for small values of $dx$ and $dy$, $f_{X|Y}(x|y)\,dx$ represents the conditional probability that $X$ is between $x$ and $x + dx$, given that $Y$ is between $y$ and $y + dy$.

The use of conditional densities allows us to define conditional probabilities of events associated with one random variable when we are given the value of a second random variable. That is, if $X$ and $Y$ are jointly continuous, then, for any set $A$,

$$P\{X \in A | Y = y\} = \int_A f_{X|Y}(x|y)\,dx$$

In particular, by letting $A = (-\infty, a]$, we can define the conditional cumulative distribution function of $X$, given that $Y = y$, by

$$F_{X|Y}(a|y) \equiv P\{X \leq a | Y = y\} = \int_{-\infty}^{a} f_{X|Y}(x|y)\,dx$$

The reader should note that, by using the ideas presented in the preceding discussion, we have been able to give workable expressions for conditional probabilities, even though the event on which we are conditioning (namely, the event $\{Y = y\}$) has probability 0.

**Example 5a.** The joint density of $X$ and $Y$ is given by

$$f(x, y) = \begin{cases} \frac{12}{5}x(2 - x - y) & 0 < x < 1, 0 < y < 1 \\ 0 & \text{otherwise} \end{cases}$$

Compute the conditional density of $X$, given that $Y = y$, where $0 < y < 1$.

***Solution:*** For $0 < x < 1$, $0 < y < 1$, we have

$$\begin{aligned} f_{X|Y}(x|y) &= \frac{f(x, y)}{f_Y(y)} \\ &= \frac{f(x, y)}{\int_{-\infty}^{\infty} f(x, y)\,dx} \\ &= \frac{x(2 - x - y)}{\int_0^1 x(2 - x - y)\,dx} \\ &= \frac{x(2 - x - y)}{\frac{2}{3} - y/2} \\ &= \frac{6x(2 - x - y)}{4 - 3y} \end{aligned}$$

■

**Example 5b.** Suppose that the joint density of $X$ and $Y$ is given by

$$f(x, y) = \begin{cases} \dfrac{e^{-x/y}e^{-y}}{y} & 0 < x < \infty, 0 < y < \infty \\ 0 & \text{otherwise} \end{cases}$$

Find $P\{X > 1 | Y = y\}$.

***Solution:*** We first obtain the conditional density of $X$, given that $Y = y$.

$$f_{X|Y}(x|y) = \frac{f(x, y)}{f_Y(y)}$$

$$= \frac{e^{-x/y}e^{-y}/y}{e^{-y}\int_0^\infty (1/y)e^{-x/y}\,dx}$$

$$= \frac{1}{y}e^{-x/y}$$

Hence

$$P\{X > 1 | Y = y\} = \int_1^\infty \frac{1}{y}e^{-x/y}\,dx$$

$$= -e^{-x/y}\Big|_1^\infty$$

$$= e^{-1/y}$$ ∎

If $X$ and $Y$ are independent continuous random variables, the conditional density of $X$, given $Y = y$, is just the unconditional density of $X$. This is so because, in the independent case,

$$f_{X|Y}(x|y) = \frac{f(x, y)}{f_Y(y)} = \frac{f_X(x)f_Y(y)}{f_Y(y)} = f_X(x)$$

We can also talk about conditional distributions when the random variables are neither jointly continuous nor jointly discrete. For example, suppose that $X$ is a continuous random variable having probability density function $f$ and $N$ is a discrete random variable, and consider the conditional distribution of $X$ given that $N = n$. Then

$$\frac{P\{x < X < x + dx | N = n\}}{dx}$$

$$= \frac{P\{N = n | x < X < x + dx\}}{P\{N = n\}} \frac{P\{x < X < x + dx\}}{dx}$$

and letting $dx$ approach 0 gives

$$\lim_{dx \to 0} \frac{P\{x < X < x + dx | N = n\}}{dx} = \frac{P\{N = n | X = x\}}{P\{N = n\}} f(x)$$

thus showing that the conditional density of $X$ given that $N = n$ is given by

$$f_{X|N}(x|n) = \frac{P\{N = n | X = x\}}{P\{N = n\}} f(x)$$

**Example 5c.** Consider $n + m$ trials having a common probability of success. Suppose, however, that this success probability is not fixed in advance but is chosen from a uniform $(0, 1)$ population. What is the conditional distribution of the success probability given that the $n + m$ trials result in $n$ successes?

***Solution:*** If we let $X$ denote the trial success probability, then $X$ is a uniform $(0, 1)$ random variable. Also, given that $X = x$, the $n + m$ trials are independent with common success probability $x$, and so $N$, the number of successes, is a binomial random variable with parameters $(n + m, x)$. Hence the conditional density of $X$ given that $N = n$ is as follows:

$$
\begin{aligned}
f_{X|N}(x|n) &= \frac{P\{N = n|X = x\}f_X(x)}{P\{N = n\}} \\
&= \frac{\binom{n+m}{n} x^n(1-x)^m}{P\{N = n\}} \qquad 0 < x < 1 \\
&= cx^n(1-x)^m
\end{aligned}
$$

where $c$ does not depend on $x$. Hence the conditional density is that of a beta random variable with parameters $n + 1$, $m + 1$.

The above result is quite interesting, for it states that if the original or *prior* (to the collection of data) distribution of a trial success probability is uniformly distributed over $(0, 1)$ [or, equivalently, is beta with parameters $(1, 1)$] then the posterior (or conditional) distribution given a total of $n$ successes in $n + m$ trials is beta with parameters $(1 + n, 1 + m)$. This is valuable, for it enhances our intuition as to what it means to assume that a random variable has a beta distribution. ∎

## 6 Order Statistics

Let $X_1, X_2, \ldots, X_n$ be $n$ independent and identically distributed, continuous random variables having a common density $f$ and distribution function $F$. Define

$$
\begin{aligned}
X_{(1)} &= \text{smallest of } X_1, X_2, \ldots, X_n \\
X_{(2)} &= \text{second smallest of } X_1, X_2, \ldots, X_n \\
&\vdots \\
X_{(j)} &= j\text{th smallest of } X_1, X_2, \ldots, X_n \\
&\vdots \\
X_{(n)} &= \text{largest of } X_1, X_2, \ldots, X_n
\end{aligned}
$$

The ordered values $X_{(1)} \leq X_{(2)} \leq \cdots \leq X_{(n)}$ are known as the *order statistics* corresponding to the random variables $X_1, X_2, \ldots, X_n$. In other words, $X_{(1)}, \ldots, X_{(n)}$ are the ordered values of $X_1, \ldots, X_n$.

The joint density function of the order statistics is obtained by noting that the order statistics $X_{(1)}, \ldots, X_{(n)}$ will take on the values $x_1 \leq x_2 \leq \cdots \leq x_n$ if and only if for some permutation $(i_1, i_2, \ldots, i_n)$ of $(1, 2, \ldots, n)$

$$X_1 = x_{i_1}, X_2 = x_{i_2}, \ldots, X_n = x_{i_n}$$

Since, for any permutation $(i_1, \ldots, i_n)$ of $(1, 2, \ldots, n)$,

$$\begin{aligned}
P&\left\{x_{i_1} - \frac{\varepsilon}{2} < X_1 < x_{i_1} + \frac{\varepsilon}{2}, \ldots, x_{i_n} - \frac{\varepsilon}{2} < X_n < x_{i_n} + \frac{\varepsilon}{2}\right\} \\
&\approx \varepsilon^n f_{X_1, \ldots, X_n}(x_{i_1}, \ldots, x_{i_n}) \\
&= \varepsilon^n f(x_{i_1}) \cdots f(x_{i_n}) \\
&= \varepsilon^n f(x_1) \cdots f(x_n)
\end{aligned}$$

we see that for $x_1 < x_2 < \cdots < x_n$

$$\begin{aligned}
P&\left\{x_1 - \frac{\varepsilon}{2} < X_{(1)} < x_1 + \frac{\varepsilon}{2}, \ldots, x_n - \frac{\varepsilon}{2} < X_{(n)} < x_n + \frac{\varepsilon}{2}\right\} \\
&\approx n!\, \varepsilon^n f(x_1) \cdots f(x_n)
\end{aligned}$$

Dividing by $\varepsilon^n$ and letting $\varepsilon \to 0$ yields

$$f_{X_{(1)}, \ldots, X_{(n)}}(x_1, x_2, \ldots, x_n) = n!\, f(x_1) \cdots f(x_n) \qquad x_1 < x_2 < \cdots < x_n \tag{6.1}$$

Equation (6.1) is most simply explained by arguing that in order for the vector $\langle X_{(1)}, \ldots, X_{(n)} \rangle$ to equal $\langle x_1, \ldots, x_n \rangle$, it is necessary and sufficient for $\langle X_1, \ldots, X_n \rangle$ to equal one of the $n!$ permutations of $\langle x_1, \ldots, x_n \rangle$. As the probability (density) that $\langle X_1, \ldots, X_n \rangle$ equals any given permutation of $\langle x_1, \ldots, x_n \rangle$ is just $f(x_1) \cdots f(x_n)$, Equation (6.1) follows.

**Example 6a.** Along a road 1 mile long are 3 people "distributed at random." Find the probability that no 2 people are less than a distance of $d$ miles apart, when $d \leq \frac{1}{2}$.

***Solution:*** Let us assume that "distributed at random" means that the positions of the 3 people are independent and uniformly distributed over the road. If $X_i$ denotes the position of the $i$th person, the desired probability is $P\{X_{(i)} > X_{(i-1)} + d, i = 2, 3\}$. As

$$f_{X_{(1)}, X_{(2)}, X_{(3)}}(x_1, x_2, x_3) = 3! \qquad 0 < x_1 < x_2 < x_3 < 1$$

it follows that

$$P\{X_{(i)} > X_{(i-1)} + d, i = 2, 3\} = \iiint\limits_{\substack{x_i > x_{i-1}+d \\ i=2,3}} f_{X_{(1)},X_{(2)},X_{(3)}}(x_1, x_2, x_3)\, dx_1\, dx_2\, dx_3$$

$$= 3! \int_0^{1-2d} \int_{x_1+d}^{1-d} \int_{x_2+d}^{1} dx_3\, dx_2\, dx_1$$

$$= 6 \int_0^{1-2d} \int_{x_1+d}^{1-d} (1 - d - x_2)\, dx_2\, dx_1$$

$$= 6 \int_0^{1-2d} \int_0^{1-2d-x_1} y_2\, dy_2\, dx_1$$

where we have made the change of variables $y_2 = 1 - d - x_2$. Hence continuing the string of equalities yields

$$= 6 \int_0^{1-2d} \frac{(1 - 2d - x_1)^2}{2}\, dx_1$$

$$= 6 \int_0^{1-2d} \frac{y_1^2}{2}\, dy_1$$

$$= (1 - 2d)^3$$

Hence the desired probability that no 2 people are within a distance $d$ of each other when 3 people are uniformly and independently distributed over an interval of size 1 is $(1 - 2d)^3$ when $d \leq \frac{1}{2}$. In fact, the same method can be used to prove that when there are $n$ people distributed at random over the unit interval the desired probability is

$$[1 - (n - 1)d]^n \qquad \text{when } d \leq \frac{1}{n - 1}$$

The proof is left as an exercise. ∎

The density function of the $j$th order statistic $X_{(j)}$ can be obtained either by integrating the joint density function (6.1) or by direct reasoning as follows: in order for $X_{(j)}$ to equal $x$, it is necessary for $j - 1$ of the $n$ values $X_1, \ldots, X_n$ to be less than $x$, $n - j$ of them to be greater than $x$, and 1 of them to equal $x$. Now, the probability density that any given set of $j - 1$ of the $X_i$'s are all less than $x$, another given set of $n - j$ are all greater than $x$, and the remaining value is equal to $x$, equals

$$[F(x)]^{j-1}[1 - F(x)]^{n-j}f(x)$$

Hence, as there are

$$\binom{n}{j-1, n-j, 1} = \frac{n!}{(n - j)!(j - 1)!}$$

different partitions of the $n$ random variables $X_1, \ldots, X_n$ into the three groups, we see that the density function of $X_{(j)}$ is given by

$$f_{X_{(j)}}(x) = \frac{n!}{(n-j)!(j-1)!}[F(x)]^{j-1}[1-F(x)]^{n-j}f(x) \qquad (6.2)$$

**Example 6b.** When a sample of $2n+1$ random variables (that is, when $2n+1$ independent and identically distributed random variables) are observed, the $(n+1)$st smallest is called the sample median. If a sample of size 3 from a uniform distribution over $(0, 1)$ is observed, find the probability that the sample median is between $\frac{1}{4}$ and $\frac{3}{4}$.

***Solution:*** From Equation (6.2) the density of $X_{(2)}$ is given by

$$f_{X_{(2)}}(x) = \frac{3!}{1!\,1!}x(1-x) \qquad 0 < x < 1$$

Hence

$$\begin{aligned} P\{\tfrac{1}{4} < X_{(2)} < \tfrac{3}{4}\} &= 6\int_{1/4}^{3/4} x(1-x)\,dx \\ &= 6\left\{\frac{x^2}{2} - \frac{x^3}{3}\right\}\Bigg|_{x=1/4}^{x=3/4} = \frac{11}{16} \end{aligned}$$ ∎

The cumulative distribution function of $X_{(j)}$ can be obtained by integrating Equation (6.2). That is

$$F_{X_{(j)}}(y) = \frac{n!}{(n-j)!(j-1)!}\int_{-\infty}^{y}[F(x)]^{j-1}[1-F(x)]^{n-j}f(x)\,dx \qquad (6.3)$$

However, $F_{X_{(j)}}(y)$ could also have been derived directly by noting that the $j$th order statistic is less than or equal to $y$ if and only if there are $j$ or more of the $X_i$'s that are less than or equal to $y$. Hence, as the number of the $X_i$'s that are less than or equal to $y$ is a binomial random variable with parameters $[n, p = F(y)]$, it follows that

$$\begin{aligned} F_{X_{(j)}}(y) = P\{X_{(j)} \le y\} &= P\{j \text{ or more of the } X_i\text{'s are} \le y\} \\ &= \sum_{k=j}^{n}\binom{n}{k}[F(y)]^k[1-F(y)]^{n-k} \qquad (6.4) \end{aligned}$$

If, in Equations (6.3) and (6.4), we take $F$ to be the uniform $(0, 1)$ distribution [that is, $F(x) = x$, $0 < x < 1$], then we obtain the interesting analytical identity

$$\sum_{k=j}^{n}\binom{n}{k}y^k(1-y)^{n-k} = \frac{n!}{(n-j)!(j-1)!}\int_0^y x^{j-1}(1-x)^{n-j}\,dx \qquad 0 \le y \le 1 \qquad (6.5)$$

By employing the same type of argument that we used in establishing Equation (6.2), we can show that the joint density function of the order statistics $X_{(i)}$ and $X_{(j)}$, when $i < j$, is

$$f_{X_{(i)},X_{(j)}}(x_i, x_j) = \frac{n!}{(i-1)!(j-i-1)!(n-j)!} \times [F(x_i)]^{i-1}[F(x_j) - F(x_i)]^{j-i-1}[1 - F(x_j)]^{n-j}f(x_i)f(x_j) \tag{6.6}$$

for all $x_i < x_j$.

**Example 6c.** Distribution of the Range of a Random Sample. Suppose that $n$ independent and identically distributed random variables $X_1, X_2, \ldots, X_n$ are observed. The random variable $R$, defined by $R = X_{(n)} - X_{(1)}$, is called the *range* of the observed random variables. If the random variables $X_i$ have distribution function $F$ and density function $f$, then the distribution of $R$ can be obtained from Equation (6.6) as follows: for $a \geq 0$,

$$\begin{aligned} P\{R \leq a\} &= P\{X_{(n)} - X_{(1)} \leq a\} \\ &= \iint_{x_n - x_1 \leq a} f_{X_{(1)},X_{(n)}}(x_1, x_n)\, dx_1\, dx_n \\ &= \int_{-\infty}^{\infty} \int_{x_1}^{x_1+a} \frac{n!}{(n-2)!}[F(x_n) - F(x_1)]^{n-2} f(x_1) f(x_n)\, dx_n\, dx_1 \end{aligned}$$

Making the change of variable $y = F(x_n) - F(x_1)$, $dy = f(x_n)\, dx_n$ yields

$$\begin{aligned} \int_{x_1}^{x_1+a} [F(x_n) - F(x_1)]^{n-2} f(x_n)\, dx_n &= \int_0^{F(x_1+a)-F(x_1)} y^{n-2}\, dy \\ &= \frac{1}{n-1}[F(x_1 + a) - F(x_1)]^{n-1} \end{aligned}$$

and thus

$$P\{R \leq a\} = n \int_{-\infty}^{\infty} [F(x_1 + a) - F(x_1)]^{n-1} f(x_1)\, dx_1 \tag{6.7}$$

Equation (6.7) can be explicitly evaluated only in a few special cases. One such case is when the $X_i$'s are all uniformly distributed on $(0, 1)$.

In this case we obtain from Equation (6.7) that, for $0 < a < 1$,

$$P\{R < a\} = n\int_0^1 [F(x_1 + a) - F(x_1)]^{n-1} f(x_1)\, dx_1$$
$$= n\int_0^{1-a} a^{n-1}\, dx_1 + n\int_{1-a}^1 (1 - x_1)^{n-1}\, dx_1$$
$$= n(1-a)a^{n-1} + a^n$$

Differentiation yields that the density function of the range is given, in this case, by

$$f_R(a) = \begin{cases} n(n-1)a^{n-2}(1-a) & 0 \le a \le 1 \\ 0 & \text{otherwise} \end{cases}$$ ∎

## 7 Joint Probability Distribution Functions of Random Variables

Let $X_1$ and $X_2$ be jointly continuous random variables with joint probability density function $f_{X_1,X_2}$. It is sometimes necessary to obtain the joint distribution of the random variables $Y_1$ and $Y_2$, which arise as functions of $X_1$ and $X_2$. Specifically, suppose that $Y_1 = g_1(X_1, X_2)$ and $Y_2 = g_2(X_1, X_2)$ for some functions $g_1$ and $g_2$.

Assume that the functions $g_1$ and $g_2$ satisfy the following conditions:

1. The equations $y_1 = g_1(x_1, x_2)$ and $y_2 = g_2(x_1, x_2)$ can be uniquely solved for $x_1$ and $x_2$ in terms of $y_1$ and $y_2$ with solutions given by, say, $x_1 = h_1(y_1, y_2)$, $x_2 = h_2(y_1, y_2)$.
2. The functions $g_1$ and $g_2$ have continuous partial derivatives at all points $(x_1, x_2)$ and are such that the following $2 \times 2$ determinant

$$J(x_1, x_2) = \begin{vmatrix} \dfrac{\partial g_1}{\partial x_1} & \dfrac{\partial g_1}{\partial x_2} \\ \dfrac{\partial g_2}{\partial x_1} & \dfrac{\partial g_2}{\partial x_2} \end{vmatrix} \equiv \frac{\partial g_1}{\partial x_1}\frac{\partial g_2}{\partial x_2} - \frac{\partial g_1}{\partial x_2}\frac{\partial g_2}{\partial x_1} \neq 0$$

at all points $(x_1, x_2)$.

Under these two conditions it can be shown that the random variables $Y_1$ and $Y_2$ are jointly continuous with joint density function given by

$$f_{Y_1,Y_2}(y_1, y_2) = f_{X_1,X_2}(x_1, x_2)|J(x_1, x_2)|^{-1} \tag{7.1}$$

where $x_1 = h_1(y_1, y_2)$, $x_2 = h_2(y_1, y_2)$.

A proof of Equation (7.1) would proceed along the following lines:

$$P\{Y_1 \leq y_1, Y_2 \leq y_2\} = \iint\limits_{\substack{(x_1,x_2):\\ g_1(x_1,x_2)\leq y_1\\ g_2(x_1,x_2)\leq y_2}} f_{X_1,X_2}(x_1, x_2)\, dx_1\, dx_2 \tag{7.2}$$

The joint density function can now be obtained by differentiating Equation (7.2) with respect to $y_1$ and $y_2$. That the result of this differentiation will be equal to the right-hand side of Equation (7.1) is an exercise in advanced calculus whose proof will not be presented in this text.

**Example 7a.** Let $X_1$ and $X_2$ be jointly continuous random variables with probability density function $f_{X_1,X_2}$. Let $Y_1 = X_1 + X_2$, $Y_2 = X_1 - X_2$. Find the joint density function of $Y_1$ and $Y_2$ in terms of $f_{X_1,X_2}$.

***Solution:*** Let $g_1(x_1, x_2) = x_1 + x_2$ and $g_2(x_1, x_2) = x_1 - x_2$. Then

$$J(x_1, x_2) = \begin{vmatrix} 1 & 1 \\ 1 & -1 \end{vmatrix} = -2$$

Also, as the equations $y_1 = x_1 + x_2$ and $y_2 = x_1 - x_2$ have as their solution $x_1 = (y_1 + y_2)/2$, $x_2 = (y_1 - y_2)/2$, it follows from Equation (7.1) that the desired density is

$$f_{Y_1,Y_2}(y_1, y_2) = \frac{1}{2} f_{X_1,X_2}\left(\frac{y_1 + y_2}{2}, \frac{y_1 - y_2}{2}\right)$$

For instance, if $X_1$ and $X_2$ are independent, uniform (0, 1) random variables, then

$$f_{Y_1,Y_2}(y_1, y_2) = \begin{cases} \frac{1}{2} & 0 \leq y_1 + y_2 \leq 2, 0 \leq y_1 - y_2 \leq 2 \\ 0 & \text{otherwise} \end{cases}$$

or if $X_1$ and $X_2$ were independent, exponential random variables with respective parameters $\lambda_1$ and $\lambda_2$, then

$$f_{Y_1,Y_2}(y_1, y_2) = \begin{cases} \dfrac{\lambda_1 \lambda_2}{2} \exp\left\{-\lambda_1\left(\dfrac{y_1 + y_2}{2}\right) - \lambda_2\left(\dfrac{y_1 - y_2}{2}\right)\right\} & y_1 + y_2 \geq 0, y_1 - y_2 \\ 0 & \text{otherwise} \end{cases}$$

Finally, if $X_1$ and $X_2$ are independent unit normal random variables,

$$f_{Y_1,Y_2}(y_1, y_2) = \frac{1}{4\pi} e^{-[(y_1+y_2)^2/8+(y_1-y_2)^2/8]}$$

$$= \frac{1}{4\pi} e^{-(y_1^2+y_2^2)/4}$$

$$= \frac{1}{\sqrt{4\pi}} e^{-y_1^2/4} \frac{1}{\sqrt{4\pi}} e^{-y_2^2/4}$$

which yields the interesting result, in this case, that $X_1 + X_2$ is independent of $X_1 - X_2$. (In fact, it can be shown that if $X_1$ and $X_2$ are independent random variables having a common distribution function $F$, then $X_1 + X_2$ will be independent of $X_1 - X_2$ if and only if $F$ is a normal distribution function.) ∎

**Example 7b.** Let $(X, Y)$ denote a random point in the plane and assume that the rectangular coordinates $X$ and $Y$ are independent unit normal random variables. We are interested in the joint distribution of $R, \Theta$, the polar coordinate representation of this point. (See Figure 6.3.)

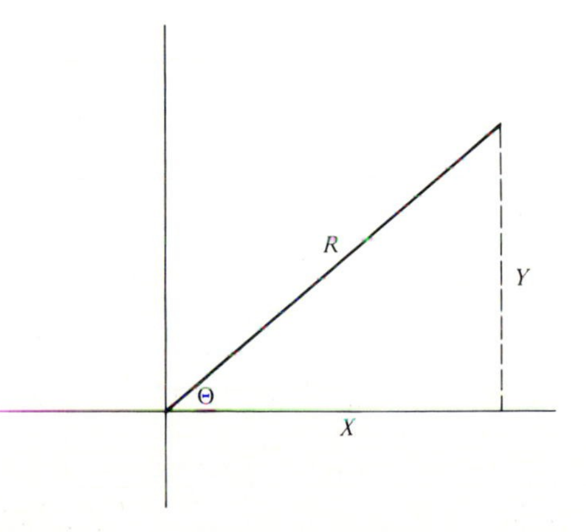


**Figure 6.3** $x$ = random point. $(X, Y) = R, \theta$

Letting $r = g_1(x, y) = \sqrt{x^2 + y^2}$ and $\theta = g_2(x, y) = \tan^{-1} y/x$, we see that

$$\frac{\partial g_1}{\partial x} = \frac{x}{\sqrt{x^2 + y^2}} \qquad \frac{\partial g_1}{\partial y} = \frac{y}{\sqrt{x^2 + y^2}}$$

$$\frac{\partial g_2}{\partial x} = \frac{1}{1 + (y/x)^2}\left(\frac{-y}{x^2}\right) = \frac{-y}{x^2 + y^2} \qquad \frac{\partial g_2}{\partial y} = \frac{1}{x[1 + (y/x)^2]} = \frac{x}{x^2 + y^2}$$

Hence

$$J(x, y) = \frac{x^2}{(x^2 + y^2)^{3/2}} + \frac{y^2}{(x^2 + y^2)^{3/2}} = \frac{1}{\sqrt{x^2 + y^2}} = \frac{1}{r}$$

As the joint density function of $X$ and $Y$ is

$$f(x, y) = \frac{1}{2\pi} e^{-(x^2+y^2)/2}$$

we see that the joint density function of $R = \sqrt{x^2 + y^2}$, $\Theta = \tan^{-1} y/x$, is given by

$$f(r, \theta) = \frac{1}{2\pi} re^{-r^2/2} \qquad 0 < \theta < 2\pi \qquad 0 < r < \infty$$

As this joint density factors into the marginal densities for $R$ and $\Theta$,we obtain that $R$ and $\Theta$ are independent random variables, with $\Theta$ being uniformly distributed over $(0, 2\pi)$ and $R$ having the Rayleigh distribution with density

$$f(r) = re^{-r^2/2} \qquad 0 < r < \infty$$

(Thus, for instance, when one is aiming at a target in the plane, if the horizontal and vertical miss distances are independent unit normals, then the absolute value of the error has the above Rayleigh distribution.)

The above result is quite interesting, for it certainly is not a priori evident that a random vector whose coordinates are independent unit normal random variables will have angle of orientation that is not only uniformly distributed, but is also independent of the vector's distance from the origin.

If we wanted the joint distribution of $R^2$ and $\Theta$, then, as the transformation $d = g_1(x, y) = x^2 + y^2$ and $\theta = g_2(x, y) = \tan^{-1} y/x$ have a Jacobian

$$J = \begin{vmatrix} 2x & 2y \\ \dfrac{-y}{x^2 + y^2} & \dfrac{x}{x^2 + y^2} \end{vmatrix} = 2$$

we see that

$$f_{R^2,\Theta}(d, \theta) = \frac{1}{2} e^{-d/2} \frac{1}{2\pi} \qquad 0 < d < \infty, 0 < \theta < 2\pi$$

Therefore, $R^2$ and $\Theta$ are independent, with $R^2$ having a exponential distribution with parameter $\frac{1}{2}$. But as $R^2 = X^2 + Y^2$, it follows, by definition, that $R^2$ has a chi-squared distribution with 2 degrees of freedom. Hence we have a verification of the result that the exponential distribution with parameter $\frac{1}{2}$ is the same as the chi-squared distribution with 2 degrees of freedom.

The above result can be used to simulate (or generate) normal random variables by making a suitable transformation on uniform random variables. Let $U_1$ and $U_2$ be independent random variables each uniformly distributed over $(0, 1)$. We will transform $U_1, U_2$ into two independent unit normal random variables $X_1$ and $X_2$ by first considering the polar coordinate representation $(R, \Theta)$ of the random vector $(X_1, X_2)$. From the above, $R^2$ and $\Theta$ will be independent, and, in addition, $R^2 = X_1^2 + X_2^2$ will have an exponential distribution with parameter $\lambda = \frac{1}{2}$. But $-2 \log U_1$ has such a distribution since, for $x > 0$,

$$\begin{aligned} P\{-2 \log U_1 < x\} &= P\left\{\log U_1 > -\frac{x}{2}\right\} \\ &= P\{U_1 > e^{-x/2}\} \\ &= 1 - e^{-x/2} \end{aligned}$$

Also, as $2\pi U_2$ is a uniform $(0, 2\pi)$ random variable, we can use it to generate $\Theta$. That is, if we let

$$\begin{aligned} R^2 &= -2 \log U_1 \\ \Theta &= 2\pi U_2 \end{aligned}$$

then $R^2$ can be taken to be the square of the distance from the origin and $\theta$ as the angle of orientation of $(X_1, X_2)$. As $X_1 = R \cos \Theta$, $X_2 = R \sin \Theta$, we obtain that

$$\begin{aligned} X_1 &= \sqrt{-2 \log U_1} \cos (2\pi U_2) \\ X_2 &= \sqrt{-2 \log U_1} \sin (2\pi U_2) \end{aligned}$$

are independent unit normal random variables. ∎

**Example 7c.** If $X$ and $Y$ are independent gamma random variables with parameters $(\alpha, \lambda)$ and $(\beta, \lambda)$, respectively, compute the joint density of $U = X + Y$ and $V = X/(X + Y)$.

***Solution:*** The joint density of $X$ and $Y$ is given by

$$\begin{aligned} f_{X,Y}(x, y) &= \frac{\lambda e^{-\lambda x}(\lambda x)^{\alpha-1}}{\Gamma(\alpha)} \frac{\lambda e^{-\lambda y}(\lambda y)^{\beta-1}}{\Gamma(\beta)} \\ &= \frac{\lambda^{\alpha+\beta}}{\Gamma(\alpha)\Gamma(\beta)} e^{-\lambda(x+y)} x^{\alpha-1} y^{\beta-1} \end{aligned}$$

Now, if $g_1(x, y) = x + y$, $g_2(x, y) = x/(x + y)$, then

$$\frac{\partial g_1}{\partial x} = \frac{\partial g_1}{\partial y} = 1 \qquad \frac{\partial g_2}{\partial x} = \frac{y}{(x + y)^2} \qquad \frac{\partial g_2}{\partial y} = -\frac{x}{(x + y)^2}$$

and so

$$J(x, y) = \begin{vmatrix} 1 & 1 \\ \dfrac{y}{(x+y)^2} & -\dfrac{x}{(x+y)^2} \end{vmatrix} = -\frac{1}{x+y}$$

Finally, as the equations $u = x + y$, $v = x/(x + y)$ have as their solutions $x = uv$, $y = u(1 - v)$, we see that

$$\begin{aligned} f_{U,V}(u, v) &= f_{X,Y}[uv, u(1-v)]u \\ &= \frac{\lambda e^{-\lambda u}(\lambda u)^{\alpha+\beta-1}}{\Gamma(\alpha+\beta)} \frac{v^{\alpha-1}(1-v)^{\beta-1}\Gamma(\alpha+\beta)}{\Gamma(\alpha)\Gamma(\beta)} \end{aligned}$$

Hence $X + Y$ and $X/(X + Y)$ are independent, with $X + Y$ having a gamma distribution with parameters $(\alpha + \beta, \lambda)$ and $X/(X + Y)$ having a beta distribution with parameters $(\alpha, b)$. The above also shows that $B(\alpha, \beta)$, the normalizing factor in the beta density, is such that

$$\begin{aligned} B(\alpha, \beta) &\equiv \int_0^1 v^{\alpha-1}(1-v)^{\beta-1}\, dv \\ &= \frac{\Gamma(\alpha)\Gamma(\beta)}{\Gamma(\alpha+\beta)} \end{aligned}$$

The above result is quite interesting. For suppose there are $n + m$ jobs to be performed, with each (independently) taking an exponential amount of time with rate $\lambda$ for performance, and suppose that we have two workers to perform these jobs. Worker I will do jobs $1, 2, \ldots, n$, and worker II will do the remaining $m$ jobs. If we let $X$ and $Y$ denote the total working times of workers I and II, respectively, then (either from the above result or from Example 3b) $X$ and $Y$ will be independent gamma random variables having parameters $(n, \lambda)$ and $(m, \lambda)$, respectively. Then the above result yields that independently of the working time needed to complete all $n + m$ jobs (that is, of $X + Y$), the proportion of this work that will be performed by worker I has a beta distribution with parameters $(n, m)$. ∎

When the joint density function of the $n$ random variables $X_1, X_2, \ldots, X_n$ is given and we want to compute the joint density function of $Y_1, Y_2, \ldots, Y_n$, where

$$Y_1 = g_1(X_1, \ldots, X_n) \qquad Y_2 = g_2(X_1, \ldots, X_n), \ldots$$
$$Y_n = g_n(X_1, \ldots, X_n)$$

the approach is the same. Namely, we assume that the functions $g_i$ have continuous partial derivatives and that the Jacobian determinant

$J(x_1, \ldots, x_n) \neq 0$ at all points $(x_1, \ldots, x_n)$, where

$$J(x_1, \ldots, x_n) = \begin{vmatrix} \frac{\partial g_1}{\partial x_1} & \frac{\partial g_1}{\partial x_2} \cdots & \frac{\partial g_1}{\partial x_n} \\ \frac{\partial g_2}{\partial x_1} & \frac{\partial g_2}{\partial x_2} \cdots & \frac{\partial g_2}{\partial x_n} \\ \frac{\partial g_n}{\partial x_1} & \frac{\partial g_n}{\partial x_2} \cdots & \frac{\partial g_n}{\partial x_n} \end{vmatrix}$$

Furthermore, we suppose that the equations $y_1 = g_1(x_1, \ldots, x_n)$, $y_2 = g_2(x_1, \ldots, x_n), \ldots, y_n = g_n(x_1, \ldots, x_n)$ have a unique solution, say, $x_1 = h_1(y_1, \ldots, y_n), \ldots, x_n = h_n(y_1, \ldots, y_n)$. Under these assumptions the joint density function of the random variables $Y_i$ is given by

$$f_{Y_1,\ldots,Y_n}(y_1, \ldots, y_n) = f_{X_1,\ldots,X_n}(x_1, \ldots, x_n)|J(x_1, \ldots, x_n)|^{-1} \quad (7.3)$$

where $x_i = h_i(y_1, \ldots, y_n)$, $i = 1, 2, \ldots, n$.

**Example 7d.** Let $X_1$, $X_2$, and $X_3$ be independent unit normal random variables. If $Y_1 = X_1 + X_2 + X_3$, $Y_2 = X_1 - X_2$, $Y_3 = X_1 - X_3$, compute the joint density function of $Y_1, Y_2, Y_3$.

***Solution:*** Letting $Y_1 = X_1 + X_2 + X_3$, $Y_2 = X_1 - X_2$, $Y_3 = X_1 - X_3$, the Jacobian of these transformations is given by

$$J = \begin{vmatrix} 1 & 1 & 1 \\ 1 & -1 & 0 \\ 1 & 0 & -1 \end{vmatrix} = 3$$

As the above transformations yield that

$$X_1 = \frac{Y_1 + Y_2 + Y_3}{3} \qquad X_2 = \frac{Y_1 - 2Y_2 + Y_3}{3} \qquad X_3 = \frac{Y_1 + Y_2 - 2Y_3}{3}$$

We see from Equation (7.3) that

$$f_{Y_1,Y_2,Y_3}(y_1, y_2, y_3) = \frac{1}{3} f_{X_1,X_2,X_3}\left(\frac{y_1 + y_2 + y_3}{3}, \frac{y_1 - 2y_2 + y_3}{3}, \frac{y_1 + y_2 - 2y_3}{3}\right)$$

Hence, as

$$f_{X_1,X_2,X_3}(x_1, x_2, x_3) = \frac{1}{(2\pi)^{3/2}} e^{-\sum_{i=1}^3 x_i^2/2}$$

we see that

$$f_{Y_1,Y_2,Y_3}(y_1, y_2, y_3) = \frac{1}{3(2\pi)^{3/2}} e^{-Q(y_1,y_2,y_3)/2}$$

where

$$\begin{aligned} Q(y_1, y_2, y_3) &= \left(\frac{y_1 + y_2 + y_3}{3}\right)^2 + \left(\frac{y_1 - 2y_2 + y_3}{3}\right)^2 + \left(\frac{y_1 + y_2 - 2y_3}{3}\right)^2 \\ &= \frac{y_1^2}{3} + \frac{2}{3} y_2^2 + \frac{2}{3} y_3^2 - \frac{2}{3} y_2 y_3. \end{aligned}$$

∎

## *Theoretical Exercises*

**1.** Verify Equation (1.2).

**2.** Suppose that the number of events that occur in a given time period is a Poisson random variable with parameter $\lambda$. If each event is classified as a type $i$ event with probability $p_i, i = 1, \ldots, n, \sum p_i = 1$, independently of other events, show that the numbers of type $i$ events that occur, $i = 1, \ldots, n$, are independent Poisson random variables with respective parameters $\lambda p_i, i = 1, \ldots, n$.

**3.** Suggest a procedure for using Buffon's needle problem to estimate $\pi$. Surprisingly enough, this was once a common method of evaluating $\pi$.

**4.** Solve Buffon's needle problem when $L > D$.

ANSWER: $\dfrac{2L}{\pi D}(1 - \sin\theta) + 2\theta/\pi$, where $\theta$ is such that $\cos\theta = D/L$.

**5.** If $X$ and $Y$ are independent continuous positive random variables, express the density function of (a) $Z = X/Y$ and (b) $Z = XY$ in terms of the density functions of $X$ and $Y$. Evaluate these expressions in the special case where $X$ and $Y$ are both exponential random variables.

**6.** Show analytically (by induction) that $X_1 + \cdots + X_n$ has a negative binomial distribution when the $X_i$, $i = 1, \ldots, n$ are independent and identically distributed geometric random variables. Also, give a second argument that verifies the above without any need for computations.

**7.** (a) If $X$ has a gamma distribution with parameters $(t, \lambda)$ what is the distribution of $cX$, $c > 0$?

(b) Show that

$$\frac{1}{2\lambda} \chi^2_{2n}$$

has a gamma distribution with parameters $n, \lambda$ when $n$ is a positive integer and $\chi^2_{2n}$ is a chi-square random variable with $2n$ degrees of freedom.

**8.** Let $X$ and $Y$ be independent continuous random variables with respective hazard rate functions $\lambda_X(t)$ and $\lambda_Y(t)$, and set $W = \text{Min}(X,Y)$.
(a) Determine the distribution function of $W$ in terms of those of $X$ and $Y$.
(b) Show that $\lambda_W(t)$, the hazard rate function of $W$, is given by

$$\lambda_W(t) = \lambda_X(t) + \lambda_Y(t)$$

**9.** Let $X_1, \ldots, X_n$ be independent exponential random variables having a common parameter $\lambda$. Determine the distribution of $\text{Min}(X_1, \ldots, X_n)$.

**10.** The lifetime of batteries are independent exponential random variables, each having parameter $\lambda$. A flashlight needs 2 batteries to work. If one has a flashlight and a stockpile of $n$ batteries, what is the distribution of time that the flashlight can operate?

**11.** Let $X_1, X_2, X_3, X_4, X_5$ be independent continuous random variables having a common distribution function $F$ and density function $f$, and set

$$I = P\{X_1 < X_2 > X_3 < X_4 > X_5\}$$

(a) Show that $I$ does not depend on $F$. *Hint:* Write $I$ as a 5-dimensional integral and make the change of variables $u_i = F(x_i)$, $i = 1, \ldots, 5$.
(b) Evaluate $I$.

**12.** Prove Proposition 3.2.

HINT: Prove it first when $n = 2$, and then use induction.

**13.** In Example 5c we computed the conditional density of a success probability for a sequence of trials when the first $n + m$ trials resulted in $n$ successes. Would the conditional density change if we actually specified which $n$ of these trials resulted in successes?

**14.** Compute the conditional probability mass function of $X$, given that $X + Y = n$ when $X$ and $Y$ are independent and identically distributed geometric random variables.

**15.** If $X$ and $Y$ are independent binomial random variables with identical parameters $n$ and $p$, show analytically that the conditional distribution of $X$, given that $X + Y = m$, is the hypergeometric distribution. Also, give a second argument that yields the result without any computations.

HINT: Suppose that $2n$ coins are flipped. Let $X$ denote the number of heads in the first $n$ flips and $Y$ the number in the second $n$ flips. Argue that given a total of $m$ heads, the number of heads in the first $n$ flips has the same distribution as the number of white balls selected when a sample of size $m$ is chosen from $n$ white and $n$ black balls.

**16.** Consider an experiment that results in one of three possible outcomes, outcome $i$ occurring with probability $p_i$, $i = 1, 2, 3$. Suppose that $n$ independent replications of this experiment are performed and let $X_i$, $i = 1, 2, 3$ denote the number of times that outcome $i$ occurs. Determine the conditional probability mass function of $X_1$, given that $X_2 = m$.

**17.** Let $X_1, X_2, X_3$ be independent and identically distributed continuous random variables. Compute
(a) $P\{X_1 > X_2 | X_1 > X_3\}$;
(b) $P\{X_1 > X_2 | X_1 < X_3\}$;
(c) $P\{X_1 > X_2 | X_2 > X_3\}$;
(d) $P\{X_1 > X_2 | X_2 < X_3\}$.

**18.** Let $U$ denote a random variable uniformly distributed over $(0, 1)$. Compute the conditional distribution of $U$ given that
(a) $U > a$;
(b) $U < a$;
where $0 < a < 1$.

**19.** Suppose that $W$, the amount of moisture in the air on a given day, is a gamma random variable with parameters $(t, \beta)$. That is, its density is $f(w) = \beta e^{-\beta w}(\beta w)^{t-1}/\Gamma(t)$, $w > 0$. Suppose also that given that $W = w$, the number of accidents during that day—call it $N$—has a Poisson distribution with mean $w$. Show that the conditional distribution of $W$ given that $N = n$ is the gamma distribution with parameters $(t + n, \beta + 1)$.

**20.** Let $W$ be a gamma random variable with parameters $(t, \beta)$, and suppose that conditional on $W = w$, $X_1, X_2, \ldots, X_n$ are independent exponential random variables with rate $w$. Show that the conditional distribution of $W$ given that $X_1 = x_1, X_2 = x_2, \ldots, X_n = x_n$ is gamma with parameters $(t + n, \beta + \sum_{i=1}^{n} x_i)$.

**21.** A rectangular array of $mn$ numbers arranged in $n$ rows, each consisting of $m$ columns, is said to contain a *saddlepoint* if there is a number that is both the minimum of its row and the maximum of its column. For instance, in the array

$$\begin{matrix} 1 & 3 & 2 \\ 0 & -2 & 6 \\ .5 & 12 & 3 \end{matrix}$$

the number 1 in the first row, first column is a saddlepoint. The existence of a saddlepoint is of significance in the theory of games. Consider a rectangular array of numbers as described above and suppose that there are two individuals—$A$ and $B$—that are playing the following game: $A$ is to choose one of the numbers $1, 2, \ldots, n$ and $B$ one of the numbers $1, 2, \ldots, m$. These choices are announced simultaneously, and if $A$ chose $i$ and $B$ chose $j$, then $A$ wins from $B$ the amount specified by the number in the $i$th row, $j$th column of the array. Now suppose the array contains a saddlepoint—say the number in the row $r$ and column $k$—call this number $x_{rk}$. Now if player $A$ chooses row $r$, then that player can guarantee herself a win at least $x_{rk}$ (since $x_{rk}$ is the minimum number in the row $r$). On the other hand, if player $B$ chooses column $k$, then he can guarantee that he will lose no more than $x_{rk}$ (since $x_{rk}$ is the maximum number in the column $k$). Hence, as $A$ has a way of playing that guarantees her a win of $x_{rk}$ and as $B$ has a way of playing that guarantees he will lose no more than $x_{rk}$, it seems reasonable to take these two strategies as being optimal and declare that the value of the game to player $A$ is $x_{rk}$.

If the $nm$ numbers in the rectangular array described above are independently chosen from an arbitrary continuous distribution, what is the probability that the resulting array will contain a saddlepoint?

**22.** The random variables $X$ and $Y$ are said to have a bivariate normal distribution if their joint density function is given by

$$f(x, y) = \frac{1}{2\pi\sigma_x\sigma_y\sqrt{1-\rho^2}}$$

$$\times \exp\left\{-\frac{1}{2(1-\rho^2)}\left[\left(\frac{x-\mu_x}{\sigma_x}\right)^2 + \left(\frac{y-\mu_y}{\sigma_y}\right)^2 - 2\rho\frac{(x-\mu_x)(y-\mu_y)}{\sigma_x\sigma_y}\right]\right\}$$

(a) Show that the conditional density of $X$, given that $Y = y$, is the normal density with parameters

$$\mu_x + \rho\frac{\sigma_x}{\sigma_y}(y-\mu_y) \qquad \text{and} \qquad \sigma_x^2(1-\rho^2)$$

(b) Show that $X$ and $Y$ are both normal random variables with respective parameters $\mu_x$, $\sigma_x^2$, and $\mu_y$, $\sigma_y^2$.

(c) Show that $X$ and $Y$ are independent when $\rho = 0$.

**23.** Suppose that $F(x)$ is a cumulative distribution function. Show that (a) $F^n(x)$, and (b) $1 - [1 - F(x)]^n$ are also cumulative distribution functions when $n$ is a positive integer.

HINT: Let $X_1, \ldots, X_n$ be independent random variables having the common distribution function $F$. Define random variables $Y$ and $Z$ in terms of the $X_i$ so that $P\{Y \le x\} = F^n(x)$, and $P\{Z \le x\} = 1 - [1 - F(x)]^n$.

**24.** Show that if $n$ people are distributed at random along a road $L$ miles long, then the probability that no 2 people are less than a distance of $D$ miles apart is, when $D \leq L/(n-1)$, $[1-(n-1)D/L]^n$. What if $D > L/(n-1)$?

**25.** Establish Equation (6.2) by differentiating Equation (6.4).

**26.** Show that the median of a sample of size $2n+1$ from a uniform distribution on $(0, 1)$ has a beta distribution with parameters $(n+1, n+1)$.

**27.** Verify Equation (6.6), which gives the joint density of $X_{(i)}$ and $X_{(j)}$.

**28.** Compute the density of the range of a sample of size $n$ from a continuous distribution having density function $f$.

**29.** Let $X_{(1)} \leq X_{(2)} \leq \cdots \leq X_{(n)}$ be the ordered values of $n$ independent uniform $(0, 1)$ random variables. Prove that for $1 \leq k \leq n+1$

$$P\{X_{(k)} - X_{(k-1)} > t\} = (1-t)^n$$

where $X_{(0)} \equiv 0$, $X_{(n+1)} \equiv 1$.

**30.** Let $X_1, \ldots X_n$ be a set of independent and identically distributed continuous random variables having distribution function $F$, and let $X_{(i)}$, $i = 1, \ldots, n$ denote their ordered values. If $X$, independent of the $X_i$, $i = 1, \ldots, n$, also has distribution $F$, determine
(a) $P\{X > X_{(n)}\}$
(b) $P\{X > X_{(1)}\}$
(c) $P\{X_{(i)} < X < X_{(j)}\}$, $1 \leq i < j \leq n$.

**31.** Let $X_1, \ldots, X_n$ be independent and identically distributed random variables having distribution function $F$ and density $f$. The quantity $M \equiv [X_{(1)} + X_{(n)}]/2$, defined to be the average of the smallest and largest value, is called the midrange. Show that its distribution function is

$$F_M(m) = n \int_{-\infty}^{m} [F(2m-x) - F(x)]^{n-1} f(x)\, dx$$

**32.** Let $X_1, \ldots, X_n$ be independent uniform $(0, 1)$ random variables. Let $R = X_{(n)} - X_{(1)}$ denote the range and $M = [X_{(n)} + X_{(1)}]/2$ the midrange. Compute the joint density function of $R$ and $M$.

## Problems

**1.** Two fair dice are rolled. Find the joint probability mass function of $X$ and $Y$ when
(a) $X$ is the largest value obtained on any die and $Y$ is the sum of the values:

(b) $X$ is the value on the first die and $Y$ is the larger of the two values;
(c) $X$ is the smallest and $Y$ the largest value obtained on the dice.

**2.** A bin of 5 transistors is known to contain 2 that are defective. The transistors are to be tested, one at a time, until the defective ones are identified. Denote by $N_1$ the number of tests made until the first defective is spotted and by $N_2$ the number of additional tests until the second defective is spotted; find the joint probability mass function of $N_1$ and $N_2$.

**3.** Consider a sequence of independent Bernoulli trials, each of which is a success with probability $p$. Let $X_1$ be the number of failures preceding the first success, and let $X_2$ be the number of failures between the first two successes. Find the joint mass function of $X_1$ and $X_2$.

**4.** The joint probability density function of $X$ and $Y$ is given by

$$f(x, y) = c(y^2 - x^2)e^{-y} \qquad -y \le x \le y, 0 < y < \infty$$

(a) Find $c$.
(b) Find the marginal densities of $X$ and $Y$.

**5.** The joint probability density function of $X$ and $Y$ is given by

$$f(x, y) = \frac{6}{7}\left(x^2 + \frac{xy}{2}\right) \qquad 0 < x < 1, 0 < y < 2$$

(a) Verify that this is indeed a joint density function.
(b) Compute the density function of $X$.
(c) Find $P\{X > Y\}$.
(d) Find $P\{Y > \frac{1}{2} | X < \frac{1}{2}\}$.

**6.** The joint probability density function of $X$ and $Y$ is given by

$$f(x, y) = e^{-(x+y)} \qquad 0 \le x < \infty, 0 \le y < \infty$$

Find (a) $P\{X < Y\}$, and (b) $P\{X < a\}$.

**7.** A television store owner figures that 45 percent of the customers entering his store will purchase an ordinary television set, 15 percent will purchase a color television set, and 40 percent will just be browsing. If 5 customers enter his store on a given day, what is the probability that he will sell exactly 2 ordinary sets and 1 color set on that day?

**8.** The number of people that enter a drugstore in a given hour is a Poisson random variable with parameter $\lambda = 10$. Compute the conditional probability that at most 3 men entered the drug store, given that 10 women entered in that hour. What assumptions have you made?

**9.** A man and a woman agree to meet at a certain location about 12:30 P.M. If the man arrives at a time uniformly distributed between 12:15 and

12:45 and if the woman independently arrives at a time uniformly distributed between 12:00 and 1 P.M., find the probability that the first to arrive waits no longer than 5 minutes. What is the probability that the man arrives first?

**10.** An ambulance travels back and forth, at a constant speed, along a road of length $L$. At a certain moment of time an accident occurs at a point uniformly distributed on the road. [That is, its distance from one of the fixed ends of the road is uniformly distributed over $(0, L)$.] Assuming that the ambulance's location at the moment of the accident is also uniformly distributed, compute, assuming independence, the distribution of its distance from the accident. What if the ambulance did not travel at a constant speed but rather traveled at different speeds when at different locations?

**11.** Suppose that a point is uniformly chosen on a square of area 1 having vertices $(0, 0)$, $(0, 1)$, $(1, 0)$, and $(1, 1)$. Let $X$ and $Y$ be the coordinates of the point chosen.
(a) Find the marginal distributions of $X$ and $Y$.
(b) Are $X$ and $Y$ independent?
(c) Find the probability that the distance from $(X, Y)$ to the center of the square is greater than $\frac{1}{4}$.

**12.** Three points $X_1, X_2, X_3$ are selected at random on a line $L$. What is the probability that $X_2$ lies between $X_1$ and $X_3$?

**13.** Two points are selected randomly on a line of length $L$ so as to be on opposite sides of the midpoint of the line. [In other words, the two points $X$ and $Y$ are independent random variables such that $X$ is uniformly distributed over $(0, L/2)$ and $Y$ is uniformly distributed over $(L/2, L)$.] Find the probability that the distance between the two points is greater than $L/3$.

**14.** In Problem 13 find the probability that the 3 line segments from 0 to $X$, from $X$ to $Y$, and from $Y$ to $L$ could be made to form the three sides of a triangle. (Note that three line segments can be made to form a triangle if the length of each of them is less than the sum of the lengths of the others.)

**15.** The joint density of $X$ and $Y$ is given by

$$f(x, y) = \begin{cases} xe^{-(x+y)} & x > 0, y > 0 \\ 0 & \text{otherwise} \end{cases}$$

Are $X$ and $Y$ independent? What if $f(x, y)$ were given by

$$f(x, y) = \begin{cases} 2 & 0 < x < y, 0 < y < 1 \\ 0 & \text{otherwise} \end{cases}$$

**16.** Suppose that $10^6$ people arrive at a service station at times that are independent random variables, each of which is uniformly distributed over $(0, 10^6)$. Let $N$ denote the number that arrive by time $i$. Find an approximation for $P\{N = i\}$.

**17.** Suppose that $A, B, C$, are independent random variables, each being uniformly distributed over $(0, 1)$.
(a) What is the joint cumulative distribution function of $A, B, C$?
(b) What is the probability that all of the roots of the equation $Ax^2 + Bx + C = 0$ are real?

**18.** If $X$ is uniformly distributed over $(0, 1)$ and $Y$ is exponentially distributed with parameter $\lambda = 1$, find the distribution of (a) $Z = X + Y$, and (b) $Z = X/Y$. Assume independence.

**19.** If $X_1$ and $X_2$ are independent exponential random variables with respective parameters $\lambda_1$ and $\lambda_2$, find the distribution of $Z = X_1/X_2$. Also compute $P\{X_1 < X_2\}$.

**20.** When a current $I$ (measured in amperes) flows through a resistance $R$ (measured in ohms), the power generated is given by $W = I^2R$ (measured in watts). Suppose that $I$ and $R$ are independent random variables with densities

$$f_I(x) = 6x(1 - x) \qquad 0 \le x \le 1$$
$$f_R(x) = 2x \qquad 0 \le x \le 1$$

Determine the density function of $W$.

**21.** Choose a number $X$ at random from the set of numbers $\{1, 2, 3, 4, 5\}$. Now choose a number at random from the subset no larger than $X$, that is, from $\{1, \ldots, X\}$. Call this second number $Y$.
(a) Find the joint mass function of $X$ and $Y$.
(b) Find the conditional mass function of $X$ given that $Y = i$. Do it for $i = 1, 2, 3, 4, 5$.
(c) Are $X$ and $Y$ independent? Why?

**22.** Two dice are rolled. Let $X$ and $Y$ denote, respectively, the largest and smallest values obtained. Compute the conditional mass function of $Y$ given $X = i$, for $i = 1, 2, \ldots, 6$. Are $X$ and $Y$ independent? Why?

**23.** The joint probability mass function of $X$ and $Y$ is given by

$$p(1, 1) = \tfrac{1}{8} \qquad p(1, 2) = \tfrac{1}{4}$$
$$p(2, 1) = \tfrac{1}{8} \qquad p(2, 2) = \tfrac{1}{2}$$

(a) Compute the conditional mass function of $X$ given $Y = i$, $i = 1, 2$.
(b) Are $X$ and $Y$ independent?
(c) Compute $P\{XY \le 3\}$, $P\{X + Y > 2\}$, $P\{X/Y > 1\}$.

**24.** The joint density function of $X$ and $Y$ is given by

$$f(x, y) = xe^{-x(y+1)} \qquad x > 0, y > 0$$

(a) Find the conditional density of $X$, given $Y = y$, and that of $Y$, given $X = x$.
(b) Find the density function of $Z = XY$.

**25.** The joint density of $X$ and $Y$ is

$$f(x, y) = c(x^2 - y^2)e^{-x} \qquad 0 \leq x < \infty, -x \leq y \leq x$$

Find the conditional distribution of $Y$, given $X = x$.

**26.** If $X_1, X_2, X_3$ are independent random variables that are uniformly distributed over $(a, b)$, compute the probability that the largest of the three is greater than the sum of the other two.

**27.** A complex machine is able to operate effectively as long as at least 3 of its 5 motors are functioning. If each motor independently functions for a random amount of time with density function $f(x) = xe^{-x}$, $x > 0$, compute the density function of the length of time that the machine functions.

**28.** If 3 trucks break down at points randomly distributed on a road of length $L$, find the probability that no 2 of the trucks are within a distance $d$ of each other when $d \leq L/2$.

**29.** Consider a sample of size 5 from a uniform distribution over $(0, 1)$. Compute the probability that the median is in the interval $(\frac{1}{4}, \frac{3}{4})$.

**30.** If $X_1, X_2, X_3, X_4, X_5$ are independent and are identically distributed exponential random variables with the parameter $\lambda$, compute (a) $P\{\min(X_1, \ldots, X_5) \leq a\}$ and (b) $P\{\max(X_1, \ldots, X_5) \leq a\}$.

**31.** Derive the distribution of the range of a sample of size 2 from a distribution having density function $f(x) = 2x$, $0 < x < 1$.

**32.** Let $X$ and $Y$ denote the coordinates of a point uniformly chosen in the circle of radius 1 centered at the origin. That is, their joint density is

$$f(x, y) = \frac{1}{\pi} \qquad x^2 + y^2 \leq 1$$

Find the joint density function of the polar coordinates $R = (X^2 + Y^2)^{1/2}$ and $\Theta = \tan^{-1} Y/X$.

**33.** If $X$ and $Y$ are independent random variables both uniformly distributed over $(0, 1)$, find the joint density function of $R = \sqrt{X^2 + Y^2}$, $\Theta = \tan^{-1} Y/X$.

**34.** If $U$ is uniform on $(0, 2\pi)$ and $Z$, independent of $U$, is exponential with rate 1, show directly (without using the results of Example 7b) that $X$ and $Y$ defined by

$$X = \sqrt{2Z} \cos U$$
$$Y = \sqrt{2Z} \sin U$$

are independent unit normal random variables.

**35.** If $X$ and $Y$ have joint density function

$$f(x, y) = \frac{1}{x^2 y^2} \qquad x \geq 1, y \geq 1$$

(a) Compute the joint density function of $U = XY$, $V = X/Y$.
(b) What are the marginal densities?

**36.** If $X$ and $Y$ are independent and identically distributed uniform random variables on $(0, 1)$, compute the joint density of
(a) $U = X + Y$, $V = X/Y$;
(b) $U = X$, $V = X/Y$;
(c) $U = X + Y$, $V = X/(X + Y)$.

**37.** Repeat Problem 36 when $X$ and $Y$ are independent exponential random variables, each with parameter $\lambda = 1$.

**38.** If $X_1$ and $X_2$ are independent exponential random variables each having parameter $\lambda$ find the joint density function of $Y_1 = X_1 + X_2$ and $Y_2 = e^{X_1}$.

**39.** If $X$, $Y$, and $Z$ are independent random variables having identical density functions $f(x) = e^{-x}$, $0 < x < \infty$, derive the joint distribution of $U = X + Y$, $V = X + Z$, $W = Y + Z$.

**40.** In Example 7d show that $Y_2$ and $Y_3$ have a joint bivariate normal distribution.

**41.** The ages of prospective parents at a certain hospital can be approximated by a bivariate normal distribution with parameters $\mu_x = 28.4$, $\sigma_x = 6.8$, $\mu_y = 31.6$, $\sigma_y = 7.4$, and $\rho = .82$. (The parameters having subscript $x$ referring to age of the pregnant woman and those with subscript $y$ to the age of the prospective father). Using the results of Theoretical Exercise 22 determine
(a) the proportion of pregnant women who are over 30.
(b) the proportion of the prospective fathers who are aged 35 that have wives over 30.

# 7

# Expectation

## 1 Introduction and Definitions

One of the most important concepts in probability theory is that of the expectation of a random variable. If $X$ is a discrete random variable having a probability mass function $p(x)$, the *expectation* or the *expected value* of $X$, denoted by $E[X]$, is defined by

$$E[X] = \sum_{x:p(x)>0} xp(x)$$

In words, the expected value of $X$ is a weighted average of the possible values that $X$ can take on, each value being weighted by the probability that $X$ assumes it. For instance, if the probability mass function of $X$ is given by

$$p(0) = \tfrac{1}{2} = p(1)$$

then

$$E[X] = 0(\tfrac{1}{2}) + 1(\tfrac{1}{2}) = \tfrac{1}{2}$$

is just the ordinary average of the two possible values 0 and 1 that $X$ can assume. On the other hand, if

$$p(0) = \tfrac{1}{2} \qquad p(1) = \tfrac{2}{3}$$

then

$$E[X] = 0(\tfrac{1}{3}) + 1(\tfrac{2}{3}) = \tfrac{2}{3}$$

is a weighted average of the two possible values 0 and 1, where the value 1 is given twice as much weight as the value 0, since $p(1) = 2p(0)$.

Another motivation of the definition of expectation is provided by the frequency interpretation of probabilities. This interpretation (partially justified by the strong law of large numbers, to be presented in the next chapter) assumes that if an infinite sequence of independent replications of an experiment is performed, then for any event $E$, the proportion of time that $E$ occurs will be $P(E)$. Now, consider a random variable $X$ that must take on one of the values $x_1, x_2, \ldots, x_n$ with respective probabilities $p(x_1), p(x_2), \ldots, p(x_n)$; and think of $X$ as representing our winnings in a single game of chance. That is, with probability $p(x_i)$ we shall win $x_i$ units $i = 1, 2, \ldots, n$. Now by the frequency interpretation, it follows that if we continually play this game, then the proportion of time that we win $x_i$ will be $p(x_i)$. As this is true for all $i$, $i = 1, 2, \ldots, n$, it follows that our average winnings per game will be

$$\sum_{i=1}^{n} x_i p(x_i) = E[X]$$

**Example 1a.** Find $E[X]$ where $X$ is the outcome when we roll a fair die.

***Solution:*** Since $p(1) = p(2) = p(3) = p(4) = p(5) = p(6) = \frac{1}{6}$, we obtain that

$$E[X] = 1(\tfrac{1}{6}) + 2(\tfrac{1}{6}) + 3(\tfrac{1}{6}) + 4(\tfrac{1}{6}) + 5(\tfrac{1}{6}) + 6(\tfrac{1}{6}) = \tfrac{7}{2}$$ ∎

**Example 1b.** Expectation of a Bernoulli Random Variable. Calculate $E[X]$ when $X$ is a Bernoulli random variable with parameter $p$.

***Solution:*** Since $p(0) = 1 - p$, $p(1) = p$, we have

$$E[X] = 0(1 - p) + 1(p) = p$$ ∎

In other words, Example 1b states that the expected number of successes in a single trial is just the probability that the trial will be a success. From this, it would seem reasonable that the expected number of successes when $n$ independent trials are performed, each trial having probability $p$ of resulting in a success, should equal $np$. This conjecture is now verified by Example 1c.

**Example 1c.** Expectation of a Binomial Random Variable. Calculate $E[X]$ when $X$ is binomially distributed, with parameters $n$ and $p$.

***Solution***

$$\begin{aligned} E[X] &= \sum_{i=0}^{n} i p(i) \\ &= \sum_{i=0}^{n} i \binom{n}{i} p^i (1 - p)^{n-i} \\ &= \sum_{i=1}^{n} \frac{i n!}{(n - i)!\, i!} p^i (1 - p)^{n-i} \end{aligned}$$

$$
\begin{aligned}
&= np \sum_{i=1}^{n} \frac{(n-1)!}{(n-i)!(i-1)!} p^{i-1}(1-p)^{n-i} \\
&= np \sum_{k=0}^{n-1} \binom{n-1}{k} p^k (1-p)^{n-1-k} \\
&= np[p + (1-p)]^{n-1} \\
&= np
\end{aligned}
$$

where the second from the last equality follows by letting $k = i - 1$. ∎

Since a Poisson random variable with parameter $\lambda$ arises as an approximation for a binomial random variable with parameters $n, p$ when $n$ is large and $\lambda = np$ moderate, it appears reasonable, from the results of Example 1c, to suppose that the expected value of a Poisson random variable with parameter $\lambda$ is equal to $\lambda$. This is now verified.

**Example 1d.** Expectation of a Poisson Random Variable. Calculate $E[X]$ when $X$ is a Poisson random variable having parameter $\lambda$.

***Solution***

$$
\begin{aligned}
E[X] &= \sum_{i=0}^{\infty} ip(i) \\
&= \sum_{i=0}^{\infty} ie^{-\lambda} \frac{\lambda^i}{i!} \\
&= \sum_{i=1}^{\infty} ie^{-\lambda} \frac{\lambda^i}{i!} \\
&= \sum_{i=1}^{\infty} e^{-\lambda} \frac{\lambda^i}{(i-1)!} \\
&= \lambda e^{-\lambda} \sum_{i=1}^{\infty} \frac{\lambda^{i-1}}{(i-1)!} \\
&= \lambda e^{-\lambda} e^{\lambda} \\
&= \lambda
\end{aligned}
$$

where we have used the identity

$$
\sum_{i=1}^{\infty} \frac{\lambda^{i-1}}{(i-1)!} = \sum_{k=0}^{\infty} \frac{\lambda^k}{k!} = e^{\lambda}
$$

∎

**Example 1e.** Expectation of a Geometric Random Variable. Calculate the expected value of a geometric random variable having parameter $p$.

***Solution:*** For a geometric random variable

$$P\{X = n\} = p(1-p)^{n-1} \qquad n \geq 1$$

Hence

$$E[X] = \sum_{n=1}^{\infty} np(1-p)^{n-1}$$
$$= p \sum_{n=1}^{\infty} nq^{n-1}$$

where $q = 1 - p$. Therefore,

$$E[X] = p \sum_{n=1}^{\infty} \frac{d}{dq}(q^n)$$
$$= p\frac{d}{dq}\left(\sum_{n=1}^{\infty} q^n\right)$$
$$= p\frac{d}{dq}\left(\frac{q}{1-q}\right)$$
$$= \frac{p}{(1-q)^2}$$
$$= \frac{1}{p}$$

In other words, if independent trials, having a common probability $p$ of being successful, are performed until the first success occurs, then the expected number of required trials equals $1/p$. Thus, for instance, we would expect to roll a fair die 6 times in order to obtain the value 1. ∎

**Example 1f.** A contestant on a quiz show is presented with two questions, questions 1 and 2, which he is to attempt to answer in some order chosen by him. If he decides to try question $i$, $i = 1, 2$ first, then he will be allowed to go on to question $j$, $j \neq i$ only if his answer to $i$ is correct. If his initial answer is incorrect, he is not allowed to answer the other question. The contestant is to receive $V_i$ dollars if he answers question $i$ correctly, $i = 1, 2$. Thus, for instance, he will receive $V_1 + V_2$ dollars if both questions are correctly answered. If the probability that he knows the answer to question $i$ is $P_i$, $i = 1, 2$, which question should he attempt first so as to maximize his expected winnings? Assume that the events $E_i$, $i = 1, 2$, that he knows the answer to question $i$, are independent events.

***Solution:*** If he attempts question 1 first, then he will win

$$\begin{array}{ll} 0 & \text{with probability } 1 - P_1 \\ V_1 & \text{with probability } P_1(1 - P_2) \\ V_1 + V_2 & \text{with probability } P_1 P_2 \end{array}$$

Hence his expected winnings in this case will be

$$V_1 P_1(1 - P_2) + (V_1 + V_2)P_1 P_2$$

On the other hand, if he attempts questions 2 first, his expected winnings will be

$$V_2 P_2(1 - P_1) + (V_1 + V_2)P_1 P_2$$

Therefore, it is better to try question 1 first if

$$V_1 P_1(1 - P_2) \geq V_2 P_2(1 - P_1)$$

or, equivalently, if

$$\frac{V_1 P_1}{1 - P_1} \geq \frac{V_2 P_2}{1 - P_2}$$

Thus, for instance, if he is 60 percent certain of answering question 1, worth \$200, correctly and he is 80 percent certain of answering question 2, worth \$100, correctly, then he should attempt question 2 first because

$$400 = \frac{(100)(.8)}{(.2)} > \frac{(200)(.6)}{(.4)} = 300$$ ∎

Although we have, up to this point, only defined expectations for discrete random variables, it is also possible to define the expected value of a continuous random variable. If $X$ is a continuous random variable having a probability density function $f(x)$, then as

$$f(x)\,dx \approx P\{x \leq X \leq x + dx\} \qquad \text{for } dx \text{ small}$$

it is reasonable to define the expected value of $X$ by

$$E[X] = \int_{-\infty}^{\infty} x f(x)\,dx$$

**Example 1g.** Expectation of a Uniform Random Variable. Calculate the expected value of a random variable that is uniformly distributed over $(a, b)$.

***Solution:*** The density of a uniform random variable over $(a, b)$ is given by

$$f(x) = \begin{cases} \dfrac{1}{b - a} & a < x < b \\ 0 & \text{otherwise} \end{cases}$$

Hence

$$E[X] = \int_a^b \frac{x}{b-a}\,dx$$

$$= \left(\frac{1}{b-a}\right)\left(\frac{b^2-a^2}{2}\right)$$

$$= \frac{(b-a)(b+a)}{(b-a)2}$$

$$= \frac{b+a}{2}$$

That is, the expected value of a random variable uniformly distributed over some interval is equal to the midpoint of the interval. ∎

**Example 1h.** Expectation of an Exponential Random Variable. Calculate the expectation of an exponentially distributed random variable having parameter $\lambda$.

***Solution:*** Since the density function is given by

$$f(x) = \begin{cases} \lambda e^{-\lambda x} & x \geq 0 \\ 0 & x < 0 \end{cases}$$

we obtain

$$E[X] = \int_0^\infty x\lambda e^{-\lambda x}\,dx$$

Integrating by parts ($\lambda e^{-\lambda x}\,dx = dv$, $u = x$) yields

$$E[X] = -xe^{-\lambda x}\Big|_0^\infty + \int_0^\infty e^{-\lambda x}\,dx$$

$$= 0 - \frac{e^{-\lambda x}}{\lambda}\Big|_0^\infty$$

$$= \frac{1}{\lambda}$$ ∎

**Example 1i.** Expectation of a Normal Random Variable. Calculate $E[X]$ when $X$ is normally distributed with parameters $\mu$ and $\sigma^2$.

***Solution***

$$E[X] = \frac{1}{\sqrt{2\pi}\sigma}\int_{-\infty}^\infty xe^{-(x-\mu)^2/2\sigma^2}\,dx$$

Writing $x$ as $(x-\mu)+\mu$ yields

$$E[X] = \frac{1}{\sqrt{2\pi}\,\sigma}\int_{-\infty}^{\infty}(x-\mu)e^{-(x-\mu)^2/2\sigma^2}\,dx + \mu\frac{1}{\sqrt{2\pi}\,\sigma}\int_{-\infty}^{\infty}e^{-(x-\mu)^2/2\sigma^2}\,dx$$

Letting $y = x-\mu$ in the first integral yields

$$E[X] = \frac{1}{\sqrt{2\pi}\,\sigma}\int_{-\infty}^{\infty} ye^{-y^2/2\sigma^2}\,dy + \mu\int_{-\infty}^{\infty} f(x)\,dx$$

where $f(x)$ is the normal density. By symmetry, the first integral must be 0, and so

$$E[X] = \mu\int_{-\infty}^{\infty} f(x)\,dx = \mu \qquad \blacksquare$$

REMARK. The concept of expectation is analogous to the physical concept of the *center of gravity* of a distribution of mass. Consider a discrete random variable $X$ having probability mass function $P(x_i)$, $i \geq 1$. If we now imagine a weightless rod in which weights with mass $P(x_i)$, $i \geq 1$, are located at the points $x_i$, $i \geq 1$ (see Figure 7.1), then the point at which the rod would be in balance is known as the center of gravity. For those readers acquainted with elementary statics it is now a simple matter to show that this point is at $E[X]$.[1]

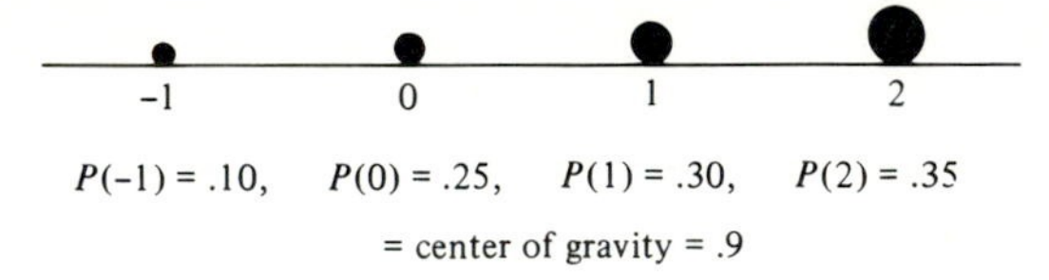


**Figure 7.1**

A TECHNICAL REMARK. We have defined the expectation of a random variable in terms of a sum in the discrete case and in terms of an integral in the continuous case. Hence the expectation is only defined when the corresponding sum or integral is defined. Now, in general integration theory, $\int_{-\infty}^{\infty} g(x)\,dx$ is defined first for nonnegative functions $g$ and then for arbitrary functions $g$ by

$$\int_{-\infty}^{\infty} g(x)\,dx = \int_{x:g(x)\geq 0} g(x)\,dx - \int_{x:g(x)<0} [-g(x)]\,dx$$

[1] To prove this, we must show that the sum of the torques tending to turn the point around $E[X]$ is equal to 0. That is, we must show that $0 = \sum_i (x_i - E[X])P(x_i)$, which is immediate.

That is, $\int_{-\infty}^{\infty} g(x)\,dx$ is defined to be the difference of $\int_{-\infty}^{\infty} g^+(x)\,dx$ and $\int_{-\infty}^{\infty} g^-(x)\,dx$, where $g^+$ and $g^-$ are nonnegative functions defined by

$$g^+(x) = \begin{cases} g(x) & \text{if } g(x) \geq 0 \\ 0 & \text{if } g(x) < 0 \end{cases} \qquad g^-(x) = \begin{cases} 0 & \text{if } g(x) \geq 0 \\ -g(x) & \text{if } g(x) < 0 \end{cases}$$

[Note that $g(x) = g^+(x) - g^-(x)$.] Hence $\int_{-\infty}^{\infty} g(x)\,dx$ is well defined as long as $\int_{-\infty}^{\infty} g^+(x)\,dx$ and $\int_{-\infty}^{\infty} g^-(x)\,dx$ are not both $+\infty$. If both integrals equal $+\infty$, we say that $\int_{-\infty}^{\infty} g(x)\,dx$ is undefined. If one of the integrals is $+\infty$ and the other one is finite, then $\int_{-\infty}^{\infty} g(x)\,dx$ equals either $+\infty$ or $-\infty$, depending on which of the integrals equals $+\infty$.

Thus, in the continuous case,

$$E[X] = \int_{x \geq 0} xf(x)\,dx - \int_{x<0} (-x)f(x)\,dx$$

and $E[X]$ is thus defined as long as both integrals are not $+\infty$. The only example of a random variable we shall come across in this text that does not have an expectation is the Cauchy random variable. In its simplest form its density is given by

$$f(x) = \frac{1}{\pi(1 + x^2)} \qquad -\infty < x < \infty$$

It can be shown that

$$\frac{1}{\pi}\int_0^{\infty} \frac{x}{1 + x^2}\,dx = \frac{1}{\pi}\int_{-\infty}^{0} \frac{(-x)}{1 + x^2}\,dx = \infty \tag{1.1}$$

and thus $E[X]$ is undefined for a Cauchy random variable.

Similarly, in the discrete case

$$E[X] = \sum_{x \geq 0} xp(x) - \sum_{x<0} (-x)p(x)$$

and $E[X]$ is defined as long as both sums are not $+\infty$.

## 2 Expectation of a Function of a Random Variable

Suppose that we are given a random variable $X$ along with its probability distribution, and that we desire to compute not the expected value of $X$ but the expected value of some function of $X$, say, $g(X)$. (For instance, we may desire $E[X^2]$ or $E[e^X]$.) How do we go about doing this? One way is as follows: Since $g(X)$ is itself a random variable, it must have a probability distribution, which should be computable from a knowledge of the distribution of $X$. Once we have obtained the distribution of $g(X)$, we can then compute $E[g(X)]$ by the definition of the expectation.

**Example 2a.** Let $X$ denote the number of heads obtained when two independent flips of a fair coin are made. Compute $E[X^2]$.

***Solution:*** The probability mass function of $X$ is given by

$$P\{X = 0\} = \tfrac{1}{4} \qquad P\{X = 1\} = \tfrac{1}{2} \qquad P\{X = 2\} = \tfrac{1}{4}$$

Hence, if we let $Y = X^2$, it follows that the probability mass function of $Y$ is given by

$$\begin{aligned} P\{Y = 0\} &= P\{X = 0\} = \tfrac{1}{4} \\ P\{Y = 1\} &= P\{X = 1\} = \tfrac{1}{2} \\ P\{Y = 4\} &= P\{X = 2\} = \tfrac{1}{4} \end{aligned}$$

Hence

$$E[X^2] = E[Y] = 0(\tfrac{1}{4}) + 1(\tfrac{1}{2}) + 4(\tfrac{1}{4}) = \tfrac{3}{2}$$

The reader should note that

$$\tfrac{3}{2} = E(X^2) \neq (E[X])^2 = 1$$

∎

**Example 2b.** If $X$ is uniformly distributed over $(0, 1)$, compute $E[e^X]$.

***Solution:*** Letting $Y = e^X$, we compute the distribution function of $Y$ as follows: For $1 \le a \le e$,

$$\begin{aligned} F_Y(a) &= P\{Y \le a\} \\ &= P\{e^X \le a\} \\ &= P\{X \le \log a\} \\ &= \log a \end{aligned}$$

where the last equality follows because $X$ is uniformly distributed over $(0, 1)$. By differentiating $F_Y(a)$, we obtain the density of $Y$,

$$f_Y(a) = \frac{1}{a} \qquad 1 \le a \le e$$

Hence

$$\begin{aligned} E[e^X] = E[Y] &= \int_{-\infty}^{\infty} a f_Y(a)\, da \\ &= \int_1^e da \\ &= e - 1 \end{aligned}$$

∎

Although the above procedure will, in theory, always enable us to compute the expectation of any function of $X$ from a knowledge of the distribution of $X$, there is, fortunately, an easier way of doing this. The following proposition shows how we can calculate the expectation of $g(X)$ without first determining its distribution.

---

*Proposition 2.1*

(a) *If $X$ is a discrete random variable with probability mass function $p(x)$, then for any real-valued function $g$,*

$$E[g(X)] = \sum_{x:p(x)>0} g(x)p(x)$$

(b) *If $X$ is a continuous random variable with probability density function $f(x)$, then for any real-valued function $g$*

$$E[g(X)] = \int_{-\infty}^{\infty} g(x)f(x)\,dx$$

---

Before attempting to prove this proposition, let us first check that it is in accord with the results of Examples 2a and 2b. Applying the proposition to Example 2a yields

$$E[X^2] = 0^2(\tfrac{1}{4}) + 1^2(\tfrac{1}{2}) + 2^2(\tfrac{1}{4}) = \tfrac{3}{2}$$

whereas an application to Example 2b yields

$$\begin{aligned} E[e^X] &= \int_0^1 e^x\,dx \qquad \text{since } f(x) = 1, 0 < x < 1 \\ &= e - 1 \end{aligned}$$

thus verifying the results previously obtained.

The proof of Proposition 2.1 is rather straightforward in the discrete case and is left as an exercise. To prove it in the continuous case, we shall need the following lemma, which is of independent interest.

---

*Lemma 2.1*

*For any random variable $Y$,*

$$E[Y] = \int_0^\infty P\{Y > y\}\,dy - \int_0^\infty P\{Y < -y\}\,dy$$

---

**Proof:** We present a proof when $Y$ is a continuous random variable with probability density function $f_Y$. We have

$$\int_0^\infty P\{Y > y\}\,dy = \int_0^\infty \int_y^\infty f_Y(x)\,dx\,dy \tag{2.1}$$

where we have used the fact that $P\{Y > y\} = \int_y^\infty f_Y(x)\,dx$. Interchanging the order of integration in Equation (2.1) yields

$$\begin{aligned} \int_0^\infty P\{Y > y\}\,dy &= \int_0^\infty \left(\int_0^x dy\right) f_Y(x)\,dx \\ &= \int_0^\infty x f_Y(x)\,dx \qquad (2.2) \end{aligned}$$

Similarly,

$$\begin{aligned} \int_0^\infty P\{Y < -y\}\,dy &= \int_0^\infty \int_{-\infty}^{-y} f_Y(x)\,dx\,dy \\ &= \int_{-\infty}^0 \left(\int_0^{-x} dy\right) f_Y(x)\,dx \\ &= -\int_{-\infty}^0 x f_Y(x)\,dx \qquad (2.3) \end{aligned}$$

Hence, from Equations (2.2) and (2.3), we see that

$$\begin{aligned} \int_0^\infty P\{Y > y\}\,dy - \int_0^\infty P\{Y < -y\}\,dy &= \int_0^\infty x f_Y(x)\,dx + \int_{-\infty}^0 x f_Y(x)\,dx \\ &= \int_{-\infty}^\infty x f_Y(x)\,dx \\ &= E[Y] \end{aligned}$$ ∎

Lemma 2.1, which is important in its own right, is most often seen in the literature as stating that for any nonnegative random variable $Y$,

$$E[Y] = \int_0^\infty P\{Y > y\}\,dy \qquad \text{when } P\{Y \geq 0\} = 1$$

We are now in a position to prove Proposition 2.1(b).

**Proof of Proposition 2.1(b):** For any function $g$ we have from Lemma 2.1 that

$$\begin{aligned} E[g(X)] &= \int_0^\infty P\{g(X) > y\}\, dy - \int_0^\infty P\{g(X) < -y\}\, dy \\ &= \int_0^\infty \int_{x:g(x)>y} f(x)\, dx\, dy - \int_0^\infty \int_{x:g(x)<-y} f(x)\, dx\, dy \\ &= \int_{x:g(x)>0} \int_0^{g(x)} dy\, f(x)\, dx - \int_{x:g(x)<0} \int_0^{-g(x)} dy\, f(x)\, dx \\ &= \int_{x:g(x)>0} g(x)f(x)\, dx + \int_{x:g(x)<0} g(x)f(x)\, dx \\ &= \int_{-\infty}^{\infty} g(x)f(x)\, dx \end{aligned}$$

which completes the proof. ∎

**Example 2c.** A product, sold seasonally, yields a net profit of $b$ dollars for each unit sold and a net loss of $\ell$ dollars for each unit left unsold when the season ends. The number of units of the product that are ordered at a specific department store during any season is a random variable having probability mass function $p(i)$, $i \geq 0$. If the store must stock this product in advance, determine the number of units the store should stock so as to maximize its expected profit.

***Solution:*** Let $X$ denote the number of units ordered. If $s$ units are stocked, then the profit, call it $P(s)$, can be expressed as

$$\begin{aligned} P(s) &= bX - (s - X)\ell \qquad && \text{if } X \leq s \\ &= sb && \text{if } X > s \end{aligned}$$

Hence the expected profit equals

$$\begin{aligned} E[P(s)] &= \sum_{i=0}^{s} [bi - (s - i)\ell]p(i) + \sum_{i=s+1}^{\infty} sbp(i) \\ &= (b + \ell) \sum_{i=0}^{s} ip(i) - s\ell \sum_{i=0}^{s} p(i) + sb\left[1 - \sum_{i=0}^{s} p(i)\right] \\ &= (b + \ell) \sum_{i=0}^{s} ip(i) - (b + \ell)s \sum_{i=0}^{s} p(i) + sb \\ &= sb + (b + \ell) \sum_{i=0}^{s} (i - s)p(i) \end{aligned}$$

To determine the optimum value of $s$, let us investigate what happens to our profit when we increase $s$ by 1 unit. By substitution we see that the expected profit in this case is given by

$$\begin{aligned} E[P(s+1)] &= b(s+1) + (b+\ell)\sum_{i=0}^{s+1}(i-s-1)p(i) \\ &= b(s+1) + (b+\ell)\sum_{i=0}^{s}(i-s-1)p(i) \end{aligned}$$

Therefore,

$$E[P(s+1)] - E[P(s)] = b - (b+\ell)\sum_{i=0}^{s} p(i)$$

Hence stocking $s+1$ units will be better than stocking $s$ units whenever

$$\sum_{i=0}^{s} p(i) < \frac{b}{b+\ell} \tag{2.4}$$

As the left-hand side of Equation (2.4) is increasing in $s$ while the right-hand side is constant, it follows that the inequality will be satisfied for all values of $s \le s^*$ where $s^*$ is the largest value of $s$ satisfying Equation (2.4). Since

$$E[P(0)] < \cdots < E[P(s^*)] < E[P(s^*+1)] > E[P(s^*+2)] > \cdots$$

it follows that stocking $s^*+1$ items will lead to a maximum expected profit. ∎

In the continuous version of Example 2c the solution is similar.

**Example 2d.** In Example 2c, suppose that the volume of demand is a continuous random variable having density $f$. Compute the optimal number of units to be stocked so as to maximize expected profit.

***Solution:*** As before, if $s$ units are stocked and the demand is $X$, then the profit, $P(s)$, is given by

$$\begin{aligned} P(s) &= bX - (s-X)\ell \qquad && \text{if } X \le s \\ &= sb && \text{if } X > s \end{aligned}$$

Hence

$$\begin{aligned} E[P(s)] &= \int_0^s (bx - (s-x)\ell)f(x)\,dx + \int_s^\infty sbf(x)\,dx \\ &= (b+\ell)\int_0^s xf(x)\,dx - s\ell\int_0^s f(x)\,dx + sb\left[1 - \int_0^s f(x)\,dx\right] \\ &= sb + (b+\ell)\int_0^s (x-s)f(x)\,dx \end{aligned} \tag{2.5}$$

That value of $s$ maximizing Equation (2.5) can now be obtained by calculus. Differentiation yields

$$\begin{aligned}\frac{d}{ds}E[P(s)] &= b + (b + \ell)\frac{d}{ds}\left[\int_0^s xf(x)\,dx - s\int_0^s f(x)\,dx\right]\\ &= b + (b + \ell)\left[sf(s) - sf(s) - \int_0^s f(x)\,dx\right]\\ &= b - (b + \ell)\int_0^s f(x)\,dx\end{aligned}$$

Equating to zero shows that the maximal expected profit is obtained when $s$ is chosen so that

$$F(s) = \frac{b}{b + \ell}$$

where $F(s) = \int_0^s f(x)\,dx$ is the cumulative distribution of demand. ∎

A simple corollary of Proposition 2.1 is Corollary 2.1.

---

*Corollary 2.1*

*If a and b are constants, then*

$$E[aX + b] = aE[X] + b$$

---

**Proof:** In the discrete case,

$$\begin{aligned}E[aX + b] &= \sum_{x:p(x)>0} (ax + b)p(x)\\ &= a\sum_{x:p(x)>0} xp(x) + b\sum_{x:p(x)>0} p(x)\\ &= aE[X] + b\end{aligned}$$

In the continuous case,

$$\begin{aligned}E[aX + b] &= \int_{-\infty}^{\infty} (ax + b)f(x)\,dx\\ &= a\int_{-\infty}^{\infty} xf(x)\,dx + b\int_{-\infty}^{\infty} f(x)\,dx\\ &= aE[X] + b\end{aligned}$$

∎

The expected value of a random variable $X$, $E[X]$ is also referred to as the mean or the first moment of $X$. The quantity $E[X^n]$, $n \geq 1$, is called the $n$th moment of $X$. By Proposition 2.1 we note that

$$E[X^n] = \begin{cases} \displaystyle\sum_{x:p(x)>0} x^n p(x) & \text{if } X \text{ is discrete} \\ \displaystyle\int_{-\infty}^{\infty} x^n f(x)\,dx & \text{if } X \text{ is continuous} \end{cases}$$

## 3 Expectation of Sums of Random Variables

A two-dimensional analog of the law of the unconscious statistician (Proposition 2.1) states that if $X$ and $Y$ are random variables and $g$ is a function of two variables, then

$$E[g(X, Y)] = \sum_y \sum_x g(x, y) p(x, y)$$

if $X$ and $Y$ have a joint probability mass function $p(x, y)$

$$= \int_{-\infty}^{\infty} \int_{-\infty}^{\infty} g(x, y) f(x, y)\,dx\,dy$$

if $X$ and $Y$ have a joint probability density function $f(x, y)$

As an application of the preceding equations, suppose that $E[X]$ and $E[Y]$ are both finite and let $g(X, Y) = X + Y$. Then, in the continuous case,

$$\begin{aligned} E[X+Y] &= \int_{-\infty}^{\infty} \int_{-\infty}^{\infty} (x+y) f(x, y)\,dx\,dy \\ &= \int_{-\infty}^{\infty} \int_{-\infty}^{\infty} x f(x, y)\,dy\,dx + \int_{-\infty}^{\infty} \int_{-\infty}^{\infty} y f(x, y)\,dx\,dy \\ &= \int_{-\infty}^{\infty} x f_X(x)\,dx + \int_{-\infty}^{\infty} y f_Y(y)\,dy \\ &= E[X] + E[Y] \end{aligned}$$

The same result holds in general; thus, whenever $E[X]$ and $E[Y]$ are finite,

$$E[X+Y] = E[X] + E[Y]$$

A simple induction proof now yields that if $E[X_i]$ is finite for all $i = 1, 2, \ldots, n$, then

$$E[X_1 + X_2 + \cdots + X_n] = E[X_1] + E[X_2] + \cdots + E[X_n] \qquad (3.1)$$

Equation (3.1) is an extremely useful formula whose utility will now be illustrated by a series of examples.

**Example 3a.** Calculate the expected sum obtained when 10 independent rolls of a fair die are made.

***Solution:*** Let $X$ denote the sum obtained. If we attempt to compute $E[X]$ by first determining the distribution of $X$, then it would take quite some time to solve this problem. However, by noting that

$$X = X_1 + X_2 + \cdots + X_{10}$$

where $X_i$ equals the value of the $i$th roll, we immediately see that

$$E[X] = E[X_1] + \cdots + E[X_{10}] = 10(\tfrac{7}{2}) = 35$$ ∎

**Example 3b.** Expectation of a Binomial Random Variable. As another example of the usefulness of Equation (3.1), let us use it to obtain the expectation of a binomial random variable having parameters $n$ and $p$. Recalling that such a random variable $X$ represents the number of successes in $n$ independent trials when each trial has probability $p$ of being a success, we have that

$$X = X_1 + X_2 + \cdots + X_n$$

where

$$X_i = \begin{cases} 1 & \text{if the } i\text{th trial is a success} \\ 0 & \text{if the } i\text{th trial is a failure} \end{cases}$$

Hence $X_i$ is a Bernoulli random variable having expectation $E[X_i] = 1(p) + 0(1 - p)$. Thus

$$E[X] = E[X_1] + E[X_2] + \cdots + E[X_n] = np$$

This derivation should be compared with the one in Example 1c. ∎

**Example 3c.** Expected Number of Matches. A group of $N$ men throw their hats into the center of a room. The hats are mixed up, and each man randomly selects one. Find the expected number of men that select their own hats.

***Solution:*** Letting $X$ denote the number of matches, we can most easily compute $E[X]$ by writing

$$X = X_1 + X_2 + \cdots + X_N$$

where

$$X_i = \begin{cases} 1 & \text{if the } i\text{th man selects his own hat} \\ 0 & \text{otherwise} \end{cases}$$

Since, for each $i$, the $i$th man is equally likely to select any of the $N$ hats,

$$E[X_i] = P\{X_i = 1\} = \frac{1}{N}$$

we see that

$$E[X] = E[X_1] + \cdots + E[X_N] = \left(\frac{1}{N}\right)N = 1$$

Hence, on the average, exactly one of the men selects his own hat. ∎

**Example 3d.** Mean of a Negative Binomial Random Variable. If independent trials, having a constant probability $p$ of being successes, are performed, determine the expected number of trials required to amass a total of $r$ successes.

***Solution:*** If $X$ denotes the number of trials needed to amass a total of $r$ successes, then $X$ is a negative binomial random variable whose mass function is given by

$$P\{X = n\} = \binom{n-1}{r-1} p^r (1-p)^{n-r}, \qquad n = r, r+1, \ldots$$

Hence it follows that

$$E[X] = \sum_{n=r}^{\infty} n \binom{n-1}{r-1} p^r (1-p)^{n-r} \tag{3.2}$$

However, a simpler expression can be obtained for $E[X]$ by noting that

$$X = X_1 + X_2 + \cdots + X_r$$

where $X_1$ is the number of trials required to obtain the first success, $X_2$ the number of additional trials until the second success is obtained, $X_3$ the number of additional trials until the third success is obtained, and so on. That is, $X_i$ represents the number of additional trials required, after the $(i - 1)$st success, until a total of $i$ successes are amassed. A little thought reveals that each of the random variables $X_i$ is a geometric random variable with parameter $p$. Hence, from the results of Example 1e, $E[X_i] = 1/p$, $i = 1, 2, \ldots, r$; and thus

$$E[X] = E[X_1] + \cdots + E[X_r] = \frac{r}{p} \tag{3.3}$$ ∎

**Example 3e.** Mean of a Hypergeometric Random Variable. If $n$ balls are randomly selected from an urn containing $N$ white and $M$ black balls, find the expected number of white balls selected.

***Solution:*** Let $X$ denote the number of white balls selected. It follows that

$$P\{X = k\} = \frac{\binom{N}{k}\binom{M}{n-k}}{\binom{N+M}{n}}$$

Hence

$$E[X] = \frac{\sum_{k=0}^{n} k\binom{N}{k}\binom{M}{n-k}}{\binom{N+M}{n}}$$

However, we can obtain a simpler expression for $E[X]$ by writing

$$X = X_1 + \cdots + X_N$$

where

$$X_i = \begin{cases} 1 & \text{if the } i\text{th white ball is selected} \\ 0 & \text{otherwise} \end{cases}$$

Now,

$$\begin{aligned} E[X_i] &= P\{X_i = 1\} \\ &= P\{i\text{th white ball is selected}\} \\ &= \frac{\binom{1}{1}\binom{M+N-1}{n-1}}{\binom{M+N}{n}} \\ &= \frac{n}{M+N} \end{aligned}$$

Hence

$$E[X] = E[X_1] + \cdots + E[X_N] = \frac{Nn}{M+N}$$

Of course, we could also have obtained this result by using the representation

$$X = Y_1 + \cdots + Y_n$$

where

$$Y_i = \begin{cases} 1 & \text{if the } i\text{th ball selected is white} \\ 0 & \text{otherwise} \end{cases}$$

Since the $i$th ball selected is equally likely to be any of the $N + M$, we have

$$E[Y_i] = \frac{N}{M + N}$$

and thus

$$E[X] = E[Y_1] + \cdots + E[Y_n] = \frac{nN}{M + N}$$ ∎

**Example 3f.** The following problem was posed and solved in the eighteenth century by Daniel Bernoulli. Suppose that a jar contains $2N$ cards, two of them marked 1, two marked 2, two marked 3, and so on. Draw out $m$ cards at random. What is the expected number of pairs that still remain in the jar? (Interestingly enough, Bernoulli proposed the above as a possible probabilistic model for determining the number of marriages that remain intact when there are a total of $m$ deaths among the $N$ married couples.)

***Solution:*** Define, for $i = 1, 2, \ldots, N$

$$X_i = \begin{cases} 1 & \text{if the } i\text{th pair remains in the jar} \\ 0 & \text{otherwise} \end{cases}$$

Now,

$$\begin{aligned} E[X_i] &= P\{X_i = 1\} \\ &= \frac{\binom{2N-2}{m}}{\binom{2N}{m}} \\ &= \frac{\dfrac{(2N-2)!}{m!(2N-2-m)!}}{\dfrac{(2N)!}{m!(2N-m)!}} \\ &= \frac{(2N-m)(2N-m-1)}{(2N)(2N-1)} \end{aligned}$$

Hence the desired result is

$$\begin{aligned} E[X_1 + X_2 + \cdots + X_N] &= E[X_1] + \cdots + E[X_N] \\ &= \frac{(2N-m)(2N-m-1)}{2(2N-1)} \end{aligned}$$ ∎

**Example 3g.** Coupon-Collecting Problems. Suppose that there are $N$ different types of coupons and each time one obtains a coupon it is equally likely to be any one of the $N$ types.

1. Find the expected number of different types of coupons that are contained in a set of $n$ coupons.
2. Find the expected number of coupons one need amass before obtaining a complete set of at least one of each type.

***Solution***

1. Let $X$ denote the number of different types of coupons in the set of $n$ coupons. We compute $E[X]$ by using the representation

$$X = X_1 + \cdots + X_N$$

where

$$X_i = \begin{cases} 1 & \text{if at least one type } i \text{ coupon is contained in the set of } n \\ 0 & \text{otherwise} \end{cases}$$

Now,

$$\begin{aligned} E[X_i] &= P\{X_i = 1\} \\ &= 1 - P\{\text{no type } i \text{ coupons are contained in the set of } n\} \\ &= 1 - \left(\frac{N-1}{N}\right)^n \end{aligned}$$

Hence

$$E[X] = E[X_1] + \cdots + E[X_n] = N\left[1 - \left(\frac{N-1}{N}\right)^n\right]$$

2. Let $Y$ denote the number of coupons collected before a complete set is attained. We compute $E[Y]$ by using the same technique as we used in computing the mean of a negative binomial random variable (Example 3d). That is, define $Y_i$, $i = 0, 1, \ldots, N-1$ to be the number of additional coupons that need be obtained after $i$ distinct types have been collected in order to obtain another distinct type, and note that

$$Y = Y_0 + Y_1 + \cdots + Y_{N-1}$$

When $i$ distinct types of coupons have already been collected, it follows that a new coupon obtained will be of a distinct type with probability $(N - i)/N$. Therefore,

$$P\{Y_i = k\} = \frac{N-i}{N}\left(\frac{i}{N}\right)^{k-1} \qquad k \geq 1$$

or in other words, $Y_i$ is a geometric random variable with parameter $(N - i)/N$. Hence, from Example 1e,

$$E[Y_i] = \frac{N}{N - i}$$

implying that

$$E[Y] = 1 + \frac{N}{N-1} + \frac{N}{N-2} + \cdots + \frac{N}{1} = N\left[1 + \cdots + \frac{1}{N-1} + \frac{1}{N}\right] \quad \blacksquare$$

**Example 3h.** Ten hunters are waiting for ducks to fly by. When a flock of ducks flies overhead, the hunters fire at the same time, but each chooses his target at random, independently of the others. If each hunter independently hits his target with probability $p$, compute the expected number of ducks that escape unhurt when a flock of size 10 flies overhead.

***Solution:*** Let $X_i$ equal 1 if the $i$th duck escapes unhurt and 0 otherwise, $i = 1, 2, \ldots, 10$. The expected number of ducks to escape can be expressed as

$$E[X_1 + \cdots + X_{10}] = E[X_1] + \cdots + E[X_{10}]$$

To compute $E[X_i] = P\{X_i = 1\}$, we note that each of the hunters will, independently, hit the $i$th duck with probability $p/10$, and so

$$P\{X_i = 1\} = \left(1 - \frac{p}{10}\right)^{10}$$

Hence

$$E[X] = 10\left(1 - \frac{p}{10}\right)^{10} \quad \blacksquare$$

**Example 3i.** Expected Number of Runs. Suppose that a sequence of $n$ 1's and $m$ 0's are randomly permuted so that each of the $(n + m)!/(n!\,m!)$ possible arrangements is equally likely. Any consecutive string of 1's is said to constitute a run of 1's—for instance, if $n = 6$, $m = 4$, and the ordering is 1, 1, 1, 0, 1, 1, 0, 0, 1, 0, then there are 3 runs of 1's—and we are interested in computing the mean number of such runs. To compute this quantity let

$$I_i = \begin{cases} 1 & \text{if a run of 1's starts at the } i\text{th position} \\ 0 & \text{otherwise} \end{cases}$$

Therefore, $R(1)$, the number of runs of 1, can be expressed as

$$R(1) = \sum_{i=1}^{n+m} I_i$$

and thus

$$E[R(1)] = \sum_{i=1}^{n+m} E[I_i]$$

Now,

$$E[I_1] = P\{\text{“1” in position } 1\}$$

$$= \frac{n}{n+m}$$

and for $1 < i \leq n + m$,

$$E[I_i] = P\{\text{“0” in position } i-1, \text{“1” in position } i\}$$

$$= \frac{m}{n+m} \frac{n}{n+m-1}$$

Hence

$$E[R(1)] = \frac{n}{n+m} + (n+m-1)\frac{nm}{(n+m)(n+m-1)}$$

Similarly, $E[R(0)]$, the expected number of runs of 0's, is

$$E[R(0)] = \frac{m}{n+m} + \frac{nm}{n+m}$$

and the expected number of runs of either type is

$$E[R(1) + R(0)] = 1 + \frac{2nm}{n+m}$$ ∎

**Example 3j.** Consider an ordinary deck of cards that is turned face up one card at a time. How many cards would one expect to turn face up in order to obtain (1) an ace and (2) a spade?

***Solution:*** Both parts (1) and (2) are special cases of the following problem: Suppose that balls are taken one by one out of an urn containing $n$ white and $m$ black balls until the first white ball is drawn. If $X$ denotes the number of balls withdrawn, compute $E[X]$.

To solve the above, imagine that the black balls in the urn have names—say, $b_1, b_2, \ldots, b_m$. If we let, for $i = 1, 2, \ldots, m$,

$$X_i = \begin{cases} 1 & \text{if } b_i \text{ is withdrawn before any of the white balls} \\ 0 & \text{otherwise} \end{cases}$$

then it is easy to see that

$$X = 1 + \sum_{i=1}^{m} X_i$$

Hence

$$E[X] = 1 + \sum_{i=1}^{m} P\{X_i = 1\}$$

However, $X_i$ will equal 1 if ball $b_i$ is withdrawn before any of the $n$ white balls. But as each of these $n + 1$ balls (the $n$ white plus ball $b_i$) has an equal probability of being the first one of this set to be withdrawn, we see that

$$E[X_i] = P\{X_i = 1\} = \frac{1}{n+1}$$

and so

$$E[X] = 1 + \frac{m}{n+1}$$ ∎

**Example 3k.** A Random Walk in the Plane. Consider a particle initially located at a given point in the plane and suppose that it undergoes a sequence of steps of fixed length but in a completely random direction. Specifically, suppose that the new position after each step is one unit of distance from the previous position and at an angle of orientation from the previous position that is uniformly distributed over $(0, 2\pi)$. (See Figure 7.2.) Compute the square of the distance from the origin after $n$ steps.

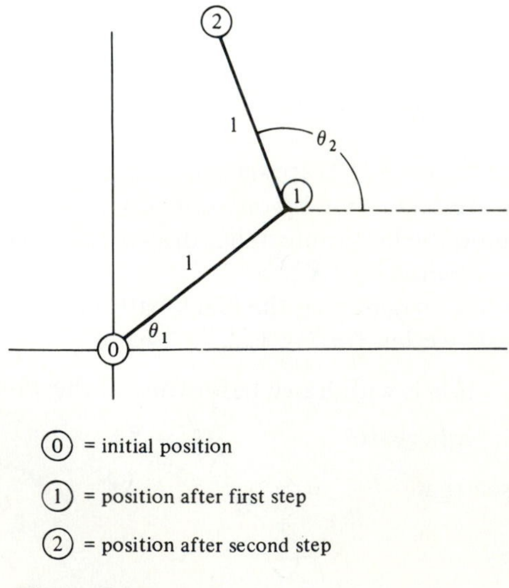


**Figure 7.2**

***Solution:*** Letting $(X_i, Y_i)$ denote the change in position at the $i$th step, $i = 1, \ldots, n$, in rectangular coordinates, we have that

$$X_i = \cos \theta_i$$
$$Y_i = \sin \theta_i$$

where $\theta_i, i = 1, \ldots, n$ are by assumption, independent, uniform $(0, 2\pi)$ random variables. As the position after $n$ steps has rectangular coordinates $\left(\sum_{i=1}^{n} X_i, \sum_{i=1}^{n} Y_i\right)$, we see that $D^2$, the square of the distance from the origin, is given by

$$\begin{aligned} D^2 &= \left(\sum_{i=1}^{n} X_i\right)^2 + \left(\sum_{i=1}^{n} Y_i\right)^2 \\ &= \sum_{i=1}^{n} (X_i^2 + Y_i^2) + \sum_{i \neq j}\sum (X_i X_j + Y_i Y_j) \\ &= n + \sum_{i \neq j}\sum (\cos \theta_i \cos \theta_j + \sin \theta_i \sin \theta_j) \end{aligned}$$

where the above uses that $\cos^2 \theta_i + \sin^2 \theta_i = 1$. Taking expectations and using the independence of $\theta_i$ and $\theta_j$ when $i \neq j$ and the fact that

$$E[\cos \theta_i] = \int_0^{2\pi} \cos u \, du = \sin 2\pi - \sin 0 = 0$$

$$E[\sin \theta_i] = \int_0^{2\pi} \sin u \, du = \cos 0 - \cos 2\pi = 0$$

we arrive at

$$E[D^2] = n$$ ∎

When one is dealing with an infinite collection of random variables $X_i, i \geq 1$, each having a finite expectation, it is not necessarily true that

$$E\left[\sum_{i=1}^{\infty} X_i\right] = \sum_{i=1}^{\infty} E[X_i] \tag{3.4}$$

To determine when (3.4) is valid, we note that $\sum_{i=1}^{\infty} X_i = \lim_{n\to\infty} \sum_{i=1}^{n} X_i$ and thus

$$\begin{aligned} E\left[\sum_{i=1}^{\infty} X_i\right] &= E\left[\lim_{n\to\infty} \sum_{i=1}^{n} X_i\right] \\ &\stackrel{?}{=} \lim_{n\to\infty} E\left[\sum_{i=1}^{n} X_i\right] \\ &= \lim_{n\to\infty} \sum_{i=1}^{n} E[X_i] \\ &= \sum_{i=1}^{\infty} E[X_i] \end{aligned} \tag{3.5}$$

Hence Equation (3.4) is valid whenever we are justified in interchanging the expectation and limit operations in Equation (3.5). Although, in general, this interchange is not justified, it can be shown to be valid in two important special cases:

1. The $X_i$ are all nonnegative random variables (that is, $P\{X_i \geq 0\} = 1$ for all $i$).
2. $\sum_{i=1}^{\infty} E[|X_i|] < \infty$.

**Example 3l.** Consider any nonnegative, integer-valued random variable $X$. If we define, for each $i \geq 1$,

$$X_i = \begin{cases} 1 & \text{if } X \geq i \\ 0 & \text{if } X < i \end{cases}$$

Then

$$\begin{aligned} \sum_{i=1}^{\infty} X_i &= \sum_{i=1}^{X} X_i + \sum_{i=X+1}^{\infty} X_i \\ &= \sum_{i=1}^{X} 1 + \sum_{i=X+1}^{\infty} 0 \\ &= X \end{aligned}$$

Hence, since the $X_i$ are all nonnegative,

$$\begin{aligned} E[X] &= \sum_{i=1}^{\infty} E(X_i) \\ &= \sum_{i=1}^{\infty} P\{X \geq i\} \end{aligned} \tag{3.6}$$

a useful identity. ∎

**Example 3m.** Suppose that $n$ elements—call them $1, 2, \ldots, n$—must be stored in a computer in the form of an ordered list. Each unit of time a request will be made for one of these elements—$i$ being requested, independently of the past, with probability $P(i), i \geq 1$, $\sum_i P(i) = 1$. Assuming these probabilities are known, what ordering minimizes the average position on the line of the element requested?

***Solution:*** Suppose that the elements are numbered so that $P(1) \geq P(2) \geq \cdots \geq P(n)$. To show that $1, 2, \ldots, n$ is the optimal ordering, let $X$ denote the position of the requested element. Now under any ordering—say,

$O = i_1, i_2, \ldots, i_n,$

$$P_O\{X \geq k\} = \sum_{j=k}^{n} P(i_j)$$

$$\geq \sum_{j=k}^{n} P(j)$$

$$= P_{1,2,\ldots,n}\{X \geq k\}$$

Summing over $k$, and using Equation (3.6), yields

$$E_O[X] \geq E_{1,2,\ldots,n}[X]$$

thus showing that ordering the elements in decreasing order of their request probabilities minimizes the expected position of the element requested. ∎

**Example 3n.** The Probability of a Union of Events. Let $A_1, \ldots A_n$ denote events and define the indicator variables $X_i$, $i = 1, \ldots, n$, by

$$X_i = \begin{matrix} 1 & \text{if } A_i \text{ occurs} \\ 0 & \text{otherwise} \end{matrix}$$

Now, note that

$$1 - \prod_{i=1}^{n} (1 - X_i) = \begin{matrix} 1 & \text{if } \cup A_i \text{ occurs} \\ 0 & \text{otherwise} \end{matrix}$$

Hence,

$$E\left[1 - \prod_{i=1}^{n} (1 - X_i)\right] = P\left(\bigcup_{i=1}^{n} A_i\right)$$

Expanding the left side of the above yields that

$$P\left(\bigcup_{i=1}^{n} A_i\right) = E\left[\sum_{i=1}^{n} X_i - \sum_{i<j}\sum X_i X_j + \sum_{i<j<k}\sum\sum X_i X_j X_k - \cdots + (-1)^{n+1} X_1 \cdots X_n\right] \tag{3.7}$$

However, as

$$X_{i_1} X_{i_2} \cdots X_{i_k} = \begin{matrix} 1 & \text{if } A_{i_1} A_{i_2} \cdots A_{i_k} \text{ occurs} \\ 0 & \text{otherwise} \end{matrix}$$

we see that

$$E[X_{i_1} \cdots X_{i_k}] = P(A_{i_1} \cdots A_{i_k})$$

and thus (3.7) is just a statement of the well-known formula for the union of events

$$P(\cup A_i) = \sum P(A_i) - \sum_{i<j}\sum P(A_iA_j) + \sum_{i<j<k}\sum\sum P(A_iA_jA_k) - \cdots + (-1)^{n+1}P(A_1 \cdots A_n)$$ ∎

## 4 Variance

Given a random variable $X$ along with its distribution function $F$, it would be extremely useful if we were able to summarize the essential properties of $F$ by certain suitably defined measures. One such measure would be $E[X]$, the expected value of $X$. However, although $E[X]$ yields the weighted average of the possible values of $X$, it does not tell us anything about the variation, or spread, of these values. For instance, although random variables $W$, $Y$, and $Z$, having probability mass functions determined by

$$W = 0 \quad \text{with probability } 1$$

$$Y = \begin{cases} -1 & \text{with probability } \frac{1}{2} \\ +1 & \text{with probability } \frac{1}{2} \end{cases}$$

$$Z = \begin{cases} -100 & \text{with probability } \frac{1}{2} \\ +100 & \text{with probability } \frac{1}{2} \end{cases}$$

all have the same expectation—namely, 0—there is much greater spread in the possible values of $Y$ than in those of $W$ (which is a constant) and in the possible values of $Z$ than in those of $Y$.

As we expect $X$ to take on values around its mean $E[X]$, it would appear that a reasonable way of measuring the possible variation of $X$ would be to look at how far apart $X$ would be from its mean on the average. One possible way to measure this would be to consider the quantity $E[|X - \mu|]$, where $\mu = E[X]$. However, it turns out to be mathematically inconvenient to deal with this quantity, and so a more tractable quantity is usually considered—namely, the expectation of the square of the difference between $X$ and its mean. We thus have the following definition.

---

*Definition*

*If $X$ is a random variable with mean $\mu$, then the* variance *of $X$, denoted by* Var $(X)$, *is defined by*

$$\text{Var}(X) = E[(X - \mu)^2]$$

---

**Example 4a.** Variance of a Normal Random Variable. Compute Var $(X)$ when $X$ is a normal random variable with parameters $\mu$ and $\sigma^2$.

***Solution:*** By recalling (see Example 1i) that $E[X] = \mu$, we obtain

$$\begin{aligned} \text{Var}(X) &= E[(X-\mu)^2] \\ &= \frac{1}{\sqrt{2\pi}\,\sigma}\int_{-\infty}^{\infty}(x-\mu)^2 e^{-(x-\mu)^2/2\sigma^2}\,dx \end{aligned} \tag{4.1}$$

Substituting $y = (x-\mu)/\sigma$ in Equation (4.1) yields

$$\begin{aligned} \text{Var}(X) &= \frac{\sigma^2}{\sqrt{2\pi}}\int_{-\infty}^{\infty} y^2 e^{-y^2/2}\,dy \\ &= \frac{\sigma^2}{\sqrt{2\pi}}\left[-ye^{-y^2/2}\Big|_{-\infty}^{\infty} + \int_{-\infty}^{\infty} e^{-y^2/2}\,dy\right] \quad \text{by integration by parts} \\ &= \sigma^2 \frac{1}{\sqrt{2\pi}}\int_{-\infty}^{\infty} e^{-y^2/2}\,dy \\ &= \sigma^2 \end{aligned}$$

An alternative formula for Var $(X)$ can be derived as follows:

$$\begin{aligned} \text{Var}(X) &= E[(X-\mu)^2] \\ &= E[X^2 - 2\mu X + \mu^2] \\ &= E[X^2] - E[2\mu X] + E[\mu^2] \\ &= E[X^2] - 2\mu E[X] + \mu^2 \\ &= E[X^2] - \mu^2 \end{aligned}$$

That is,

$$\text{Var}(X) = E[X^2] - (E[X])^2 \tag{4.2}$$

or, in words, the variance of $X$ is equal to the expected value of $X^2$ minus the square of the expected value of $X$. This is, in practice, usually the easiest way to compute Var $(X)$.

**Example 4b.** Calculate Var $(X)$ if $X$ represents the outcome when a fair die is rolled.

***Solution:*** It was shown in Example 1a that $E[X] = \frac{7}{2}$. Also

$$\begin{aligned} E[X^2] &= 1^2(\tfrac{1}{6}) + 2^2(\tfrac{1}{6}) + 3^2(\tfrac{1}{6}) + 4^2(\tfrac{1}{6}) + 5^2(\tfrac{1}{6}) + 6^2(\tfrac{1}{6}) \\ &= (\tfrac{1}{6})(91) \end{aligned}$$

Hence

$$\text{Var}(X) = \tfrac{91}{6} - (\tfrac{7}{2})^2 = \tfrac{35}{12}$$

**Example 4c.** Compute the variance of a binomial random variable $X$ with parameters $n$ and $p$.

***Solution:*** We first compute $E[X^2]$ as follows:

$$E[X^2] = \sum_{i=0}^{n} i^2 \binom{n}{i} p^i (1-p)^{n-i}$$

To evaluate the above, we use the identity $i^2 = i(i-1) + i$ to obtain

$$\begin{aligned}
E[X^2] &= \sum_{i=0}^{n} i(i-1)\frac{n!}{(n-i)!\,i!}p^i(1-p)^{n-i} + \sum_{i=0}^{n} i\binom{n}{i}p^i(1-p)^{n-i} \\
&= \sum_{i=2}^{n} \frac{n!}{(n-i)!\,(i-2)!}p^i(1-p)^{n-i} + E[X] \\
&= n(n-1)p^2 \sum_{i=2}^{n} \binom{n-2}{i-2} p^{i-2}(1-p)^{n-i} + E[X] \\
&= n(n-1)p^2[p + (1-p)]^{n-2} + E[X] \\
&= n(n-1)p^2 + E[X]
\end{aligned}$$

Since $E[X] = np$ (from Example 1c), we obtain

$$\begin{aligned}
\text{Var}(X) &= n(n-1)p^2 + np - n^2p^2 \\
&= np(1-p)
\end{aligned}$$ ∎

A useful identity is that for any constants $a$ and $b$

$$\text{Var}(aX + b) = a^2\,\text{Var}(X) \tag{4.3}$$

To prove Equation (4.3), we note from Corollary 2.1 that $E[aX + b] = aE[X] + b$. Hence

$$\begin{aligned}
\text{Var}(aX + b) &= E[(aX + b - (aE[X] + b))^2] \\
&= E[(aX - aE[X])^2] \\
&= E[a^2(X - E[X])^2) \\
&= a^2 E[(X - E[X])^2] \\
&= a^2\,\text{Var}(X)
\end{aligned}$$

REMARK. Analogous to the mean's being the center of gravity of a distribution of mass, the variance represents, in the terminology of mechanics, the moment of inertia.

## 5 Covariance, Variance of Sums, and Correlations

We start with the following proposition, which shows that the expectation of a product of independent random variables is just the product of their expectations.

*Proposition 5.1*

*If X and Y are independent, then for any functions h and g*

$$E[g(X)h(Y)] = E[g(X)]E[h(Y)]$$

**Proof:** Suppose that $X$ and $Y$ are jointly continuous with joint density $f(x, y)$. Then

$$\begin{aligned} E[g(X)h(Y)] &= \int_{-\infty}^{\infty}\int_{-\infty}^{\infty} g(x)h(y)f(x, y)\,dx\,dy \\ &= \int_{-\infty}^{\infty}\int_{-\infty}^{\infty} g(x)h(y)f_X(x)f_Y(y)\,dx\,dy \\ &= \int_{-\infty}^{\infty} h(y)f_Y(y)\,dy \int_{-\infty}^{\infty} g(x)f_X(x)\,dx \\ &= E[h(Y)]E[g(X)] \end{aligned}$$

The proof in the discrete case is similar. ∎

The covariance of any two random variables $X$ and $Y$, denoted by $\text{Cov}(X, Y)$, is defined by

$$\text{Cov}(X, Y) = E[(X - E[X])(Y - E[Y])]$$

Expanding the right side of the preceding equation yields

$$\begin{aligned} \text{Cov}(X, Y) &= E[XY - E[X]Y - XE[Y] + E[Y]E[X]] \\ &= E[XY] - E[X]E[Y] - E[X]E[Y] + E[X]E[Y] \\ &= E[XY] - E[X]E[Y] \end{aligned}$$

Note that if $X$ and $Y$ are independent then, by Proposition 5.1, it follows that $\text{Cov}(X, Y) = 0$. However, the converse is not true. A simple example of two dependent random variables $X$ and $Y$ having zero covariance is obtained by letting $X$ be a random variable such that

$$P\{X = 0\} = P\{X = 1\} = P\{X = -1\} = \tfrac{1}{3}$$

and define

$$Y = \begin{cases} 0 & \text{if } X \neq 0 \\ 1 & \text{if } X = 0 \end{cases}$$

Now, $XY = 0$, and so $E[XY] = 0$. Also, $E[X] = 0$ and thus

$$\text{Cov}(X, Y) = E[XY] - E[X]E[Y] = 0$$

However, $X$ and $Y$ are clearly not independent.

A useful expression for the variance of the sum of two random variables may be obtained in terms of the covariance in the following manner:

$$\begin{aligned}
\text{Var}(X+Y) &= E[(X+Y-E[X+Y])^2] \\
&= E[(X+Y-EX-EY)^2] \\
&= E[((X-EX)+(Y-EY))^2] \\
&= E[(X-EX)^2+(Y-EY)^2+2(X-EX)(Y-EY)] \\
&= E[(X-EX)^2]+E[(Y-EY)^2] \\
&\quad +2E[(X-EX)(Y-EY)] \\
&= \text{Var}(X)+\text{Var}(Y)+2\,\text{Cov}(X,Y)
\end{aligned}$$

In fact, a similar argument can be used to prove

$$\text{Var}\left(\sum_{i=1}^{n} X_i\right) = \sum_{i=1}^{n} \text{Var}(X_i) + 2\sum_{i<j}\sum \text{Cov}(X_i, X_j) \tag{5.1}$$

If $X_1, \ldots, X_n$ are all pairwise independent, then Equation (5.1) reduces to

$$\text{Var}\left(\sum_{i=1}^{n} X_i\right) = \sum_{i=1}^{n} \text{Var}(X_i) \tag{5.2}$$

**Example 5a.** Variance of a Binomial Random Variable. Compute the variance of a binomial random variable $X$ with parameters $n$ and $p$.

***Solution:*** Since such a random variable represents the number of successes in $n$ independent trials when each trial has a common probability $p$ of being a success, we may write

$$X = X_1 + \cdots + X_n$$

where the $X_i$ are independent Bernoulli random variables such that

$$X_i = \begin{cases} 1 & \text{if the } i\text{th trial is a success} \\ 0 & \text{otherwise} \end{cases}$$

Hence, from Equation (5.2) we obtain

$$\text{Var}(X) = \text{Var}(X_1) + \cdots + \text{Var}(X_n)$$

But

$$\begin{aligned}
\text{Var}(X_i) &= E[X_i^2] - (E[X_i])^2 \\
&= E[X_i] - (E[X_i])^2 \qquad \text{since } X_i^2 = X_i \\
&= p - p^2
\end{aligned}$$

and thus

$$\text{Var}(X) = np(1-p)$$ ∎

**Example 5b.** Variance of the Number of Matches. Compute the variance of $X$, the number of men that select their own hats in Example 3c.

***Solution:*** Using the same representation for $X$ as we do in Example 3c, namely,

$$X = X_1 + \cdots + X_N$$

where

$$X_i = \begin{cases} 1 & \text{if } i\text{th man selects his own hat} \\ 0 & \text{otherwise} \end{cases}$$

we obtain from Equation (5.1) that

$$\text{Var}(X) = \sum_{i=1}^{N} \text{Var}(X_i) + 2 \sum_{i<j}\sum \text{Cov}(X_i, X_j) \tag{5.3}$$

Since $P\{X_i = 1\} = 1/N$, we see from the previous example that

$$\text{Var}(X_i) = \frac{1}{N}\left(1 - \frac{1}{N}\right) = \frac{N-1}{N^2}$$

Also

$$\text{Cov}(X_i, X_j) = E[X_i X_j] - E[X_i]E[X_j]$$

Now,

$$X_i X_j = \begin{cases} 1 & \text{if the } i\text{th and } j\text{th men both select their own hats} \\ 0 & \text{otherwise} \end{cases}$$

and thus

$$\begin{aligned} E[X_i X_j] &= P\{X_i = 1, X_j = 1\} \\ &= P\{X_i = 1\}P\{X_j = 1 \mid X_i = 1\} \\ &= \frac{1}{N}\frac{1}{N-1} \end{aligned}$$

Hence

$$\text{Cov}(X_i, X_j) = \frac{1}{N(N-1)} - \left(\frac{1}{N}\right)^2 = \frac{1}{N^2(N-1)}$$

and from Equation (5.3),

$$\begin{aligned} \text{Var}(X) &= \frac{N-1}{N} + 2\binom{N}{2}\frac{1}{N^2(N-1)} \\ &= \frac{N-1}{N} + \frac{1}{N} \\ &= 1 \end{aligned}$$

Thus both the mean and variance of the number of matches are equal to 1. In a way, this result is not unexpected because, as shown in Section

5 of Chapter 2, when $N$ is large, the probability of $i$ matches is approximately $e^{-1}/i!$. That is, when $N$ is large, the number of matches is approximately distributed as a Poisson random variable with mean 1. Hence, as the mean and variance of a Poisson random variable are equal (see Theoretical Exercise 10), the result obtained in this example is not surprising. ∎

**Example 5c.** Sampling from a Finite Population. Consider a set of $N$ individuals each of whom has an opinion about a certain subject that is measured by a real number $v$, which represents the individual's "strength of feeling" about the subject. Let $v_i$ represent the strength of feeling of individual $i$, $i = 1, \ldots, N$. Suppose that these quantities $v_i$, $i = 1, \ldots, N$ are unknown and to gather information a group of $n$ of the $N$ individuals is "randomly chosen" in the sense that all of the $\binom{N}{n}$ subsets of size $n$ are equally likely to be chosen. These $n$ individuals are then questioned and their feelings determined. If $S$ denotes the sum of the $n$ sampled values, determine its mean and variance.

An important application of the above is to a forthcoming election in which each individual in the population is either for or against a certain candidate or proposition. If we take $v_i$ to equal 1 if individual $i$ is in favor and 0 if he or she is against, then $\bar{v} = \sum_{i=1}^{N} v_i/N$ represents the proportion of the population that is in favor. To estimate $\bar{v}$, a random sample of $n$ individuals is chosen, and these individuals are polled. The proportion of those polled that are in favor—that is, $S/n$—is often used as an estimate of $\bar{v}$.

***Solution:*** For each individual $i$, $i = 1, \ldots, N$, define an indicator variable $I_i$ to indicate whether or not that person is included in the sample. That is,

$$I_i = \begin{cases} 1 & \text{if individual } i \text{ is in the random sample} \\ 0 & \text{otherwise} \end{cases}$$

Now $S$ can be expressed by

$$S = \sum_{i=1}^{N} v_i I_i$$

and so

$$E[S] = \sum_{i=1}^{N} v_i E[I_i]$$

$$\begin{aligned} \text{Var}(S) &= \sum_{i=1}^{N} \text{Var}(v_i I_i) + 2 \sum_{i<j}\sum \text{Cov}(v_i I_i, v_j I_j) \\ &= \sum_{i=1}^{N} v_i^2 \, \text{Var}(I_i) + 2 \sum_{i<j}\sum v_i v_j \, \text{Cov}(I_i, I_j) \end{aligned}$$

As

$$E[I_i] = \frac{n}{N}$$

$$E[I_iI_j] = \frac{n}{N}\frac{n-1}{N-1}$$

we see that

$$\text{Var}(I_i) = \frac{n}{N}\left(1 - \frac{n}{N}\right)$$

$$\text{Cov}(I_i, I_j) = \frac{n(n-1)}{N(N-1)} - \left(\frac{n}{N}\right)^2$$

$$= \frac{-n(N-n)}{N^2(N-1)}$$

Hence

$$E[S] = n\sum_{i=1}^{N}\frac{v_i}{N} = n\bar{v}$$

$$\text{Var}(S) = \frac{n}{N}\left(\frac{N-n}{N}\right)\sum_{i=1}^{N} v_i^2 - \frac{2n(N-n)}{N^2(N-1)}\sum_{i<j}\sum v_iv_j$$

The expression for Var $(S)$ can be simplified somewhat by using the identity $(v_1 + \cdots + v_N)^2 = \sum_{i=1}^{N} v_i^2 + 2\sum_{i<j}\sum v_iv_j$ to give, after some simplification,

$$\text{Var}(S) = \frac{n(N-n)}{N-1}\left(\frac{\sum_{i=1}^{N} v_i^2}{N} - \bar{v}^2\right)$$

Consider now the special case in which $Np$ of the $v$'s are equal to 1 and the remainder equal to 0. Then in this case $S$ is a hypergeometric random variable and has mean and variance given by

$$E[S] = n\bar{v} = np \qquad \text{since } \bar{v} = \frac{Np}{N} = p$$

$$\text{Var}(S) = \frac{n(N-n)}{(N-1)}\left(\frac{Np}{N} - p^2\right)$$

$$= \frac{n(N-n)}{N-1}p(1-p)$$

The quantity $S/n$, equal to the proportion of those sampled that have values equal to 1, is such that

$$E\left[\frac{S}{n}\right] = p$$

$$\text{Var}\left(\frac{S}{n}\right) = \frac{(N-n)}{n(N-1)}p(1-p) \qquad \blacksquare$$

The correlation of two random variables $X$ and $Y$, denoted by $\rho(X, Y)$, is defined, as long as $\text{Var}(X)\,\text{Var}(Y)$ is positive, by

$$\rho(X, Y) = \frac{\text{Cov}(X, Y)}{\sqrt{\text{Var}(X)\,\text{Var}(Y)}}$$

It can be shown that

$$-1 \le \rho(X, Y) \le 1 \tag{5.4}$$

To prove Equation (5.4), suppose that $X$ and $Y$ have variances given by $\sigma_x^2$ and $\sigma_y^2$, respectively. Then

$$\begin{aligned} 0 &\le \text{Var}\left(\frac{X}{\sigma_x} + \frac{Y}{\sigma_y}\right) \\ &= \frac{\text{Var}(X)}{\sigma_x^2} + \frac{\text{Var}(Y)}{\sigma_y^2} + \frac{2\,\text{Cov}(X, Y)}{\sigma_x \sigma_y} \\ &= 2[1 + \rho(X, Y)] \end{aligned}$$

implying that

$$-1 \le \rho(X, Y)$$

On the other hand,

$$\begin{aligned} 0 &\le \text{Var}\left(\frac{X}{\sigma_x} - \frac{Y}{\sigma_y}\right) \\ &= \frac{\text{Var}(X)}{\sigma_x^2} + \frac{\text{Var}\,Y}{(-\sigma_y)^2} - \frac{2\,\text{Cov}(X, Y)}{\sigma_x \sigma_y} \\ &= 2[1 - \rho(X, Y)] \end{aligned}$$

implying that

$$\rho(X, Y) \le 1$$

which completes the proof of Equation (5.4).

In fact, since $\text{Var}(Z) = 0$ implies that $Z$ is constant with probability 1 (this intuitive fact will be rigorously proved in Chapter 8), we see from the proof of (5.4) that $\rho(X, Y) = 1$ implies that $Y = a + bX$, where $b = \sigma_y/\sigma_x > 0$ and $\rho(X, Y) = -1$ implies that $Y = a + bX$, where $b = -\sigma_y/\sigma_x < 0$. We leave it as an exercise for the reader to show that the reverse is also true: that if $Y = a + bX$, then $\rho(X, Y)$ is either $+1$ or $-1$, depending on the sign of $b$.

The correlation coefficient is a measure of the degree of linearity between $X$ and $Y$. A value of $\rho(X, Y)$ near $+1$ or $-1$ indicates a high degree of linearity between $X$ and $Y$, whereas a value near 0 indicates a lack of such linearity. A positive value of $\rho(X, Y)$ indicates that $Y$ tends to increase when $X$ does, whereas a negative value indicates that $Y$ tends to decrease when $X$ increases. If $\rho(X, Y) = 0$, then $X$ and $Y$ are said to be *uncorrelated*.

**Example 5d.** Let $I_A$ and $I_B$ be indicator variables for the events $A$ and $B$. That is,

$$I_A = \begin{cases} 1 & \text{if } A \text{ occurs} \\ 0 & \text{otherwise} \end{cases}$$

$$I_B = \begin{cases} 1 & \text{if } B \text{ occurs} \\ 0 & \text{otherwise} \end{cases}$$

Then

$$\begin{aligned} E[I_A] &= P(A) \\ E[I_B] &= P(B) \\ E[I_A I_B] &= P(AB) \end{aligned}$$

and so

$$\begin{aligned} \text{Cov}\,(I_A, I_B) &= P(AB) - P(A)P(B) \\ &= P(B)[P(A|B) - P(A)] \end{aligned}$$

Thus we obtain the quite intuitive result that the indicator variables for $A$ and $B$ are either positively correlated, uncorrelated, or negatively correlated depending on whether $P(A|B)$ is greater than, equal to, or less than $P(A)$. ∎

A useful result concerning covariances is that

$$\text{Cov}\left(\sum_{i=1}^{n} X_i, \sum_{i=1}^{m} Y_i\right) = \sum_{j=1}^{m} \sum_{i=1}^{n} \text{Cov}\,(X_i, Y_j) \tag{5.5}$$

We will leave the proof of the above as an exercise. We illustrate its use by the following example.

**Example 5e.** Consider $m$ independent trials, each of which results in any of $r$ possible outcomes with probabilities $P_1, P_2, \ldots, P_r, \sum_1^r P_i = 1$. If we let $N_i, i = 1, \ldots, r$, denote the number of the $m$ trials that result in outcome $i$, then $N_1, N_2, \ldots, N_r$ has the multinomial distribution

$$P\{N_1 = n_1, N_2 = n_2, \ldots, N_r = n_r\}$$
$$= \frac{m!}{n_1!\, n_2! \ldots n_r!} P_1^{n_1} P_2^{n_2} \cdots P_r^{n_r} \qquad \sum_{i=1}^{r} n_i = m$$

For $i \neq j$ it seems likely that, when $N_i$ is large, $N_j$ would tend to be small, and hence it is intuitive that they should be negatively correlated. Let us compute their covariance by using the identity (5.5) and the representation

$$N_i = \sum_{k=1}^{m} I_i(k) \qquad \text{and} \qquad N_j = \sum_{k=1}^{m} I_j(k)$$

where

$$I_i(k) = \begin{cases} 1 & \text{if trial } k \text{ results in outcome } i \\ 0 & \text{otherwise} \end{cases}$$

$$I_j(k) = \begin{cases} 1 & \text{if trial } k \text{ results in outcome } j \\ 0 & \text{otherwise} \end{cases}$$

From Equation (5.5) we have

$$\text{Cov}\,(N_i, N_j) = \sum_{\ell=1}^{m} \sum_{k=1}^{m} \text{Cov}\,(I_i(k), I_j(\ell))$$

Now, when $k \neq \ell$,

$$\text{Cov}\,(I_i(k), I_j(\ell)) = 0$$

since the outcome of trial $k$ is independent of the outcome of trial $\ell$. On the other hand,

$$\text{Cov}\,(I_i(\ell), I_j(\ell)) = E[I_i(\ell)I_j(\ell)] - E[I_i(\ell)]E[I_j(\ell)]$$
$$= 0 - P_iP_j = -P_iP_j$$

where the above uses that $I_i(\ell)I_j(\ell) = 0$ since trial $\ell$ cannot result in both outcome $i$ and outcome $j$. Hence we obtain that

$$\text{Cov}\,(N_i, N_j) = -mP_iP_j$$

which is in accord with our intuition that $N_i$ and $N_j$ are negatively correlated. ∎

## 6 Conditional Expectation

### 6.1 *Definitions*

Recall that if $X$ and $Y$ are jointly discrete random variables, the conditional probability mass function of $X$, given that $Y = y$, is defined for all $y$ such that $P\{Y = y\} > 0$, by

$$p_{X|Y}(x|y) = P\{X = x \mid Y = y\} = \frac{p(x, y)}{p_Y(y)}$$

It is therefore natural to define, in this case, the conditional expectation of $X$, given that $Y = y$, for all values of $y$ such that $p_Y(y) > 0$ by

$$\begin{aligned} E[X \mid Y = y] &= \sum_x xP\{X = x \mid Y = y\} \\ &= \sum_x x p_{X|Y}(x|y) \end{aligned}$$

**Example 6a.** If $X$ and $Y$ are independent binomial random variables with identical parameters $n$ and $p$, calculate the conditional expected value of $X$, given that $X + Y = m$.

***Solution:*** Let us first calculate the conditional probability mass function of $X$, given that $X + Y = m$. For $k \le \min(n, m)$,

$$\begin{aligned} P\{X = k \mid X + Y = m\} &= \frac{P\{X = k, X + Y = m\}}{P\{X + Y = m\}} \\ &= \frac{P\{X = k, Y = m - k\}}{P\{X + Y = m\}} \\ &= \frac{P\{X = k\}P\{Y = m - k\}}{P\{X + Y = m\}} \\ &= \frac{\binom{n}{k} p^k (1-p)^{n-k} \binom{n}{m-k} p^{m-k} (1-p)^{n-m+k}}{\binom{2n}{m} p^m (1-p)^{2n-m}} \\ &= \frac{\binom{n}{k}\binom{n}{m-k}}{\binom{2n}{m}} \end{aligned}$$

where we have used the fact (see Example 3d of Chapter 6) that $X + Y$ is a binomial random variable with parameters $2n$ and $p$. Hence the conditional distribution of $X$, given that $X + Y = m$, is the hypergeometric distribution; thus, from Example 3e, we obtain

$$E[X \mid X + Y = m] = \frac{m}{2} \quad \blacksquare$$

Similarly, let us recall that if $X$ and $Y$ are jointly continuous, with a joint probability density function $f(x, y)$, the conditional probability density of $X$, given that $Y = y$, is defined for all values of $y$ such that $f_Y(y) > 0$ by

$$f_{X|Y}(x \mid y) = \frac{f(x, y)}{f_Y(y)}$$

It is natural, in this case, to define the conditional expectation of $X$, given that $Y = y$, by

continuous case

$$E[X \mid Y = y] = \int_{-\infty}^{\infty} x f_{X|Y}(x \mid y)\, dx$$

provided that $f_Y(y) > 0$.

**Example 6b.** Suppose the joint density of $X$ and $Y$ is given by

$$f(x, y) = \frac{e^{-x/y} e^{-y}}{y} \qquad 0 < x < \infty, 0 < y < \infty$$

Compute $E[X \mid Y = y]$.

***Solution:*** We start by computing the conditional density

$$\begin{aligned} f_{X|Y}(x \mid y) &= \frac{f(x, y)}{f_Y(y)} \\ &= \frac{f(x, y)}{\int_{-\infty}^{\infty} f(x, y)\, dx} \\ &= \frac{(1/y) e^{-x/y} e^{-y}}{\int_0^{\infty} (1/y) e^{-x/y} e^{-y}\, dx} \\ &= \frac{(1/y) e^{-x/y}}{\int_0^{\infty} (1/y) e^{-x/y}\, dx} \\ &= \frac{(1/y) e^{-x/y}}{-e^{-x/y} \big|_{x=0}^{x=\infty}} \\ &= \left(\frac{1}{y}\right) e^{-x/y} \end{aligned}$$

Hence the conditional distribution of $X$, given that $Y = y$, is just the exponential distribution with mean $y$. Thus

$$E[X \mid Y = y] = \int_0^\infty \frac{x}{y} e^{-x/y}\, dx = y \qquad \blacksquare$$

REMARK. Just as conditional probabilities satisfy all of the properties of ordinary probabilities, so do conditional expectations satisfy the properties of ordinary expectations. For instance, such formulas as

$$E[g(X) \mid Y = y] = \begin{cases} \sum_x g(x) p_{X|Y}(x \mid y) & \text{in the discrete case} \\ \int_{-\infty}^{\infty} g(x) f_{X|Y}(x \mid y)\, dx & \text{in the continuous case} \end{cases}$$

and

$$E\left[\sum_{i=1}^n X_i \mid Y = y\right] = \sum_{i=1}^n E[X_i \mid Y = y]$$

remain valid. As a matter of fact, conditional expectation given $Y = y$ can be thought of as being an ordinary expectation on a reduced sample space consisting only of outcomes for which $Y = y$.

## 6.2 *Computing Expectations by Conditioning*

Let us denote by $E[X \mid Y]$ that function of the random variable $Y$ whose value at $Y = y$ is $E[X \mid Y = y]$. Note that $E[X \mid Y]$ is itself a random variable. An extremely important property of conditional expectation is given by the following proposition.

---

*Proposition 6.1*

$$E[X] = E[E[X \mid Y]] \tag{6.1}$$

---

If $Y$ is a discrete random variable, then Equation (6.1) states that

$$E[X] = \sum_y E[X \mid Y = y] P\{Y = y\} \tag{6.1a}$$

whereas if $Y$ is continuous with density $f_Y(y)$, then Equation (6.1) states

$$E[X] = \int_{-\infty}^{\infty} E[X \mid Y = y] f_Y(y)\, dy \tag{6.1b}$$

We now give a proof of Equation (6.1) in the case where $X$ and $Y$ are both discrete random variables.

**Proof of Equation (6.1) when $X$ and $Y$ are discrete:** We must show that

$$E[X] = \sum_y E[X \mid Y = y]P\{Y = y\} \tag{6.2}$$

Now, the right-hand-side of Equation (6.2) can be written as

$$\begin{aligned}
\sum_y E[X \mid Y = y]P\{Y = y\} &= \sum_y \sum_x xP\{X = x \mid Y = y\}P\{Y = y\} \\
&= \sum_y \sum_x x \frac{P\{X = x, Y = y\}}{P\{Y = y\}} P\{Y = y\} \\
&= \sum_y \sum_x xP\{X = x, Y = y\} \\
&= \sum_x x \sum_y P\{X = x, Y = y\} \\
&= \sum_x xP\{X = x\} \\
&= E[X]
\end{aligned}$$

and the result is proved. ∎

One way to understand Equation (6.2) is to interpret it as follows: To calculate $E[X]$, we may take a weighted average of the conditional expected value of $X$, given that $Y = y$, each of the terms $E[X \mid Y = y]$ being weighted by the probability of the event on which it is conditioned. (Of what does this remind you?) This is an extremely useful result that often enables us to easily compute expectations by first conditioning on some appropriate random value. The following examples illustrate its use.

**Example 6c.** A miner is trapped in a mine containing 3 doors. The first door leads to a tunnel that will take him to safety after 3 hours of travel. The second door leads to a tunnel that will return him to the mine after 5 hours of travel. The third door leads to a tunnel that will return him to the mine after 7 hours. If we assume that the miner is at all times equally likely to choose any one of the doors, what is the expected length of time until he reaches safety?

***Solution:*** Let $X$ denote the amount of time (in hours) until the miner reaches safety, and let $Y$ denote the door he initially chooses. Now

$$\begin{aligned}
E[X] &= E[X \mid Y = 1]P\{Y = 1\} + E[X \mid Y = 2]P\{Y = 2\} \\
&\quad + E[X \mid Y = 3]P\{Y = 3\} \\
&= \tfrac{1}{3}(E[X \mid Y = 1] + E[X \mid Y = 2] + E[X \mid Y = 3])
\end{aligned}$$

However,

$$\begin{aligned} E[X \mid Y = 1] &= 3 \\ E[X \mid Y = 2] &= 5 + E[X] \\ E[X \mid Y = 3] &= 7 + E[X] \end{aligned} \tag{6.3}$$

To understand why Equation (6.3) is correct, consider, for instance, $E[X \mid Y = 2]$ and reason as follows: If the miner chooses the second door, he spends 5 hours in the tunnel and then returns to his cell. But once he returns to his cell the problem is as before; thus his expected additional time until safety is just $E[X]$. Hence $E[X \mid Y = 2] = 5 + E[X]$. The argument behind the other equalities in Equation (6.3) is similar. Hence

$$E[X] = \tfrac{1}{3}(3 + 5 + E[X] + 7 + E[X])$$

or

$$E[X] = 15$$ ∎

**Example 6d.** The Expectation of a Random Number of Random Variables. Suppose that the number of people entering a department store on a given day is a random variable with mean 50. Suppose further that the amounts of money spent by these customers are independent random variables having a common mean of $8. Assume also that the amount of money spent by a customer is also independent of the total number of customers to enter the store. What is the expected amount of money spent in the store on a given day?

***Solution:*** If we let $N$ denote the number of customers that enter the store and $X_i$ the amount spent by the $i$th such customer, then the total amount of money spent can be expressed as $\sum_{i=1}^{N} X_i$. Now,

$$E\left[\sum_1^N X_i\right] = E\left[E\left[\sum_1^N X_i \mid N\right]\right]$$

But

$$\begin{aligned} E\left[\sum_1^N X_i \mid N = n\right] &= E\left[\sum_1^n X_i \mid N = n\right] \\ &= E\left[\sum_1^n X_i\right] \qquad \text{by the independence of the } X_i \text{ and } N \\ &= nE[X] \qquad \text{where } E[X] = E[X_i] \end{aligned}$$

which implies that

$$E\left[\sum_{1}^{N} X_i | N\right] = NE[X]$$

and thus

$$E\left[\sum_{i=1}^{N} X_i\right] = E[NE[X]] = E[N]E[X]$$

Hence, in our example, the expected amount of money spent in the store is 50 × 8 or $400. ∎

**Example 6e.** An urn contains $a$ white and $b$ black balls. One ball at a time is randomly withdrawn until the first white ball is drawn. Find the expected number of black balls that are withdrawn.

***Solution:*** This problem was previously treated in Example 3j. Here we present a solution using conditioning. Let $X$ denote the number of black balls withdrawn and, to make explicit the dependence on $a$ and $b$, let $M_{a,b} = E[X]$. We obtain an expression for $M_{a,b}$ by conditioning on the initial ball that is withdrawn. That is, define

$$Y = \begin{cases} 1 & \text{if the first ball selected is white} \\ 0 & \text{if the first ball selected is black} \end{cases}$$

Conditioning on $Y$ yields

$$M_{a,b} = E[X] = E[X | Y = 1]P\{Y = 1\} + E[X | Y = 0]P\{Y = 0\}$$

However,

$$E[X | Y = 1] = 0 \tag{6.4}$$

$$E[X | Y = 0] = 1 + M_{a,b-1} \tag{6.5}$$

To understand Equations (6.4) and (6.5), suppose, for instance, that the first ball withdrawn is black. Then, after the first withdrawal, the situation is exactly the same as if we had started with $a$ white balls and $b - 1$ black balls, which establishes Equation (6.5).

Since $P\{Y = 0\} = b/(a + b)$, we see that

$$M_{a,b} = \frac{b}{a + b}[1 + M_{a,b-1}]$$

Now, $M_{a,0}$ is clearly equal to 0, and we obtain

$$M_{a,1} = \frac{1}{a+1}[1 + M_{a,0}] = \frac{1}{a+1}$$

$$M_{a,2} = \frac{2}{a+2}[1 + M_{a,1}] = \frac{2}{a+2}\left[1 + \frac{1}{a+1}\right] = \frac{2}{a+1}$$

$$M_{a,3} = \frac{3}{a+3}[1 + M_{a,2}] = \frac{3}{a+3}\left[1 + \frac{2}{a+1}\right] = \frac{3}{a+1}$$

By using induction, it is easy to verify that

$$M_{a,b} = \frac{b}{a+1}$$ ∎

It is also possible to obtain the variance of a random variable by conditioning. We illustrate this by the following example.

**Example 6f.** Variance of the Geometric Distribution. Independent trials each resulting in a success with probability $p$ are successively performed. Let $N$ be the time of the first success. Find Var $(N)$.

***Solution:*** Let $Y = 1$ if the first trial results in a success and $Y = 0$ otherwise. Now,

$$\text{Var}(N) = E[N^2] - (E[N])^2$$

To calculate $E[N^2]$, we condition on $Y$ as follows:

$$E[N^2] = E[E[N^2 | Y]]$$

However,

$$E[N^2 | Y = 1] = 1$$
$$E[N^2 | Y = 0] = E[(1 + N)^2]$$

These two equations follow because, if the first trial results in a success, then clearly $N = 1$; thus $N^2 = 1$. On the other hand, if the first trial results in a failure, then the total number of trials necessary for the first success will have the same distribution as one (the first trial that results in failure) plus the necessary number of additional trials. Since this latter quantity has the same distribution as $N$, we obtain that $E[N^2 | Y = 0] = E[(1 + N)^2]$. Hence we see that

$$\begin{aligned} E[N^2] &= E[N^2 | Y = 1]P\{Y = 1\} + E[N^2 | Y = 0]P\{Y = 0\} \\ &= p + (1-p)E[(1+N)^2] \\ &= 1 + (1-p)E[2N + N^2] \end{aligned}$$

However, as was shown in Example 1e, $E[N] = 1/p$; therefore, this yields

$$E[N^2] = 1 + \frac{2(1-p)}{p} + (1-p)E[N^2]$$

or

$$E[N^2] = \frac{2-p}{p^2}$$

Therefore,

$$\begin{aligned}\text{Var}(N) &= E[N^2] - (E[N])^2 \\ &= \frac{2-p}{p^2} - \left(\frac{1}{p}\right)^2 \\ &= \frac{1-p}{p^2}\end{aligned}$$

▮

## 6.3 *Computing Probabilities by Conditioning*

Not only can we obtain expectations by first conditioning on an appropriate random variable, but we may also use this approach to compute probabilities. To see this, let $E$ denote an arbitrary event and define the indicator random variable $X$ by

$$X = \begin{cases} 1 & \text{if } E \text{ occurs} \\ 0 & \text{if } E \text{ does not occur} \end{cases}$$

It follows from the definition of $X$ that

$$\begin{aligned} E[X] &= P(E) \\ E[X \mid Y = y] &= P(E \mid Y = y) \qquad \text{for any random variable } Y \end{aligned}$$

Therefore, from Equations (5.1a) and (5.1b) we obtain

$$\begin{aligned} P(E) &= \sum_y P(E \mid Y = y)P(Y = y) && \text{if } Y \text{ is discrete} \\ &= \int_{-\infty}^{\infty} P(E \mid Y = y) f_Y(y)\, dy && \text{if } Y \text{ is continuous} \end{aligned} \tag{6.6}$$

Note that if $Y$ is a discrete random variable taking on one of the values $y_1, \ldots, y_n$, then, by defining the events $F_i$, $i = 1, \ldots, n$ by $F_i = \{Y = y_i\}$,

Equation (6.6) reduces to the familiar equation

$$P(E) = \sum_{i=1}^{n} P(E \mid F_i)P(F_i)$$

where $F_1, \ldots, F_n$ are mutually exclusive events whose union is the sample space.

**Example 6g.** Suppose that $X$ and $Y$ are independent continuous random variables having densities $f_X$ and $f_Y$, respectively. Compute $P\{X < Y\}$.

***Solution:*** Conditioning on the value of $Y$ yields

$$\begin{aligned} P\{X < Y\} &= \int_{-\infty}^{\infty} P\{X < Y \mid Y = y\} f_Y(y)\, dy \\ &= \int_{-\infty}^{\infty} P\{X < y \mid Y = y\} f_Y(y)\, dy \\ &= \int_{-\infty}^{\infty} P\{X < y\} f_Y(y)\, dy \qquad \text{by independence} \\ &= \int_{-\infty}^{\infty} F_X(y) f_Y(y)\, dy \end{aligned}$$

where

$$F_X(y) = \int_{-\infty}^{y} f_X(x)\, dx$$

∎

**Example 6h.** Suppose that $X$ and $Y$ are independent continuous random variables. Find the distribution of $X + Y$.

***Solution:*** By conditioning on the value of $Y$, we obtain

$$\begin{aligned} P\{X + Y < a\} &= \int_{-\infty}^{\infty} P\{X + Y < a \mid Y = y\} f_Y(y)\, dy \\ &= \int_{-\infty}^{\infty} P\{X + y < a \mid Y = y\} f_Y(y)\, dy \\ &= \int_{-\infty}^{\infty} P\{X < a - y\} f_Y(y)\, dy \\ &= \int_{-\infty}^{\infty} F_X(a - y) f_Y(y)\, dy \end{aligned}$$

∎

## 6.4 *Conditional Variance*

Just as we have defined the conditional expectation of $X$ given the value of $Y$, we can also define the conditional variance of $X$ given that $Y = y$, which is defined as follows:

$$\text{Var}(X \mid Y) \equiv E[[X - E(X \mid Y)]^2 \mid Y]$$

That is, $\text{Var}(X \mid Y)$ is equal to the (conditional) expected square of the difference between $X$ and its (conditional) mean when the value of $Y$ is given. Or, in other words, $\text{Var}(X \mid Y)$ is exactly analogous to the usual definition of variance, but now all expectations are conditional on the fact that $Y$ is known.

There is a very useful relationship between $\text{Var}(X)$, the unconditional variance of $X$, and $\text{Var}(X \mid Y)$, the conditional variance of $X$ given $Y$, that can often be applied to compute $\text{Var}(X)$. To obtain this relationship, note first that by the same reasoning that yields $\text{Var}(X) = E[X^2] - (E[X])^2$ we have that

$$\text{Var}(X \mid Y) = E[X^2 \mid Y] - (E[X \mid Y])^2$$

and so

$$\begin{aligned} E[\text{Var}(X \mid Y)] &= E[E[X^2 \mid Y]] - E[(E[X \mid Y])^2] \\ &= E[X^2] - E[(E[X \mid Y])^2] \end{aligned} \tag{6.7}$$

Also, as $E[E[X \mid Y]] = E[X]$, we have that

$$\text{Var}(E[X \mid Y]) = E[(E[X \mid Y])^2] - (E[X])^2 \tag{6.8}$$

Hence, by adding Equations (6.7) and (6.8), we arrive at the following proposition.

---

*Proposition 6.2 The Conditional Variance Formula*

$$\text{Var}(X) = E[\text{Var}(X \mid Y)] + \text{Var}(E[X \mid Y])$$

---

**Example 6i.** Suppose that by any time $t$ the number of people that have arrived at a train depot is a Poisson random variable with mean $\lambda t$. If the initial train arrives at the depot at a time (independent of when the passengers arrive) that is uniformly distributed over $(0, T)$, what is the mean and variance of the number of passengers that enter the train?

***Solution:*** Let, for each $t \geq 0$, $N(t)$ denote the number of arrivals by $t$, and let $Y$ denote the time at which the train arrives. The random variable of interest is then $N(Y)$. Conditioning on $Y$ gives:

$$\begin{aligned} E[N(Y)\,|\,Y=t] &= E[N(t)\,|\,Y=t] \\ &= E[N(t)] \qquad \text{by the independence of } Y \text{ and } N(t) \\ &= \lambda t \qquad \text{since } N(t) \text{ is Poisson with mean } \lambda t \end{aligned}$$

Hence

$$E[N(Y)\,|\,Y] = \lambda Y$$

and so taking expectations gives

$$E[N(Y)] = \lambda E[Y] = \frac{\lambda T}{2}$$

To obtain Var $(N(Y))$, we use the conditional variance formula:

$$\begin{aligned} \text{Var}\,(N(Y)\,|\,Y=t) &= \text{Var}\,(N(t)\,|\,Y=t) \\ &= \text{Var}\,(N(t)) \qquad \text{by independence} \\ &= \lambda t \end{aligned}$$

and so

$$\begin{aligned} \text{Var}\,(N(Y)\,|\,Y) &= \lambda Y \\ E[N(Y)\,|\,Y] &= \lambda Y \end{aligned}$$

Hence, from the conditional variance formula,

$$\begin{aligned} \text{Var}\,(N(Y)) &= E[\lambda Y] + \text{Var}\,(\lambda Y) \\ &= \lambda\frac{T}{2} + \lambda^2\frac{T^2}{12} \end{aligned}$$

where the above uses that Var $(Y) = T^2/12$. ∎

**Example 6j.** The Variance of a Random Number of Random Variables. Let $X_1, X_2, \ldots$ be a sequence of independent and identically distributed random variables and let $N$ be a nonnegative integer valued random variable that is independent of the sequence $X_i$, $i \geq 1$. To compute $\text{Var}\left(\sum_{i=1}^{N} X_i\right)$, we condition on $N$:

$$E\left[\sum_{i=1}^{N} X_i \,|\, N\right] = NE[X]$$

$$\text{Var}\left(\sum_{i=1}^{N} X_i \,|\, N\right) = N\,\text{Var}\,(X)$$

The above follows, since given $N$, $\sum_{i=1}^{N} X_i$ is just the sum of a fixed number of independent random variables, and so its expectation and variance is just the sum of the individual means and variances. Hence, from the conditional variance formula,

$$\mathrm{Var}\left(\sum_{i=1}^{N} X_i\right) = E[N]\,\mathrm{Var}\,(X) + (E[X])^2\,\mathrm{Var}\,(N) \quad \blacksquare$$

## 7 Conditional Expectation and Prediction

Sometimes a situation arises where the value of a random variable $X$ is observed and then, based on the observed value, an attempt is made to predict the value of a second random variable $Y$. Let $g(X)$ denote the predictor, that is, if $X$ is observed to equal $x$, then $g(x)$ is our prediction for the value of $Y$. Clearly, we would like to choose $g$ so that $g(X)$ tends to be close to $Y$. One possible criterion for closeness is to choose $g$ so as to minimize $E[(Y - g(X))^2]$. We now show that, under this criterion, the best possible predictor of $Y$ is $g(X) = E[Y|X]$.

---

*Proposition 7.1*

$$E[(Y - g(X))^2] \geq E[(Y - E[Y|X])^2]$$

---

**Proof**

$$\begin{aligned} E[(Y - g(X))^2|X] &= E[(Y - E[Y|X] + E[Y|X] - g(X))^2|X] \\ &= E[(Y - E[Y|X])^2|X] \\ &\quad + E[(E[Y|X] - g(X))^2|X] \\ &\quad + 2E[(Y - E[Y|X])(E[Y|X] - g(X))|X] \end{aligned} \tag{7.1}$$

However, given $X$, $E[Y|X] - g(X)$, being a function of $X$, can be treated as a constant. Thus

$$\begin{aligned} &E[(Y - E[Y|X])(E[Y|X] - g(X))|X] \\ &\quad = (E[Y|X] - g(X))E[Y - E[Y|X]|X] \\ &\quad = (E[Y|X] - g(X))(E[Y|X] - E[Y|X]) \\ &\quad = 0 \end{aligned} \tag{7.2}$$

Thus, from Equations (7.1) and (7.2), we obtain

$$E[(Y - g(X))^2|X] \geq E[(Y - E[Y|X])^2|X]$$

and the result follows by taking expectations of both sides of the above. $\blacksquare$

REMARK. A second, more intuitive although less rigorous argument verifying Proposition 7.1 is as follows. It is straightforward to verify that $E[(Y - c)^2]$ is minimized at $c = E[Y]$. (See Theoretical Exercise 5). Thus, if we want to predict the value of $Y$ when there are no data available to use, the best possible prediction, in the sense of minimizing the mean square error, is to predict that $Y$ will equal its mean. On the other hand, if the value of the random variable $X$ is observed to be $x$, then the prediction problem remains exactly as in the previous (no data) case with the exception that all probabilities and expectations are now conditional on the event that $X = x$. Hence it follows that the best prediction in this situation is to predict that $Y$ will equal its conditional expected value given that $X = x$, thus establishing Proposition 7.1.

**Example 7a.** Suppose that the son of a man of height $x$ (in inches) attains a height that is normally distributed with mean $x + 1$ and variance 4. What is the best prediction of the height at full growth of the son of a man who is 6 feet tall?

***Solution:*** Formally, this model can be written as

$$Y = X + 1 + e$$

where $e$ is a normal random variable, independent of $X$, having mean 0 and variance 4. The $X$ and $Y$, of course, represent the heights of the man and his son, respectively. The best prediction $E[Y|X = 72]$ is thus equal to

$$\begin{aligned} E[Y|X = 72] &= E[X + 1 + e|X = 72] \\ &= 73 + E[e|X = 72] \\ &= 73 + E(e) \qquad \text{by independence} \\ &= 73 \end{aligned}$$

∎

**Example 7b.** Suppose that if a signal value $s$ is sent from location $A$, then the signal value received at location $B$ is normally distributed with parameters $(s, 1)$. If $S$, the value of the signal sent at $A$, is normally distributed with parameters $(\mu, \sigma^2)$, what is the "best" estimate of the signal sent if $R$, the value received at $B$, is equal to $r$?

***Solution:*** Let us start by computing the conditional density of $S$ given $R$ as follows:

$$\begin{aligned} f_{S|R}(s|r) &= \frac{f_{S,R}(s, r)}{f_R(r)} \\ &= \frac{f_S(s)f_{R|S}(r|s)}{f_R(r)} \\ &= Ke^{-(s-\mu)^2/2\sigma^2}e^{-(r-s)^2/2} \end{aligned}$$

where $K$ does not depend on $s$. Now,

$$\frac{(s-\mu)^2}{2\sigma^2}+\frac{(r-s)^2}{2}=s^2\left(\frac{1}{2\sigma^2}+\frac{1}{2}\right)-\left(\frac{\mu}{\sigma^2}+r\right)s+C_1$$

$$=\frac{1+\sigma^2}{2\sigma^2}\left[s^2-2\left(\frac{\mu+r\sigma^2}{1+\sigma^2}\right)s\right]+C_1$$

$$=\frac{1+\sigma^2}{2\sigma^2}\left(s-\frac{(\mu+r\sigma^2)}{1+\sigma^2}\right)^2+C_2$$

where $C_1$ and $C_2$ do not depend on $s$. Hence

$$f_{S|R}(s|r)=C\exp\left\{\frac{-\left[s-\frac{(\mu+r\sigma^2)}{1+\sigma^2}\right]^2}{2\left(\frac{\sigma^2}{1+\sigma^2}\right)}\right\}$$

where $C$ does not depend on $s$. Hence we may conclude that the conditional distribution of $S$, the signal sent, given that $r$ is received, is normal with mean and variance now given by

$$E[S|R=r]=\frac{\mu+r\sigma^2}{1+\sigma^2}$$

$$\text{Var}\,(S|R=r)=\frac{\sigma^2}{1+\sigma^2}$$

Hence, from Proposition 7.1, given that the value received is $r$, the best estimate, in the sense of minimizing the mean square error, for the signal sent is

$$E[S|R=r]=\frac{1}{1+\sigma^2}\mu+\frac{\sigma^2}{1+\sigma^2}r$$

Writing the conditional mean as we did above is informative, for it shows that it equals a weighted average of $\mu$, the a priori expected value of the signal and $r$, the value received. The relative weights given to $\mu$ and $r^2$ are in the same proportion to each other as 1 (the conditional variance of the received signal when $s$ is sent) is to $\sigma^2$ (the variance of the signal to be sent). ∎

**Example 7c.** In digital signal processing raw continuous analog data $X$ must be quantized, or discretized, in order to obtain a digital representation. In order to quantize the raw data $X$, an increasing set of numbers $a_i$, $i=0,\pm1,\pm2,\ldots$, such that $\lim_{i\to+\infty} a_i=\infty$, $\lim_{i\to-\infty} a_i=-\infty$, is fixed and the raw data are then quantized according to the interval $(a_i, a_{i+1})$ in which $X$ lies. Let us denote by $y_i$ the discretized value when $X \in (a_i, a_{i+1})$, and let $Y$ denote the observed discretized value—that is,

$$Y = y_i \qquad \text{if } a_i < X \leq a_{i+1}$$

The distribution of $Y$ is given by

$$P\{Y = y_i\} = F_X(a_{i+1}) - F_X(a_i)$$

Suppose now that we want to choose the values $y_i$, $i = 0, \pm 1, \pm 2, \ldots$ so as to minimize $E[(X - Y)^2]$, the expected mean square difference between the raw data and their quantized version.

(i) Find the optimal values $y_i$, $i = 0, \pm 1, \ldots$

For the optimal quantizer $Y$ show that

(ii) $E[Y] = E[X]$, and so the mean square error quantizer preserves the input mean, and
(iii) $\text{Var}(Y) = \text{Var}(X) - E[(X - Y)^2]$

***Solution:*** (i) For any quantizer $Y$, upon conditioning on the value of $Y$ we obtain

$$E[(X - Y)^2] = \sum_i E[(X - y_i)^2 \,|\, a_i < X \leq a_{i+1}] P\{a_i < X \leq a_{i+1}\}$$

Now, if we let

$$I = i \qquad \text{if } a_i < X \leq a_{i+1}$$

then

$$E[(X - y_i)^2 \,|\, a_i < X \leq a_{i+1}] = E[(X - y_i)^2 \,|\, I = i]$$

and by Proposition 7.1 this quantity is minimized when

$$\begin{aligned} y_i &= E[X \,|\, I = i] \\ &= E[X \,|\, a_i < X \leq a_{i+1}] \\ &= \int_{a_i}^{a_{i+1}} \frac{x f_X(x)\, dx}{F_X(a_{i+1}) - F_X(a_i)} \end{aligned}$$

Now, since the optimal quantizer is given by $Y = E[X \,|\, I]$, it follows that

(ii) $E[Y] = E[X]$
(iii) $\begin{aligned}[t] \text{Var}(X) &= E[\text{Var}(X \,|\, I)] + \text{Var}(E[X \,|\, I]) \\ &= E[E[(X - Y)^2 \,|\, I]] + \text{Var}(Y) \\ &= E[(X - Y)^2] + \text{Var}(Y) \end{aligned}$ ∎

It sometimes happens that the joint probability distribution of $X$ and $Y$ is not completely known; or if it is known, it is such that the calculation of $E[Y|X = x]$ is mathematically intractable. If, however, the means and variances of $X$ and $Y$ and the correlation of $X$ and $Y$ are known, then we can at least determine the best *linear* predictor of $Y$ with respect to $X$.

To obtain the best linear predictor of $Y$ with respect to $X$, we need to choose $a$ and $b$ so as to minimize $E[(Y - (a + bX))^2]$. Now,

$$\begin{aligned} E[(Y - (a + bX))^2] &= E[Y^2 - 2aY - 2bXY + a^2 + 2abX + b^2X^2] \\ &= E[Y^2] - 2aE[Y] - 2bE[XY] + a^2 \\ &\quad + 2abE[X] + b^2E[X^2] \end{aligned}$$

Taking partial derivatives, we obtain

$$\frac{\partial}{\partial a} E[(Y - a - bX)^2] = -2E[Y] + 2a + 2bE[X]$$

$$\frac{\partial}{\partial b} E[(Y - a - bX)^2] = -2E[XY] + 2aE[X] + 2bE[X^2] \tag{7.3}$$

Equating Equations (7.3) to 0 and solving for $a$ and $b$ yields the solutions

$$b = \frac{E[XY] - E[X]E[Y]}{E[X^2] - (E[X])^2} = \frac{\text{Cov}(X, Y)}{\sigma_x^2} = \rho\frac{\sigma_y}{\sigma_x}$$

$$a = E[Y] - bE[X] = E[Y] - \frac{\rho\sigma_y E[X]}{\sigma_x} \tag{7.4}$$

where $\rho = \text{Correlation}(X, Y)$, $\sigma_y^2 = \text{Var}(Y)$, and $\sigma_x^2 = \text{Var}(X)$. It is easy to verify that the values of $a$ and $b$ from Equation (7.4) minimize $E[(Y - a - bX)^2]$, and thus the best (in the sense of mean square error) linear predictor $Y$ with respect to $X$ is

$$\mu_y + \frac{\rho\sigma_y}{\sigma_x}(X - \mu_x)$$

where $\mu_y = E[Y]$ and $\mu_x = E[X]$.

The mean square error of this predictor is given by

$$\begin{aligned} &E\left[\left(Y - \mu_y - \rho\frac{\sigma_y}{\sigma_x}(X - \mu_x)\right)^2\right] \\ &= E[(Y - \mu_y)^2] + \rho^2\frac{\sigma_y^2}{\sigma_x^2}E[(X - \mu_x)^2] - 2\rho\frac{\sigma_y}{\sigma_x}E[(Y - \mu_y)(X - \mu_x)] \\ &= \sigma_y^2 + \rho^2\sigma_y^2 - 2\rho^2\sigma_y^2 \\ &= \sigma_y^2(1 - \rho^2) \end{aligned} \tag{7.5}$$

We note from Equation (7.5) that if $\rho$ is near +1 or −1, then the mean square error of the best linear predictor is near 0.

**Example 7d.** An example in which the conditional expectation of $Y$ given $X$ is linear in $X$, and hence the best linear predictor of $Y$ with respect to $X$ is the best overall predictor, is when $X$ and $Y$ have a bivariate normal distribution. In this case their joint density is given by

$$f(x, y) = \frac{1}{2\pi\sigma_x\sigma_y\sqrt{1-\rho^2}} \exp\left\{-\frac{1}{2(1-\rho^2)}\left[\left(\frac{x-\mu_x}{\sigma_x}\right)^2 - \frac{2\rho(x-\mu_x)(y-\mu_y)}{\sigma_x\sigma_y} + \left(\frac{y-\mu_y}{\sigma_y}\right)^2\right]\right\}$$

We leave it for the reader to verify that the conditional density of $Y$, given $X = x$, is given by

$$f_{Y|X}(y|x) = \frac{1}{\sqrt{2\pi}\sigma_y\sqrt{1-\rho^2}} \exp\left\{-\frac{1}{2\sigma_y^2(1-\rho^2)}\left(y - \mu_y - \frac{\rho\sigma_y}{\sigma_x}(x-\mu_x)\right)^2\right\}$$

Hence the conditional distribution of $Y$, given $X = x$, is the normal distribution with mean

$$E[Y|X = x] = \mu_y + \rho\frac{\sigma_y}{\sigma_x}(x - \mu_x)$$

and variance $\sigma_y^2(1-\rho^2)$.

## 8 Moment Generating Functions

The moment generating function $\phi(t)$ of the random variable $X$ is defined for all real values of $t$ by

$$\phi(t) = E[e^{tX}] = \begin{cases} \sum_x e^{tx}p(x) & \text{if } X \text{ is discrete with mass function } p(x) \\ \int_{-\infty}^{\infty} e^{tx}f(x)\,dx & \text{if } X \text{ is continuous with density } f(x) \end{cases}$$

We call $\phi(t)$ the moment generating function because all of the moments of $X$ can be obtained by successively differentiating $\phi(t)$ and then evaluating the result at $t = 0$. For example,

$$\begin{aligned}\phi'(t) &= \frac{d}{dt} E[e^{tX}] \\ &= E\left[\frac{d}{dt}(e^{tX})\right] \\ &= E[Xe^{tX}] \qquad (8.1)\end{aligned}$$

where we have assumed that the interchange of the differentiation and expectation operators is legitimate. That is, we have assumed that

$$\frac{d}{dt}\left[\sum_x e^{tx}p(x)\right] = \sum_x \frac{d}{dt}[e^{tx}p(x)]$$

in the discrete case, and

$$\frac{d}{dt}\left[\int e^{tx}f(x)\,dx\right] = \int \frac{d}{dt}[e^{tx}f(x)]\,dx$$

in the continuous case. This assumption can almost always be justified and, indeed, is valid for all of the distributions considered in this text. Hence, from Equation (8.1) we obtain, by evaluating at $t = 0$, that

$$\phi'(0) = E[X]$$

Similarly,

$$\begin{aligned}\phi''(t) &= \frac{d}{dt}\phi'(t) \\ &= \frac{d}{dt} E[Xe^{tX}] \\ &= E\left[\frac{d}{dt}(Xe^{tX})\right] \\ &= E[X^2 e^{tX}]\end{aligned}$$

and thus

$$\phi''(0) = E[X^2]$$

In general, the $n$th derivative of $\phi(t)$ is given by

$$\phi^n(t) = E[X^n e^{tX}] \qquad n \geq 1$$

implying that

$$\phi^n(0) = E[X^n] \qquad n \geq 1$$

We now compute $\phi(t)$ for some common distributions.

**Example 8a.** The Binomial Distribution with Parameters $n$ and $p$. If $X$ is a binomial random variable with parameters $n$ and $p$, then

$$\begin{aligned}\phi(t) &= E[e^{tX}]\\ &= \sum_{k=0}^{n} e^{tk}\binom{n}{k}p^k(1-p)^{n-k}\\ &= \sum_{k=0}^{n}\binom{n}{k}(pe^t)^k(1-p)^{n-k}\\ &= (pe^t+1-p)^n\end{aligned}$$

where the last equality follows from the binomial theorem.

$$\phi'(t) = n(pe^t+1-p)^{n-1}pe^t$$

and thus

$$E[X] = \phi'(0) = np$$

which checks with the result first obtained in Example 1c. Differentiating a second time yields

$$\phi''(t) = n(n-1)(pe^t+1-p)^{n-2}(pe^t)^2 + n(pe^t+1-p)^{n-1}pe^t$$

and so

$$E[X^2] = \phi''(0) = n(n-1)p^2+np$$

The variance of $X$ is given by

$$\begin{aligned}\text{Var}(X) &= E[X^2]-(E[X])^2\\ &= n(n-1)p^2+np-n^2p^2\\ &= np(1-p)\end{aligned}$$

verifying the result of Example 5a. ∎

**Example 8b.** The Poisson Distribution with Mean $\lambda$. If $X$ is a Poisson random variable with parameter $\lambda$, then

$$\begin{aligned}\phi(t) &= E[e^{tX}]\\ &= \sum_{n=0}^{\infty}\frac{e^{tn}e^{-\lambda}\lambda^n}{n!}\\ &= e^{-\lambda}\sum_{n=0}^{\infty}\frac{(\lambda e^t)^n}{n!}\\ &= e^{-\lambda}e^{\lambda e^t}\\ &= \exp\{\lambda(e^t-1)\}\end{aligned}$$

Differentiation yields

$$\phi'(t) = \lambda e^t \exp\{\lambda(e^t - 1)\}$$
$$\phi''(t) = (\lambda e^t)^2 \exp\{\lambda(e^t - 1)\} + \lambda e^t \exp\{\lambda(e^t - 1)\}$$

and thus

$$\begin{aligned} E[X] &= \phi'(0) = \lambda \\ E[X^2] &= \phi''(0) = \lambda^2 + \lambda \\ \text{Var}(X) &= E[X^2] - (E[X])^2 \\ &= \lambda \end{aligned}$$

Hence both the mean and the variance of the Poisson random variable equal $\lambda$. ▮

**Example 8c.** The Exponential Distribution with Parameter $\lambda$

$$\begin{aligned} \phi(t) &= E[e^{tX}] \\ &= \int_0^\infty e^{tx}\lambda e^{-\lambda x}\,dx \\ &= \lambda \int_0^\infty e^{-(\lambda - t)x}\,dx \\ &= \frac{\lambda}{\lambda - t} \qquad \text{for } t < \lambda \end{aligned}$$

We note from this derivation that for the exponential distribution, $\phi(t)$ is only defined for values of $t$ less than $\lambda$. Differentiation of $\phi(t)$ yields

$$\phi'(t) = \frac{\lambda}{(\lambda - t)^2} \qquad \phi''(t) = \frac{2\lambda}{(\lambda - t)^3}$$

Hence

$$E[X] = \phi'(0) = \frac{1}{\lambda} \qquad E[X^2] = \phi''(0) = \frac{2}{\lambda^2}$$

The variance of $X$ is given by

$$\begin{aligned} \text{Var}(X) &= E[X^2] - (E[X])^2 \\ &= \frac{1}{\lambda^2} \end{aligned}$$ ▮

**Example 8d.** The Normal Distribution. We first compute the moment generating function of a unit normal random variable with parameters

0 and 1. Letting $Z$ be such a random variable, we have

$$\begin{aligned}
\phi_Z(t) &= E[e^{tZ}] \\
&= \frac{1}{\sqrt{2\pi}} \int_{-\infty}^{\infty} e^{tx} e^{-x^2/2}\, dx \\
&= \frac{1}{\sqrt{2\pi}} \int_{-\infty}^{\infty} \exp\left\{ -\frac{(x^2 - 2tx)}{2} \right\} dx \\
&= \frac{1}{\sqrt{2\pi}} \int_{-\infty}^{\infty} \exp\left\{ -\frac{(x-t)^2}{2} + \frac{t^2}{2} \right\} dx \\
&= e^{t^2/2} \frac{1}{\sqrt{2\pi}} \int_{-\infty}^{\infty} e^{-(x-t)^2/2}\, dx \\
&= e^{t^2/2} \frac{1}{\sqrt{2\pi}} \int_{-\infty}^{\infty} e^{-y^2/2}\, dy \qquad \text{by the substitution } y = x - t \\
&= e^{t^2/2}
\end{aligned}$$

Hence the moment generating function of the unit normal random variable $Z$ is given by $\phi_Z(t) = e^{t^2/2}$. To obtain the moment generating function of an arbitrary normal random variable, we recall (see Section 3 of Chapter 5) that $X = \mu + \sigma Z$ will have a normal distribution with parameters $\mu$ and $\sigma^2$ whenever $Z$ is a unit normal random variable. Hence the moment generating function of such a random variable is given by

$$\begin{aligned}
\phi_X(t) &= E[e^{tX}] \\
&= E[e^{t(u+\sigma Z)}] \\
&= E[e^{t\mu} e^{t\sigma Z}] \\
&= e^{t\mu} E[e^{t\sigma Z}] \\
&= e^{t\mu} \phi_Z(t\sigma) \\
&= e^{t\mu} e^{(t\sigma)^2/2} \\
&= \exp\left\{ \frac{\sigma^2 t^2}{2} + \mu t \right\}
\end{aligned}$$

By differentiating, we obtain

$$\phi_X'(t) = (\mu + t\sigma^2) \exp\left\{ \frac{\sigma^2 t^2}{2} + \mu t \right\}$$

$$\phi_X''(t) = (\mu + t\sigma^2)^2 \exp\left\{ \frac{\sigma^2 t^2}{2} + \mu t \right\} + \sigma^2 \exp\left\{ \frac{\sigma^2 t^2}{2} + \mu t \right\}$$

and thus

$$\begin{aligned}
E[X] &= \phi'(0) = \mu \\
E[X^2] &= \phi''(0) = \mu^2 + \sigma^2
\end{aligned}$$

implying that

$$\begin{aligned}\text{Var}(X) &= E[X^2] - E([X])^2 \\ &= \sigma^2\end{aligned}$$ ▮

Tables 7.1 and 7.2 give the moment generating function for some common discrete and continuous distributions.

An important property of moment generating functions is that the moment generating function of the sum of independent random variables equals the product of the individual moment generating functions. To prove this, suppose that $X$ and $Y$ are independent and have moment generating functions $\phi_X(t)$ and $\phi_Y(t)$, respectively. Then $\phi_{X+Y}(t)$, the moment generating function of $X + Y$, is given by

$$\begin{aligned}\phi_{X+Y}(t) &= E[e^{t(X+Y)}] \\ &= E[e^{tX}e^{tY}] \\ &= E[e^{tX}]E[e^{tY}] \\ &= \phi_X(t)\phi_Y(t)\end{aligned}$$

where the next to the last equality follows from Proposition 5.1, since $X$ and $Y$ are independent.

Another important result is that the moment generating function uniquely determines the distribution. That is, if $\phi_X(t)$ exists and is finite in some region about $t = 0$, then the distribution of $X$ is uniquely determined. For instance, if $\phi_X(t) = (\frac{1}{2})^{10}(e^t + 1)^{10}$, then it follows from Table 7.1 that $X$ is a binomial random variable with parameters 10 and $\frac{1}{2}$.

**Example 8e.** Suppose the moment generating function of a random variable $X$ is given by $\phi(t) = e^{3(e^t-1)}$. What is $P\{X = 0\}$?

***Solution:*** We see from Table 7.1 that $\phi(t) = e^{3(e^t-1)}$ is the moment generating function of a Poisson random variable with mean 3. Hence, by the one-to-one correspondence between moment generating functions and distribution functions, it follows that $X$ must be a Poisson random variable with mean 3. Thus $P\{X = 0\} = e^{-3}$. ▮

**Example 8f.** Sums of Independent Binomial Random Variables. If $X$ and $Y$ are independent binomial random variables with parameters $(n, p)$ and $(m, p)$, respectively, what is the distribution of $X + Y$?

***Solution:*** The moment generating function of $X + Y$ is given by

$$\begin{aligned}\phi_{X+Y}(t) = \phi_X(t)\phi_Y(t) &= (pe^t + 1 - p)^n(pe^t + 1 - p)^m \\ &= (pe^t + 1 - p)^{m+n}\end{aligned}$$

**Table 7.1**

| *Discrete probability distribution* | *Probability mass function, $p(x)$* | *Moment generating function, $\phi(t)$* | *Mean* | *Variance* |
|---|---|---|---|---|
| Binomial with parameters $n$, $p$ | $\binom{n}{x} p^x(1-p)^{n-x}$ | $(pe^t + 1 - p)^n$ | $np$ | $np(1-p)$ |
| $0 \leq p \leq 1$ | $x = 0, 1, \ldots, n$ | | | |
| Poisson with parameter | $e^{-\lambda}\frac{\lambda^x}{x!}$ | $\exp\{\lambda(e^t - 1)\}$ | $\lambda$ | $\lambda$ |
| $\lambda > 0$ | $x = 0, 1, 2, \ldots$ | | | |
| Geometric with parameter | $p(1-p)^{x-1}$ | $\frac{pe^t}{1-(1-p)e^t}$ | $\frac{1}{p}$ | $\frac{1-p}{p^2}$ |
| $0 \leq p \leq 1$ | $x = 1, 2, \ldots$ | | | |
| Negative binomial with parameters $r$, $p$ | $\binom{n-1}{r-1} p^r(1-p)^{n-r}$ | $\left[\frac{pe^t}{1-(1-p)e^t}\right]^r$ | $\frac{r}{p}$ | $\frac{r(1-p)}{p^2}$ |
| $0 \leq p \leq 1$ | $n = r, r+1, \ldots$ | | | |

**Table 7.2**

| *Continuous probability distribution* | *Probability density function f(x)* | *Moment generating function, $\phi(t)$* | *Mean* | *Variance* |
|---|---|---|---|---|
| Uniform over $(a, b)$ | $f(x) = \begin{cases} \dfrac{1}{b-a} & a < x < b \\ 0 & \text{otherwise} \end{cases}$ | $\dfrac{e^{tb} - e^{ta}}{t(b-a)}$ | $\dfrac{a+b}{2}$ | $\dfrac{(b-a)^2}{12}$ |
| Exponential with parameter $\lambda > 0$ | $f(x) = \begin{cases} \lambda e^{-\lambda x} & x \geq 0 \\ 0 & x < 0 \end{cases}$ | $\dfrac{\lambda}{\lambda - t}$ | $\dfrac{1}{\lambda}$ | $\dfrac{1}{\lambda^2}$ |
| Gamma with parameters $(s, \lambda)$ $\lambda > 0$ | $f(x) = \begin{cases} \dfrac{\lambda e^{-\lambda x}(\lambda x)^{s-1}}{\Gamma(s)} & x \geq 0 \\ 0 & x < 0 \end{cases}$ | $\left(\dfrac{\lambda}{\lambda - t}\right)^s$ | $\dfrac{s}{\lambda}$ | $\dfrac{s}{\lambda^2}$ |
| Normal with parameters $(\mu, \sigma^2)$ | $f(x) = \dfrac{1}{\sqrt{2\pi}\sigma} e^{-(x-\mu)^2/2\sigma^2} \quad -\infty < x < \infty$ | $\exp\left\{\mu t + \dfrac{\sigma^2 t^2}{2}\right\}$ | $\mu$ | $\sigma^2$ |

However, $(pe^t + 1 - p)^{m+n}$ is the moment generating function of a binomial random variable having parameters $m + n$ and $p$. Thus this must be the distribution of $X + Y$. ∎

**Example 8g.** Sums of Independent Poisson Random Variables. Calculate the distribution of $X + Y$ when $X$ and $Y$ are independent Poisson random variables with means $\lambda_1$ and $\lambda_2$, respectively.

***Solution***

$$\begin{aligned}\phi_{X+Y}(t) &= \phi_X(t)\phi_Y(t)\\ &= \exp\{\lambda_1(e^t - 1)\}\exp\{\lambda_2(e^t - 1)\}\\ &= \exp\{(\lambda_1 + \lambda_2)(e^t - 1)\}\end{aligned}$$

Hence $X + Y$ is Poisson distributed with mean $\lambda_1 + \lambda_2$, verifying the result given in Example 3c of Chapter 6. ∎

**Example 8h.** Sums of Independent Normal Random Variables. Show that if $X$ and $Y$ are independent normal random variables with parameters $(\mu_1, \sigma_1^2)$ and $(\mu_2, \sigma_2^2)$, respectively, then $X + Y$ is normal with mean $\mu_1 + \mu_2$ and variance $\sigma_1^2 + \sigma_2^2$.

***Solution***

$$\begin{aligned}\phi_{X+Y}(t) &= \phi_X(t)\phi_Y(t)\\ &= \exp\left\{\frac{\sigma_1^2 t^2}{2} + \mu_1 t\right\}\exp\left\{\frac{\sigma_2^2 t^2}{2} + \mu_2 t\right\}\\ &= \exp\left\{\frac{(\sigma_1^2 + \sigma_2^2)t^2}{2} + (\mu_1 + \mu_2)t\right\}\end{aligned}$$

which is the moment generating function of a normal random variable with mean $\mu_1 + \mu_2$ and variance $\sigma_1^2 + \sigma_2^2$. Hence the result follows because the moment generating function uniquely determines the distribution. ∎

**Example 8i.** The Moment Generating Function of the Sum of a Random Number of Random Variables. Let $X_1, X_2, \ldots$ be a sequence of independent and identically distributed random variables, and let $N$ be a nonnegative, integer-valued random variable that is independent of the sequence $X_i$, $i \geq 1$. We want to compute the moment generating function of

$$Y = \sum_{i=1}^{N} X_i$$

(In Example 6d, $Y$ was interpreted as the amount of money spent in a store on a given day when both the amount spent by a customer and the number of such customers are random variables.)

To compute the moment generating function of $Y$, we first condition on $N$ as follows:

$$\begin{aligned} E[e^{t\sum_1^N X_i} | N = n] &= E[e^{t\sum_1^n X_i} | N = n] \\ &= E[e^{t\sum_1^n X_i}] \\ &= (\phi_X(t))^n \end{aligned}$$

where

$$\phi_X(t) = E[e^{tX_i}]$$

Hence

$$E[e^{tY} | N] = (\phi_X(t))^N$$

and thus

$$\phi_Y(t) = E[(\phi_X(t))^N]$$

The moments of $Y$ can now be obtained upon differentiation, as follows:

$$\phi'_Y(t) = E[N(\phi_X(t))^{N-1}\phi'_X(t)]$$

and so

$$\begin{aligned} E[Y] &= \phi'_Y(0) \\ &= E[N(\phi_X(0))^{N-1}\phi'_X(0)] \\ &= E[NEX] \\ &= E[N]E[X] \end{aligned} \tag{8.2}$$

verifying the result of Example 6d. (In this last set of equalities we have used the fact that $\phi_X(0) = E[e^{0X}] = 1$.)

Also,

$$\phi''_Y(t) = E[N(N-1)(\phi_X(t))^{N-2}(\phi'_X(t))^2 + N(\phi_X(t))^{N-1}\phi''_X(t)]$$

and so

$$\begin{aligned} E[Y^2] &= \phi''_Y(0) \\ &= E[N(N-1)(E[X])^2 + NE[X^2]] \\ &= (E[X])^2(E[N^2] - E[N]) + E[N]E[X^2] \\ &= E[N](E[X^2] - (E[X])^2) + (E[X])^2E[N^2] \\ &= E[N] \operatorname{Var}(X) + (E[X])^2E[N^2] \end{aligned} \tag{8.3}$$

Hence, from Equations (8.2) and (8.3), we see that

$$\begin{aligned}\operatorname{Var}(Y) &= E[N]\operatorname{Var}(X) + (E[X])^2(E[N^2] - (E[N])^2) \\ &= E[N]\operatorname{Var}(X) + (E[X])^2 \operatorname{Var}(N)\end{aligned}$$ ∎

It is also possible to define the joint moment generating function of two or more random variables. This is done as follows. For any $n$ random variables $X_1, \ldots, X_n$, the joint moment generating function, $\phi(t_1, \ldots, t_n)$, is defined for all real values of $t_1, \ldots, t_n$ by

$$\phi(t_1, \ldots, t_n) = E[e^{t_1X_1 + \cdots + t_nX_n}]$$

The individual moment generating functions can be obtained from $\phi(t_1, \ldots, t_n)$ by letting all but one of the $t_j$ be 0. That is,

$$\phi_{X_i}(t) = E[e^{tX_i}] = \phi(0, \ldots, 0, t, 0, \ldots, 0)$$

where the $t$ is in the $i$th place.

It can be proved (although the proof is too advanced for this text) that $\phi(t_1, \ldots, t_n)$ uniquely determines the joint distribution of $X_1, \ldots, X_n$. This result can then be used to prove that the $n$ random variables $X_1, \ldots, X_n$ are independent if and only if

$$\phi(t_1, \ldots, t_n) = \phi_{X_1}(t_1) \cdots \phi_{X_n}(t_n) \tag{8.4}$$

This follows because, if the $n$ random variables are independent, then

$$\begin{aligned}\phi(t_1, \ldots, t_n) &= E[e^{(t_1X_1 + \cdots + t_nX_n)}] \\ &= E[e^{t_1X_1} \cdots e^{t_nX_n}] \\ &= E[e^{t_1X_1}] \cdots E[e^{t_nX_n}] \qquad \text{by independence} \\ &= \phi_{X_1}(t_1) \cdots \phi_{X_n}(t_n)\end{aligned}$$

On the other hand, if Equation (8.4) is satisfied, then the joint moment generating function $\phi(t_1, \ldots, t_n)$ is the same as the joint moment generating function of $n$ independent random variables, the $i$th of which has the same distribution as $X_i$. As the joint moment generating function uniquely determines the joint distribution, this must be the joint distribution; hence the random variables are independent.

**Example 8j.** The Multivariate Normal Distribution. Let $Z_1, \ldots, Z_n$ be a set of $n$ independent unit normal random variables. If, for some constants $a_{ij}$, $1 \leq i \leq m$, $1 \leq j \leq n$, and $\mu_i$, $1 \leq i \leq m$,

$$\begin{aligned}X_1 &= a_{11}Z_1 + \cdots + a_{1n}Z_n + \mu_1 \\ X_2 &= a_{21}Z_1 + \cdots + a_{2n}Z_n + \mu_2 \\ &\vdots \\ X_i &= a_{i1}Z_1 + \cdots + a_{in}Z_n + \mu_i \\ &\vdots \\ X_m &= a_{m1}Z_1 + \cdots + a_{mn}Z_n + \mu_m\end{aligned}$$

then the random variables $X_1, \ldots, X_m$ are said to have a multivariate normal distribution.

It follows from the fact that the sum of independent normal random variables is itself a normal random variable (see Example 8h) that each $X_i$ is a normal random variable with mean and variance given by

$$E[X_i] = \mu_i$$

$$\text{Var}(X_i) = \sum_{j=1}^{n} a_{ij}^2$$

The covariance of $X_i$ and $X_j$ is given by

$$\begin{aligned}
\text{Cov}(X_i, X_j) &= \text{Cov}\left(\mu_i + \sum_{k=1}^{n} a_{ik} Z_k, \mu_j + \sum_{\ell=1}^{n} a_{j\ell} Z_\ell\right) \\
&= \text{Cov}\left(\sum_{k=1}^{n} a_{ik} Z_k, \sum_{\ell=1}^{n} a_{j\ell} Z_\ell\right) \\
&= \sum_{k,\ell} a_{ik} a_{j\ell} \text{Cov}(Z_k, Z_\ell) \\
&= \sum_{k=1}^{n} a_{ik} a_{jk}
\end{aligned}$$

since

$$\text{Cov}(Z_k, Z_\ell) = \begin{cases} 1 & \text{if } k = \ell \\ 0 & \text{if } k \neq \ell \end{cases}$$

The joint moment generating function is given by

$$\phi(t_1, \ldots, t_m) = E[e^{(t_1 X_1 + \cdots + t_m X_m)}]$$

Now,

$$\begin{aligned}
t_1 X_1 + \cdots + t_m X_m = &(a_{11} t_1 + a_{21} t_2 + \cdots + a_{m1} t_m) Z_1 \\
&+ (a_{12} t_1 + a_{22} t_2 + \cdots + a_{m2} t_m) Z_2 \\
&+ \\
&\vdots \\
&+ (a_{1n} t_1 + a_{2n} t_2 + \cdots + a_{mn} t_m) Z_n \\
&+ \mu_1 t_1 + \mu_2 t_2 + \cdots + \mu_m t_m
\end{aligned}$$

and thus

$$\sum_{i=1}^{m} t_i X_i$$

has a normal distribution with mean

$$E\left[\sum_{i=1}^{m} t_i X_i\right] = \sum_{i=1}^{m} t_i \mu_i$$

and variance

$$\text{Var}\left(\sum_{i=1}^{m} t_1 X_i\right) = \sum_{k=1}^{n}\left(\sum_{i=1}^{m} a_{ik} t_i\right)^2$$

Hence, by using the fact that if $Y$ is a normal random variable with mean $\mu$ and variance $\sigma^2$, then

$$E[e^Y] = \phi_Y(t)\big|_{t=1} = e^{\mu+\sigma^2/2}$$

we obtain that

$$\begin{aligned}\phi(t_1, \ldots, t_m) &= E\left[\exp\left\{\sum_1^m t_i X_i\right\}\right] \\ &= \exp\left\{\sum_{i=1}^{m} t_i \mu_i + \frac{1}{2}\sum_{k=1}^{n}\left(\sum_{i=1}^{m} a_{ik} t_i\right)^2\right\}\end{aligned}$$

Now,

$$\begin{aligned}\sum_{k=1}^{n}\left(\sum_{i=1}^{m} a_{ik} t_i\right)^2 &= \sum_{k=1}^{n}\sum_{i=1}^{m} a_{ik} t_i \sum_{j=1}^{m} a_{jk} t_j \\ &= \sum_{j=1}^{m}\sum_{i=1}^{m} t_i t_j \sum_{k=1}^{n} a_{ik} a_{jk} \\ &= \sum_{j=1}^{m}\sum_{i=1}^{m} t_i t_j \,\text{Cov}(X_i, X_j)\end{aligned}$$

and thus $\phi(t_1, \ldots, t_m)$ can be expressed as

$$\phi(t_1, \ldots, t_m) = \exp\left\{\sum_{i=1}^{m} t_i \mu_i + \frac{1}{2}\sum_{j=1}^{m}\sum_{i=1}^{m} t_i t_j \,\text{Cov}(X_i, X_j)\right\}$$

which shows that the joint distribution of $X_1, \ldots, X_m$ is completely determined from a knowledge of the values of $\mu_i = E[X_i]$ and $\text{Cov}(X_i, X_j)$, $i, j = 1, \ldots, m$.

## 9 General Definition of Expectation

Up to this point we have defined expectations only for discrete and continuous random variables. However, there also exist random variables that are neither discrete nor continuous, and they too may possess an expectation. As an example of such a random variable, let $X$ be a Bernoulli random variable with parameter $p = \frac{1}{2}$, and let $Y$ be a uniformly distributed random variable over the interval [0, 1]. Furthermore, suppose that $X$ and

$Y$ are independent and define the new random variable $W$ by

$$W = \begin{cases} X & \text{if } X = 1 \\ Y & \text{if } X \neq 1 \end{cases}$$

Clearly, $W$ is neither a discrete (since its set of possible values $[0, 1]$ is uncountable) nor a continuous (since $P\{W = 1\} = \frac{1}{2}$) random variable.

In order to define the expectation of an arbitrary random variable, we require the notion of a Stieltjes integral. Before defining the Stieltjes integral, let us recall that for any function $g$, $\int_b^a g(x)\,dx$ is defined by

$$\int_a^b g(x)\,dx = \lim \sum_{i=1}^{n} g(x_i)(x_i - x_{i-1})$$

where the limit is taken over all $a = x_0 < x_1 < x_2 \cdots < x_n = b$ as $n \to \infty$ and $\max_{i=1,\cdots,n} (x_i - x_{i-1}) \to 0$.

For any distribution function $F$, we define the Stieltjes integral of the nonnegative function $g$ over the interval $[a, b]$ by

$$\int_a^b g(x)\,dF(x) = \lim \sum_{i=1}^{n} g(x_i)[F(x_i) - F(x_{i-1})]$$

where, as before, the limit is taken over all $a = x_0 < x_1 < \cdots < x_n = b$ as $n \to \infty$ and $\max_{i=1,\ldots,n} (x_i - x_{i-1}) \to 0$. Further, we define the Stieltjes integral over the whole real line by

$$\int_{-\infty}^{\infty} g(x)\,dF(x) = \lim_{\substack{a \to -\infty \\ b \to +\infty}} \int_a^b g(x)\,dF(x)$$

Finally, if $g$ is not a nonnegative function, we define $g^+$ and $g^-$ by

$$g^+(x) = \begin{cases} g(x) & \text{if } g(x) \geq 0 \\ 0 & \text{if } g(x) < 0 \end{cases}$$

$$g^-(x) = \begin{cases} 0 & \text{if } g(x) \geq 0 \\ -g(x) & \text{if } g(x) < 0 \end{cases}$$

As $g(x) = g^+(x) - g^-(x)$ and $g^+$ and $g^-$ are both nonnegative functions, it is natural to define

$$\int_{-\infty}^{\infty} g(x)\,dF(x) = \int_{-\infty}^{\infty} g^+(x)\,dF(x) - \int_{-\infty}^{\infty} g^-(x)\,dF(x)$$

and we say that $\int_{-\infty}^{\infty} g(x)\,dF(x)$ exists as long as $\int_{-\infty}^{\infty} g^+(x)\,dF(x)$ and $\int_{-\infty}^{\infty} g^-(x)\,dF(x)$ are not both equal to $+\infty$.

If $X$ is an arbitrary random variable having cumulative distribution $F$, we define the expected value of $X$ by

$$E[X] = \int_{-\infty}^{\infty} x\, dF(x) \tag{9.1}$$

It can be shown that if $X$ is a discrete random variable with mass function $p(x)$, then

$$\int_{-\infty}^{\infty} x\, dF(x) = \sum_{x:p(x)>0} xp(x)$$

whereas, if $X$ is a continuous random variable with density function $f(x)$, then

$$\int_{-\infty}^{\infty} x\, dF(x) = \int_{-\infty}^{\infty} xf(x)\, dx$$

The reader should note that Equation (9.1) yields an intuitive definition of $E[X]$, for consider the approximating sum

$$\sum_{i=1}^{n} x_i[F(x_i) - F(x_{i-1})]$$

of $E[X]$. As $F(x_i) - F(x_{i-1})$ is just the probability that $X$ will be in the interval $(x_{i-1}, x_i]$, the approximating sum multiplies the approximate value of $X$ when it is in the interval $(x_{i-1}, x_i]$ by the probability that it will be in that interval and then sums over all the intervals. Clearly, as these intervals get smaller and smaller in length, we obtain the "expected value" of $X$.

Stieltjes integrals are mainly of theoretical interest because they yield a compact way of defining and dealing with the properties of expectation. For instance, use of Stieltjes integrals avoids the necessity of having to give separate statements and proofs of theorems for the continuous and the discrete cases. However, their properties are very much the same as those of ordinary integrals, and all of the proofs presented in this chapter can easily be translated into proofs in the general case.

## Theoretical Exercises

1. Prove Proposition 2.1(a).
2. There are some textbooks that "define" $E[g(X)] = \int_{-\infty}^{\infty} g(x) f_X(x)\, dx$. However, as $g(X)$ is itself a random variable, it is not clear that the above "definition" is logically consistent. For, it also follows from this "definition" that $E[g(X)] = \int_{-\infty}^{\infty} y f_{g(X)}(y)\, dy$, and it is not a priori

evident that the two expressions for $E[g(X)]$ need be identical. That they are identical is, of course, the substance of the law of the unconscious statistician. Suppose now that we "define" for any continuous random variable $X$ and real-valued function $g$, the $S$-pectation of $g(X)$ by

$$S[g(X)] = \int_{-\infty}^{\infty} g(x)(f_X(x))^2 \, dx$$

Show, by letting $X$ be a uniform (0, 1) random variable and by computing $S(X^2)$ in two different ways, that this "definition" of $S$-pectation is logically inconsistent (that is, it leads to contradictions).

**3.** For a nonnegative integer-valued random variable $N$ show that

$$\sum_{i=0}^{\infty} iP\{N > i\} = \tfrac{1}{2}(E[N^2] - E[N])$$

HINT: $\sum_i iP\{N > i\} = \sum_i i \sum_{k=i+1}^{\infty} P\{N = k\}$. Now interchange the order of summation.

**4.** Prove, for nonnegative random variables $X$, that

$$E[X^n] = \int_0^{\infty} nx^{n-1}(1 - F(x)) \, dx$$

**5.** Show that $E[(X - a)^2]$ is minimized at $a = E[X]$.

**6.** Suppose that $X$ is a continuous random variable with density function $f$. Show that $E[|X - a|]$ is minimized when $a$ is equal to the median of $F$.

HINT: Write

$$E[|X - a|] = \int |x - a| f(x) \, dx$$

Now break up the integral into the regions where $x < a$ and where $x > a$ and differentiate.

**7.** Prove the two-dimensional analog of Proposition 2.1 when (a) $X$ and $Y$ have a joint probability mass function, and (b) $X$ and $Y$ have a joint probability density function and $g(x, y) \geq 0$ for all $x, y$.

**8.** Let $X$ be a random variable having finite expectation $\mu$ and variance $\sigma^2$, and let $g(\cdot)$ be a twice differentiable function. Show that

$$E[g(X)] \approx g(\mu) + \frac{g''(\mu)}{2}\sigma^2$$

HINT: Expand $g(\cdot)$ in a Taylor series about $\mu$. Use the first three terms and ignore the remainder.

**9.** A coin having probability $p$ of landing heads is flipped $n$ times. Compute the expected number of runs of heads of size 1, of size 2, of size $k$, $1 \leq k \leq n$.

**10.** Let $X_1, X_2, \ldots, X_n$ be independent and identically distributed positive random varialbes. Find, for $k$   $n$,

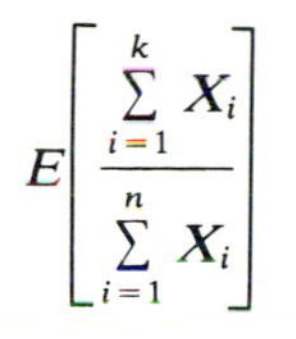

$$E\left[\frac{\sum_{i=1}^{k} X_i}{\sum_{i=1}^{n} X_i}\right]$$

**11.** Consider $n$ independent trials each resulting in any one of $r$ possible outcomes with probabilities $P_1, P_2, \ldots, P_r$. Let $X$ denote the number of outcomes that never occur in any of the trials. Find $E[X]$ and show that among all probability vectors $P_1, \ldots, P_r$ $E[X]$ is minimized when $P_i = 1/r$, $i = 1, \ldots, r$.

**12.** Based on your knowledge of the variance of a binomial random variable and the relationship between Poisson and binomial random variables, what would you guess would be the variance of a Poisson random variable with parameter $\lambda$? Verify this analytically.

**13.** Compute the variance of a geometric random variable having mean $1/p$.

**14.** Compute the variance of an exponential random variable having mean $1/\lambda$.

**15.** Compute the variance of a random variable uniformly distributed over $(a, b)$.

**16.** Compute the mean and variance of a beta random variable with parameters $(a, b)$.

**17.** Compute the mean and variance of a gamma random variable with parameters $(t, \lambda)$.

**18.** From a set of $n$ elements a nonempty subset is chosen at random in the sense that all of the nonempty subsets are equally likely to be selected. Let $X$ denote the number of elements in the chosen subset. Using the identities given in Theoretical Exercise 13 of Chapter 1, show that

$$E[X] = \frac{n}{2 - \left(\frac{1}{2}\right)^{n-1}}$$

$$\text{Var}(X) = \frac{n2^{2n-2} - n(n+1)2^{n-2}}{(2^n - 1)^2}$$

Show also that for $n$ large

$$\text{Var}(X) \sim \frac{n}{4}$$

in the sense that the ratio of the above approaches 1 as $n$ approaches $\infty$. Compare this with the limiting form of Var ($Y$) when $P\{Y = i\} = 1/n$, $i = 1, \ldots, n$.

**19.** Independent trials are performed. If the $i$th such trial results in a success with probability $P_i$, compute (a) the expected number, and (b) the variance, of the number of successes that occur in the first $n$ trials. Does independence make a difference in part (a)? In part (b)?

**20.** Let $X_1, \ldots, X_n$ be independent and identically distributed random variables having mean $\mu$ and variance $\sigma^2$, and let $\bar{X} = \sum_{i=1}^{n} X_i/n$. Show that:

(a) $E[\bar{X}] = \mu$; (b) $\text{Var}(\bar{X}) = \dfrac{\sigma^2}{n}$;

(c) $E\left[\sum_{i=1}^{n} (X_i - \bar{X})^2\right] = (n-1)\sigma^2$.

**21.** Let $X_1, \ldots, X_n$ be independent and identically distributed, continuous random variables. We say that a record value occurs at time $j$, $j \leq n$, if $X_j \geq X_i$ for all $1 \leq i \leq j$. Show that

(a) $E[\text{number of record values}] = \sum_{j=1}^{n} 1/j$.

(b) $\text{Var}(\text{number of record values}) = \sum_{j=1}^{n} (j-1)/j^2$

**22.** For Example 3g show that the variance of the number of coupons needed to amass a full set is equal to

$$\sum_{i=1}^{N-1} \frac{iN}{(N-i)} 2$$

When $N$ is large, this can be shown to be approximately equal (in the sense that their ratio approaches 1 as $n \to \infty$) to $N^2\pi^2/6$.

**23.** Consider $n$ independent trials, the $i$th of which results in a success with probability $P_i$.

(a) Compute the expected number of successes in the $n$ trials—call it $\mu$.
(b) For fixed value of $\mu$, what choice of $P_1, \ldots, P_n$ maximizes the variance of the number of successes?
(c) What choice minimizes the variance?

**24.** Verify Equation (5.1).

**25.** Suppose that balls are randomly removed from an urn initially containing $n$ white and $m$ black balls. It was shown in Example 3j that $E[X] = 1 + m/(n + 1)$, when $X$ is the number of draws needed to obtain a white ball.

(a) Compute Var $(X)$.

(b) Show that the expected number of balls that need be drawn to amass a total of $k$ white balls is $k[1 + m/(n + 1)]$.

HINT: Let $Y_i$, $i = 1, \ldots, n + 1$, denote the number of black balls withdrawn after the $(i - 1)$st white ball and before the $i$th white ball. Argue that the $Y_i$, $i = 1, \ldots, n + 1$, are identically distributed.

**26.** Compute the variance of a negative binomial random variable.

**27.** Suppose that $X_1$ and $X_2$ are independent random variables having a common mean $\mu$. Suppose also that Var $(X_1) = \sigma_1^2$ and Var $(X_2) = \sigma_2^2$. The value of $\mu$ is unknown and it is proposed to estimate $\mu$ by a weighted average of $X_1$ and $X_2$. That is, $\lambda X_1 + (1 - \lambda)X_2$ will be used as an estimate of $\mu$, for some appropriate value of $\lambda$. Which value of $\lambda$ yields the estimate having the lowest possible variance? Explain why it is desirable to use this value of $\lambda$.

**28.** Prove

(a) $\text{Cov}(a + bX, c + dY) = bd\,\text{Cov}(X, Y)$;

(b) $\text{Cov}(X + Y, Z) = \text{Cov}(X, Z) + \text{Cov}(Y, Z)$;

(c) $\text{Cov}(\sum_1^n X_i, \sum_1^m Y_i) = \sum_{j=1}^m \sum_{i=1}^n \text{Cov}(X_i, Y_j)$.

**29.** In Example 5e we showed that the covariance of the multinomial random variables $N_i$ and $N_j$ is equal to $-mP_iP_j$ by expressing $N_i$ and $N_j$ as the sum of indicator variables. This result could also have been obtained by using the formula

$$\text{Var}(N_i + N_j) = \text{Var}(N_i) + \text{Var}(N_j) + 2\text{Cov}(N_i,N_j)$$

(a) What is the distribution of $N_i + N_j$?

(b) Use the above identity to show that $\text{Cov}(N_i,N_j) = -mP_iP_j$

**30.** If $X$ and $Y$ are identically distributed, not necessarily independent, show

$$\text{Cov}(X + Y, X - Y) = 0$$

**31.** *The Conditional Covariance Formula.* The conditional covariance of $X$ and $Y$, given $Z$, is defined by

$$\text{Cov}(X,Y|Z) = E[(X - E[X|Z])(Y - E[Y|Z])|Z]$$

(a) Show that

$$\text{Cov}(X,Y|Z) = E[XY|Z] - E[X|Z]E[Y|Z]$$

(b) Prove the conditional covariance formula

$$\text{Cov}(X,Y) = E[\text{Cov}(X,Y|Z)] + \text{Cov}(E[X|Z], E[Y|Z])$$

(c) Set $X = Y$ in (b) and obtain the conditional variance formula.

**32.** Let $X_{(i)}$, $i = 1, \ldots, n$ denote the order statistics from a set of $n$ uniform (0,1) random variables and note that the density function of $X_{(i)}$ is given by

$$f(x) = \frac{n!}{(i-1)!(n-i)!} x^{i-1}(1-x)^{n-i}, \; 0 < x < 1$$

(a) Compute $\text{Var}(X_{(i)})$, $i = 1, \ldots, n$.
(b) Which value of $i$ minimizes and which value maximizes $\text{Var}(X_{(i)})$?

**33.** If $Y = a + bX$, show that

$$\rho(X, Y) = \begin{cases} +1 & \text{if } b > 0 \\ -1 & \text{if } b < 0 \end{cases}$$

**34.** If $Z$ is a unit normal random variable and if $Y$ is defined by $Y = a + bZ + cZ^2$, show that

$$\rho(Y, Z) = \frac{b}{\sqrt{b^2 + 2c^2}}$$

**35.** Prove the Cauchy-Schwarz inequality, namely, that

$$(E[XY])^2 \le E[X^2]E[Y^2]$$

HINT: Unless $Y = -tX$ for some constant, in which case this inequality holds with equality, it follows that for all $t$

$$0 < E[(tX + Y)^2] = E[X^2]t^2 + 2E[XY]t + E[Y^2]$$

Hence the roots of the quadratic equation

$$E[X^2]t^2 + 2E[XY]t + E[Y^2] = 0$$

must be imaginary, which implies that the discriminant of this quadratic equation must be negative.

**36.** Show that if $X$ and $Y$ are independent, then

$$E[X \mid Y = y] = E[X] \qquad \text{for all } y$$

(a) in the discrete case, and (b) in the continuous case.

**37.** Prove that $E[g(X)Y \mid X] = g(X)E[Y \mid X]$.

**38.** Prove that if $E[Y|X = x] = E[Y]$ for all $x$, then $X$ and $Y$ are uncorrelated, and give a counterexample to show that the converse is not true.

HINT: Prove and use the fact that $E[XY] = E[XE[Y|X]]$.

**39.** Show $\text{Cov}(X, E[Y|X]) = \text{Cov}(X, Y)$.

**40.** Let $X_1, \ldots, X_n$ be independent and identically distributed random variables. Find

$$E[X_1 | X_1 + \cdots + X_n = x]$$

**41.** Consider Example 5e, which is concerned with the multinomial distribution. Use conditional expectation to compute $E[N_i N_j]$ and then use this to verify the formula for $\text{Cov}(N_i, N_j)$ given in Example 5e.

**42.** An urn initially contains $b$ black and $w$ white balls. At each stage we add $r$ black balls and then withdraw, at random, $r$ from the $b + w + r$. Show that $E[\text{number of white balls after stage } t] = \left(\dfrac{b + w}{b + w + r}\right)^t w$.

**43.** Prove Equation (6.1b).

**44.** A coin, which lands on heads with probability $p$, is continually flipped. Compute the expected number of flips that are made until a string of $r$ heads in a row is obtained.

HINT: Condition on the time of the first occurrence of tails, to obtain the equation

$$E[X] = (1 - p) \sum_{i=1}^{r} p^{i-1}(i + E[X]) + (1 - p) \sum_{i=r+1}^{\infty} p^{i-1} r$$

Simplify and solve for $E[X]$.

**45.** One ball at a time is randomly selected from an urn containing $a$ white and $b$ black balls until all of the remaining balls are of the same color. Let $M_{a,b}$ denote the expected number of balls left in the urn when the experiment ends. Compute a recursive formula for $M_{a,b}$ and solve when $a = 3$, $b = 5$.

**46.** An urn contains $a$ white and $b$ black balls. After a ball is drawn, it is returned to the urn if it is white; but if it is black, it is replaced by a white ball from another urn. Let $M_n$ denote the expected number of white balls in the urn after the foregoing operation has been repeated $n$ times.

(a) Derive the recursive equation

$$M_{n+1} = \left(1 - \frac{1}{a + b}\right) M_n + 1$$

(b) Use part (a) to prove that

$$M_n = a + b - b\left(1 - \frac{1}{a+b}\right)^n$$

(c) What is the probability that the $(n + 1)$st ball drawn is white?

**47.** The best linear predictor of $Y$ with respect to $X_1$ and $X_2$ is equal to $a + bX_1 + cX_2$ where $a$, $b$, and $c$ are chosen to minimize

$$E[(Y - (a + bX_1 + cX_2))^2].$$

Determine $a$, $b$, and $c$.

**48.** The best quadratic predictor of $Y$ with respect to $X$ is $a + bX + cX^2$, where $a$, $b$, and $c$ are chosen to minimize $E[(Y - (a + bX + cX^2))^2]$. Determine $a$, $b$, and $c$.

**49.** If $X$ and $Y$ are jointly normally distributed with joint density function given by

$$f(x, y) = \frac{1}{2\pi\sigma_x\sigma_y\sqrt{1-\rho^2}} \times \exp\left\{-\frac{1}{2(1-\rho^2)}\left[\left(\frac{x-\mu_x}{\sigma_x}\right)^2 + \left(\frac{y-\mu_y}{\sigma_y}\right)^2 - 2\rho\frac{(x-\mu_x)(y-\mu_y)}{\sigma_x\sigma_y}\right]\right\}$$

(a) show that the conditional distribution of $Y$, given $X = x$, is normal with mean $\mu_y + \rho\dfrac{\sigma_y}{\sigma_x}(x - \mu_x)$ and variance $\sigma_2^2(1 - \rho^2)$;
(b) show that Corr $(X, Y) = \rho$;
(c) argue that $X$ and $Y$ are independent if and only if $\rho = 0$.

**50.** Let $X$ be a normal random variable with parameters $\mu = 0$ and $\sigma^2 = 1$ and let $I$, independent of $X$, be such that $P\{I = 1\} = \frac{1}{2} = P\{I = 0\}$. Now define $Y$ by

$$Y = \begin{cases} X & \text{if } I = 1 \\ -X & \text{if } I = 0 \end{cases}$$

In words, $Y$ is equally likely to equal either $X$ or $-X$.
(a) Are $X$ and $Y$ independent?
(b) Are $I$ and $Y$ independent?
(c) Show that $Y$ is normal with mean 0 and variance 1.
(d) Show that Cov $(X, Y) = 0$.

(e) Do (a), (c), and (d) contradict the results of Theoretical Exercise 42, which implies that uncorrelated jointly normal random variables are independent?

**51.** It follows from Proposition 7.1 and the fact that the best linear predictor of $Y$ with respect to $X$ is $\mu_y + \rho\frac{\sigma_y}{\sigma_x}(X - \mu_x)$ that if

$$E[Y|X] = a + bX$$

then

$$a = \mu_y - \rho\frac{\sigma_y}{\sigma_x}\mu_x \qquad b = \rho\frac{\sigma_y}{\sigma_x}$$

(Why?) Verify this directly.

**52.** For random variables $X$ and $Z$ show that

$$E[(X - Y)^2] = E[X^2] - E[Y^2]$$

where

$$Y = E[X|Z]$$

**53.** Consider a population consisting of individuals able to produce offspring of the same kind. Suppose that each individual will, by the end of its lifetime have produced $j$ new offspring with probability $P_j$, $j \geq 0$, independently of the number produced by any other individual. The number of individuals initially present, denoted by $X_0$, is called the size of the 0th generation. All offspring of the 0th generation constitute the first generation, and their number is denoted by $X_1$. In general, let $X_n$ denote the size of the $n$th generation. Let $\mu = \sum_{j=0}^{\infty} jP_j$ and $\sigma^2 = \sum_{j=0}^{\infty} (j - \mu)^2 P_j$ denote, respectively, the mean and the variance of the number of offspring produced by a single individual.

Suppose that $X_0 = 1$—that is, initially there is a single individual in the population.

(a) Show that

$$E[X_n] = \mu E[X_{n-1}]$$

(b) Use (a) to conclude that

$$E[X_n] = \mu^n$$

(c) Show that

$$\text{Var}(X_n) = \sigma^2\mu^{n-1} + \mu^2\,\text{Var}(X_{n-1})$$

(d) Use (c) to conclude that

$$\text{Var}(X_n) = \begin{cases} \sigma^2\mu^{n-1}\left(\dfrac{\mu^n - 1}{\mu - 1}\right) & \text{if } \mu \neq 1 \\ n\sigma^2 & \text{if } \mu = 1 \end{cases}$$

The above is known as a "branching process," and an important question for a population that evolves along such lines is the probability that the population will eventually die out. Let $\pi$ denote this probability when the population starts with a single individual. That is,

$$\pi = P\{\text{population eventually dies out} \mid X_0 = 1)$$

(e) Argue that $\pi$ satisfies

$$\pi = \sum_{j=0}^{\infty} P_j \pi^j$$

HINT: Condition on the number of offspring of the initial member of the population.

**54.** Verify the formula for the moment generating function of a uniform random variable that is given in Table 7.2. Also, differentiate to verify the formulas for the mean and variance.

**55.** If $Y = aX + b$, where $a$ and $b$ are constants, express the moment generating function of $Y$ in terms of the moment generating function of $X$.

**56.** Let $X$ have moment generating function $\phi(t)$, and define $\Psi(t) = \log \phi(t)$. Show that

$$\Psi''(t)|_{t=0} = \text{Var}(X)$$

**57.** Use Table 7.2 to determine the distribution of $\sum_{i=1}^{n} X_i$ when $X_1, \ldots, X_n$ are independent and identically distributed exponential random variables, each having mean $1/\lambda$.

**58.** Show how to compute $\text{Cov}(X, Y)$ from the joint moment generating function of $X$ and $Y$.

**59.** Suppose that $X_1, \ldots, X_n$ have a multivariate normal distribution. Show that $X_1, \ldots, X_n$ are independent random variables if and only if

$$\text{Cov}(X_i, X_j) = 0 \qquad \text{when } i \neq j$$

**60.** If $Z$ is a unit normal random variable what is $\text{Cov}(Z, Z^2)$?

## Problems

**1.** A man aiming at a target receives 10 points if his shot is within 1 inch of the target, 5 points if it is between 1 and 3 inches from the target, and 3 points if it is between 3 and 5 inches from the target. Find the expected number of points scored if:

(a) The man's shot is uniformly distributed in a circle of radius 8 inches centered at the target.

(b) The vertical and horizontal distances of the man's shot from the target are (in inches) independent and identically distributed normal random variables with parameters $\mu = 0$ and $\sigma^2 = 4$.

**2.** A popular dice game in British pubs is the game of chuck-a-luck, which is played by the house rolling 3 dice. A player may wager on any of the outcomes 1 through 6. If exactly 1 of the three dice shows that outcome, the player wins at even odds; if 2 dice show that outcome, the payoff is 2 to 1; if all 3 dice show the player's choice, the payoff is 3 to 1; if none of the dice shows the players choice, the player loses. Compute the expected winnings if a player wagers 1 unit on this game.

**3.** A bin of 5 electrical components is known to contain 2 that are defective. If the components are to be tested one at a time, in random order, until the defectives are discovered, find the expected number of tests that are made.

**4.** $A$ and $B$ play the following game: $A$ writes down either number 1 or number 2 and $B$ must guess which one. If the number that $A$ has written down is $i$ and $B$ has guessed correctly, $B$ receives $i$ units from $A$. If $B$ makes a wrong guess, $B$ pays $\frac{3}{4}$ units to $A$. If $B$ randomizes his decision by guessing 1 with probability $p$ and 2 with probability $1 - p$, determine his expected gain if (a) $A$ has written down number 1, and (b) $A$ has written down number 2.

What value of $p$ maximizes the minimum possible value of $B$'s expected gain and what is this maximin value? (Note that $B$'s expected gain depends not only on $p$, but also on what $A$ does.)

Consider now player $A$. Suppose that he also randomizes his decision, writing down number 1 with probability $q$. What is $A$'s expected loss if (c) $B$ chooses number 1, and (d) $B$ chooses number 2.

What value of $q$ minimizes $A$'s maximum expected loss? Show that the minimum of $A$'s maximum expected loss is equal to the maximum of $B$'s minimum expected gain. This result, known as the minimax theorem, was first established in generality by the mathematician John Van Neumann and is the fundamental result in the mathematical discipline known as the theory of games. The common value is called the value of the game to player $B$.

**5.** A typical slot machine has 3 dials, each with 20 symbols (cherries, lemons, plums, oranges, bells, and bars). A typical set of dials is set up as follows:

| | *Dial 1* | *Dial 2* | *Dial 3* |
|---|---|---|---|
| Cherries | 7 | 7 | 0 |
| Oranges | 3 | 7 | 6 |
| Lemons | 3 | 0 | 4 |
| Plums | 4 | 1 | 6 |
| Bells | 2 | 2 | 3 |
| Bars | 1 | 3 | 1 |
| | 20 | 20 | 20 |

According to this table, of the 20 slots on dial 1, 7 are cherries, 3 are oranges, and so on. A typical payoff on a 1-unit bet is as shown in the following table.

| *Dial 1* | *Dial 2* | *Dial 3* | *Payoff* |
|---|---|---|---|
| Bar | Bar | Bar | 60 |
| Bell | Bell | Bell | 20 |
| Bell | Bell | Bar | 18 |
| Plum | Plum | Plum | 14 |
| Orange | Orange | Orange | 10 |
| Orange | Orange | Bar | 8 |
| Cherry | Cherry | Anything | 4 |
| Cherry | No cherry | Anything | 2 |
| | Anything else | | −1 |

Compute the player's expected winnings on a single play of the slot machine. Assume that each dial acts independently.

**6.** One of the numbers 1 through 10 is randomly chosen. You are to try to guess the number chosen by asking questions with "yes-no" answers. Compute the expected number of questions you will need to ask in each of the two cases:

(a) Your $i$th question is to be "Is it $i$?", $i = 1, 2, 3, 4, 5, 6, 7, 8, 9, 10$.

(b) With each question you try to eliminate one half of the remaining numbers, as nearly as possible.

**7.** An insurance company writes a policy to the effect that an amount of money $A$ must be paid if some event $E$ occurs within a year. If the company estimates that $E$ will occur within a year with probability $p$, what should it charge the customer in order that its expected profit will be 10 percent of $A$?

**8.** A sample of 3 items is selected at random from a box containing 20 items of which 4 are defective. Find the expected number of defective items in the sample.

**9.** There are two possible causes for a breakdown of a machine. To check the first possibility would cost $C_1$ dollars, and, if that were the cause of the breakdown, the trouble could be repaired at a cost of $R_1$ dollars. Similarly, there are costs $C_2$ and $R_2$ associated with the second possibility. Let $p$ and $1 - p$ denote, respectively, the probabilities that the breakdown is caused by the first and second possibilities. Under what conditions on $p$, $C_i$, $R_i$, $i = 1, 2$, should we check the first possible cause of breakdown and then the second, as opposed to reversing the checking order, so as to minimize the expected cost involved in returning the machine to working order?
NOTE: If the first check is negative, we must still check the other possibility.

**10.** An individual tosses a fair coin until a tail appears for the first time. If the tail appears on the $n$th flip, the individual wins $2^n$ dollars. Let $X$ denote the player's winnings. Show that $E[X] = +\infty$. This problem is known as the St. Petersburg paradox.
(a) Would you be willing to pay \$1 million to play this game once?
(b) Would you be willing to pay \$1 million for each game if you could play for as long as you liked and only had to settle up when you stopped playing?

**11.** Let $X_1, X_2, \ldots$ be a sequence of independent and identically distributed continuous random variables. Let $N \geq 2$ be such that

$$X_1 \geq X_2 \geq \cdots \geq X_{N-1} < X_N$$

That is, $N$ is the point at which the sequence stops decreasing. Show that $E[N] = e$.

**12.** Compute $E[X]$ if $X$ has a density function given by

(a) $f(x) = \begin{cases} \frac{1}{4}xe^{-x/2} & x > 0 \\ 0 & \text{otherwise} \end{cases}$

(b) $f(x) = \begin{cases} c(1 - x^2) & -1 < x < 1 \\ 0 & \text{otherwise} \end{cases}$

(c) $f(x) = \begin{cases} \dfrac{5}{x^2} & x > 5 \\ 0 & x \leq 5 \end{cases}$

**13.** The density function of $X$ is given by

$$f(x) = \begin{cases} a + bx^2 & 0 \le x \le 1 \\ 0 & \text{otherwise} \end{cases}$$

If $E[X] = \frac{3}{5}$, find $a$, $b$.

**14.** The lifetime in hours of electronic tubes is a random variable having a probability density function given by

$$f(x) = \alpha^2 x e^{-\alpha x} \qquad x \ge 0$$

Compute the expected lifetime of such a tube.

**15.** If $X_1, X_2, \ldots, X_n$ are independent and identically distributed random variables having uniform distributions over $(0, 1)$, find (a) $E[\max(X_1, \ldots, X_n)]$, and (b) $E[\min(X_1, \ldots, X_n]$.

**16.** Let $X$ be a uniform $(0, 1)$ random variable. Compute $E[X^n]$ by using Proposition 2.1 and then check the result by using the definition of expectation.

**17.** Each night different meteorologists give us the "probability" that it will rain the next day. To judge how well these people predict, we will score each of them as follows: If a meteorologist says that it will rain with probability $p$, then he or she will receive a score of

$$\begin{array}{rl} 1 - (1-p)^2 & \text{if it does rain} \\ 1 - p^2 & \text{if it does not rain} \end{array}$$

We will then keep track of scores over a certain time span and conclude that the meteorologist with the highest average score is the best predictor of weather. Suppose now that a given meteorologist is aware of this and so wants to maximize his or her expected score. If this individual truly believes that it will rain tomorrow with probability $p^*$, what value of $p$ should he or she assert so as to maximize the expected score?

**18.** (a) A fire station is to be located along a road of length $A$, $A < \infty$. If fires will occur at points uniformly chosen on $(0, A)$, where should the station be located so as to minimize the expected distance from the fire? That is, choose $a$ so as to

$$\text{minimize } E[|X - a|]$$

when $X$ is uniformly distributed over $(0, A)$.

(b) Now suppose the road is of infinite length—stretching from point 0 outward to $\infty$. If the distance of a fire from point 0 is exponentially distributed with rate $\lambda$, where should the fire station now be located? That is, we want to minimize $E[|X - a|]$ where $X$ is now exponential with rate $\lambda$.

(c) For an arbitrary distribution $F$, we say that $m$ is a median of this distribution if

$$F(m) \geq \tfrac{1}{2} \qquad \text{and} \qquad 1 - F(m) \geq \tfrac{1}{2}$$

(Hence, for a continuous distribution the median is the unique value $m$ such that $F(m) = \frac{1}{2}$.) What can we say about that value of $a$ which minimizes $E[|X - a|]$ when $X$ has distribution function $F$?

**19.** A newsboy purchases papers at 10 cents and sells them at 15 cents. However, he is not allowed to return unsold papers. If his daily demand is a binomial random variable with $n = 300$, $p = \frac{1}{3}$, approximately how many papers should he purchase so as to maximize his expected profit?

**20.** In Example 2c, suppose that the department store incurs an additional cost of $c$ for each unit of unmet demand. (This is often referred to as a goodwill cost because the store loses the goodwill of those customers whose demands it cannot meet.) Compute the expected profit when the store stocks $s$ units, and determine the value of $s$ that maximizes the expected profit.

**21.** Repeat Example 2d where the store incurs an additional cost at a rate of $c$ per each unit of unmet demand.

**22.** The game of Red-Dog is a two-person game played with a deck of cards representing all values from 0 to 1. Each player antes 1 unit; then player $A$ deals a card $X$ to his opponent, player $B$, and a card $Y$ to himself. Player $B$ is then given the option of betting any amount of money between 1 and $M$ units, or of folding. If $B$ folds, $A$ wins the ante; if $B$ bets, $A$ must cover the wager. Whoever has the higher valued card wins the pot. Suppose that $X$ and $Y$ are independent random variables, each being uniformly distributed over $(0, 1)$. A strategy for player $B$ is a function $f(x)$, $0 < x < 1$, where for $x \in (0, 1)$ $f(x)$ is either 0 or between 1 and $M$. The strategy $f(x)$ is the strategy that tells player $B$ that, when his card has value $x$, player $B$ should fold if $f(x) = 0$ and bet $f(x)$ if $f(x) \neq 0$.
(a) Compute the expected winnings of player $B$ in a single play of Red-Dog if $B$ employs strategy $f(x)$.
(b) Show that the optimal strategy for player $B$, that is, the strategy maximizing his expected winnings, is given by

$$f(x) = \begin{cases} 0 \ (B \text{ folds}) & \text{if } x < \frac{1}{4} \\ 1 & \text{if } \frac{1}{4} \leq x < \frac{1}{2} \\ M & \text{if } \frac{1}{2} \leq x < 1 \end{cases}$$

(c) Compute the expected winnings of $B$ if he employs the optimal strategy.

**23.** A binary message—either 0 or 1—is to be communicated from location $A$ to location $B$. Because of transmission "noise," the value 2 is to be transmitted if the message is 1 and the value $-2$ if the message is 0. If $x$ is the value transmitted, then $R = x + N$ is the value received, where $N$, representing the transmission error or noise, is a unit normal random variable. The rule used at location $B$ to decipher the message is of the following form:

| | |
|---|---|
| if $R > C$ | conclude that the message is 1 |
| if $R < C$ | conclude that the message is 0 |

Suppose that a cost of 10 units is incurred when message 0 is sent and message 1 is incorrectly concluded and a cost a 20 is incurred if message 0 is concluded when the message sent was actually 1. If 25 percent of the time message 1 is the one sent, what value of $C$ minimizes the expected cost incurred?

**24.** A binary message is to be communicated from location $A$ to location $B$. Because of errors in transmission, the message 0 will be coded as $-x$ and the message 1 as $+x$. If the value $y$ is sent from location $A$, then the value received at location $B$ is $R = y + N$, where $N$ is a unit normal random variable. If $R$ is received at location $B$, then the decoding rule employed is to

| | |
|---|---|
| conclude 1 | if $R > 0$ |
| conclude 0 | if $R < 0$ |

Suppose that the cost of sending the value $x$ is $2 + 10|x|$ and the cost of concluding the incorrect message is 60. What value of $x$ minimizes the expected total cost when the message to be sent is equally likely to be either 0 or 1?

**25.** To determine whether or not they have a certain disease, 100 people are to have their blood tested. However, rather than testing each individual separately, it has been decided first to group the people in groups of 10. The blood samples of the 10 people in each group will be pooled and analyzed together. If the test is negative, one test will suffice for the 10 people; whereas, if the test is positive each of the 10 people will also be individually tested and, in all, 11 tests will be made on this group. Assume the probability that a person has the disease is .1 for all people, independently of each other, and compute the expected number of tests necessary for the 100 people. (Note that we are assuming that the pooled test will be positive if at least one person in the pool has the disease.)

**26.** A ball is chosen, at random, from each of 5 urns. The urns contain, respectively, 1 white, 5 black; 3 white, 3 black; 6 white, 4 black; 2

white, 6 black; and 3 white, 7 black balls. Compute the expected number of white balls selected.

**27.** Let $Z$ be a unit normal random variable, and for a fixed $x$, set

$$X = \begin{cases} Z & \text{if } Z > x \\ 0 & \text{otherwise} \end{cases}$$

Show that $E[X] = \dfrac{1}{\sqrt{2\pi}} e^{-x^2/2}$

**28.** A deck of $n$ cards, numbered 1 through $n$, is thoroughly shuffled so that all possible $n!$ orderings can be assumed to be equally likely. Suppose you are to sequentially make $n$ guesses, where the $i^{\text{th}}$ one is a guess of the card in position $i$. Let $N$ denote the number of correct guesses.

(a) If you are not given any information about your earlier guesses show that, for any strategy, $E[N] = 1$.

(b) Suppose that after each guess you are shown the card that was in the position in question. What do you think is the best strategy? Show that under this strategy

$$\begin{aligned} E[N] &= 1/n + 1/(n-1) + \cdots + 1 \\ &\approx \int_1^n \frac{1}{x}\,dx = \log n \end{aligned}$$

(c) Suppose that you are told after each guess whether you are right or wrong. In this case it can be shown that the strategy that maximizes $E[N]$ is one which keeps on guessing the same card until you are told you are correct and then changes to a new card. For this strategy show that

$$\begin{aligned} E[N] &= 1 + 1/2! + 1/3! + \cdots + 1/n! \\ &\approx e - 1 \end{aligned}$$

HINT: For all parts express $N$ as the sum of indicator (that is, Bernoulli) random variables.

**29.** Cards from an ordinary deck of 52 playing cards are turned face up one at a time. If the first card is an ace, or the second a deuce, or the third a three, or . . . , or the thirteenth a king, or the fourteenth an ace, and so on, we say that a match occurs. Note that we do not require that the $(13n + 1)$th card be any particular ace for a match to occur but only that it be an ace. Compute the expected number of matches that occur.

**30.** A certain region is inhabited by $r$ distinct types of a certain kind of insect species, and each insect caught will, independently of the types of the previous catches, be of type $i$ with probability

$$P_i, i = 1, \ldots, r \qquad \sum_1^r P_i = 1$$

(a) Compute the mean number of insects that are caught before the first type 1 catch.
(b) Compute the mean number of types of insects that are caught before the first type 1 catch.

**31.** An urn contains $n$ balls—the $i$th having weight $W(i)$, $i = 1, \ldots, n$. The balls are removed without replacement one at a time according to the following rule: At each selection the probability that a given ball in the urn is chosen is equal to its weight divided by the sum of the weights remaining in the urn. Thus, for instance, if at some time $i_1, \ldots i_r$ is the set of balls remaining in the urn, then the next selection will be $i_j$ with probability $W(i_j) \Big/ \sum_{k=1}^r W(i_k)$, $j = 1, \ldots, r$. Compute the expected number of balls that are withdrawn before ball number 1.

**32.** For a group of 100 people compute (a) the expected number of days of the year that are birthdays of exactly 3 people, and (b) the expected number of distinct birthdays.

**33.** How many times would you expect to roll a fair die before all 6 sides appeared at least once?

**34.** Urn 1 contains 5 white and 6 black balls, while urn 2 contains 8 white and 10 black balls. Two balls are randomly selected from urn 1 and are then put in urn 2. If 3 balls are then randomly selected from urn 2, compute the expected number of white balls in the trio.

HINT: Let $X_i = 1$ if the $i$th white ball initially in urn 1 is one of the three selected, and let $X_i = 0$ otherwise. Similarly, let $Y_i = 1$ if the $i$th white ball from urn 2 is one of the three selected, and let $Y_i = 0$ otherwise. The number of white balls in the trio can now be written as $\sum_1^5 X_i + \sum_1^8 Y_i$.

**35.** If $E[X] = 1$ and $\text{Var}(X) = 5$ find
(a) $E[(2 + X)^2]$
(b) $\text{Var}(4 + 3X)$

**36.** If 10 married couples are randomly seated at a round table, compute (a) the expected number, and (b) the variance of the number of wives who are seated next to their husbands.

**37.** Cards from an ordinary deck are turned face up one at a time. Compute the expected number of cards that need be turned face up in order to obtain (a) 2 aces, (b) 5 spades, and (c) all 13 hearts.

**38.** Let $X$ be the numbers of 1's and $Y$ the number of 2's that occur in $n$ rolls of a fair die. Compute Cov $(X, Y)$.

HINT: Use part (c) of Theoretical Exercise 26.

**39.** A die is rolled twice. Let $X$ equal the sum of the outcomes, and let $Y$ equal the first outcome minus the second. Compute Cov $(X, Y)$.

**40.** The random variables $X$ and $Y$ have a joint density function given by

$$f(x, y) = \begin{cases} 2e^{-2x}/x & 0 \le x < \infty, 0 \le y \le x \\ 0 & \text{otherwise} \end{cases}$$

Compute Cov $(X, Y)$.

**41.** Let $X_1, \ldots$ be independent with common mean $\mu$ and common variance $\sigma^2$, and set $Y_n = X_n + X_{n+1} + X_{n+2}$. Find, for $j \ge 0$, $\text{Cov}(Y_n, Y_{n+j})$.

**42.** The joint density function of $X$ and $Y$ is given by

$$f(x,y) = \frac{1}{y} e^{-(y+x/y)}, \; x > 0, \; y > 0$$

Find $E[X]$, $E[Y]$, and show that $\text{Cov}(X,Y) = 1$.

**43.** A pond contains 100 fish, of which 30 are carp. If 20 fish are caught, what are the mean and variance of the number of carp among these 20? What assumptions are you making?

**44.** A group of 20 people—consisting of 10 men and 10 women—are randomly arranged into 10 pairs of 2 each. Compute the expectation and variance of the number of pairs that consist of a man and a woman. Now suppose the 20 people consisted of 10 married couples. Compute the mean and variance of the number of married couples that are paired together.

**45.** Let $X_1, X_2, \ldots, X_n$ be independent random variables having an unknown continuous distribution function $F$, and let $Y_1, Y_2, \ldots, Y_m$ be independent random variables having an unknown continuous distribution function $G$. Now order those $n + m$ variables and let

$$I_i = \begin{cases} 1 & \text{if the } i\text{th smallest of the } n+m \text{ variables} \\ & \text{is from the } X \text{ sample} \\ 0 & \text{otherwise} \end{cases}$$

The random variable $R = \sum_{i=1}^{n+m} iI_i$ is the sum of the ranks of the $X$

sample and is the basis of a standard statistical procedure (called the Wilcoxon sum of ranks test) for testing whether $F$ and $G$ are identical distributions. This test accepts the hypothesis that $F = G$ when $R$ is neither too large nor too small. Assuming that the hypothesis of equality is in fact correct, compute the mean and variance of $R$.

HINT: Use the results of Example 5c.

**46.** There are two distinct methods for manufacturing a certain goods, the quality of a goods produced by method $i$ being a continuous random variable having distribution $F_i$, $i = 1, 2$. Suppose that $n$ goods are produced by method 1 and $m$ by method 2. Rank the $n + m$ goods according to quality and let

$$X_j = \begin{cases} 1 & \text{if the } j\text{th best was produced from method 1} \\ 2 & \text{otherwise} \end{cases}$$

For the vector $X_1, X_2, \ldots, X_{n+m}$, which consists of $n$ 1's and $m$ 2's, let $R$ denote the number of runs of 1. For instance, if $n = 5$, $m = 2$, and $X = 1, 2, 1, 1, 1, 1, 2$, then $R = 2$. If $F_1 = F_2$ (that is, if the two methods produce identically distributed goods), what are the mean and variance of $R$?

**47.** If $X_1$, $X_2$, $X_3$, $X_4$ are (pairwise) uncorrelated random variables each having mean 0 and variance 1, compute the correlations of (a) $X_1 + X_2$ and $X_2 + X_3$, and (b) $X_1 + X_2$ and $X_3 + X_4$.

**48.** Consider the following dice game, as played at a certain gambling casino: Players 1 and 2 roll in turn a pair of dice. The bank then rolls the dice to determine the outcome according to the following: player $i$, $i = 1, 2$, wins if the roll is strictly greater than the bank's. Let for $i = 1, 2$,

$$I_i = \begin{cases} 1 & \text{if } i \text{ wins} \\ 0 & \text{otherwise} \end{cases}$$

and show that $I_1$ and $I_2$ are positively correlated. Explain why this result was to be expected.

**49.** A fair die is successively rolled. Let $X$ and $Y$ denote, respectively, the number of rolls necessary to obtain a 6 and a 5. Find (a) $E[X]$, (b) $E[X \mid Y = 1]$, and (c) $E[X \mid Y = 5]$.

**50.** An urn contains 4 white and 6 black balls. Two successive random samples of sizes 3 and 5, respectively, are drawn from the urn without replacement. Let $X$ and $Y$ denote the number of white balls in the two samples, and compute $E[X | Y = i]$, for $i = 1, 2, 3, 4$.

**51.** The joint density of $X$ and $Y$ is given by

$$f(x, y) = \frac{e^{-x/y}e^{-y}}{y} \qquad 0 < x < \infty; 0 < y < \infty$$

Compute $E[X^2 | Y = y]$.

**52.** The joint density of $X$ and $Y$ is given by

$$f(x, y) = \frac{e^{-y}}{y} \qquad 0 < x < y, 0 < y < \infty$$

Compute $E[X^3 | Y = y]$.

**53.** A prisoner is trapped in a cell containing 3 doors. The first door leads to a tunnel that returns him to his cell after 2 days travel. The second to a tunnel that returns him to his cell after 4 days travel. The third door leads to freedom after 1 day of travel. If it is assumed that the prisoner will always select doors 1, 2, and 3 with respective probabilities .5, .3, and .2, what is the expected number of days until the prisoner reaches freedom?

**54.** Ten hunters are waiting for ducks to fly by. When a flock of ducks flies overhead, the hunters fire at the same time, but each chooses his target at random, independently of the others. If each hunter independently hits his target with probability .6, compute the expected number of ducks that are hit. Assume that the number of ducks in a flock is a Poisson random variable with mean 6.

**55.** The number of people that enter an elevator on the ground floor is a Poisson random variable with mean 10. If there are $N$ floors above the ground floor and if each person is equally likely to get off at any one of these $N$ floors, independently of where the others get off, compute the expected number of stops that the elevator will make before discharging all of its passengers.

**56.** Suppose that the expected number of accidents per week at an industrial plant is 5. Suppose also that the numbers of workers injured in each accident are independent random variables with a common mean of 2.5. If the number of workers injured in each accident is independent of the number of accidents that occur, compute the expected number of workers injured in a week.

**57.** In Example 6c compute the variance of the length of time until the miner reaches safety.

**58.** The dice game of craps was defined in Problem 10 of Chapter 2. Compute (a) the mean, and (b) the variance, of the number of rolls of the dice that it takes to complete one game of craps.

**59.** Consider a gambler who at each gamble either wins or loses his bet with probabilities $p$ and $1 - p$. When $p > \frac{1}{2}$, a popular gambling system, known as the Kelley strategy, is to always bet the fraction $2p - 1$ of your current fortune. Compute the expected fortune after $n$ gambles of a gambler who starts with $x$ units and employs the Kelley strategy.

**60.** The number of accidents that an individual has in a given year is a Poisson random variable with mean $\lambda$. However, suppose that the value of $\lambda$ changes from person to person, being equal to 2 for 60 percent of the population and 3 for the other 40 percent. If a person is chosen at random, what is the probability that he will have (a) 0 accidents, and (b) exactly 3 accidents, in a year? What is the conditional probability that he will have 3 accidents in a given year, given that he had no accidents the previous year?

**61.** Repeat Problem 60 when the proportion of the population having a value of $\lambda$ less than $x$ is equal to $1 - e^{-x}$.

**62.** Consider an urn containing a large number of coins and suppose that each of the coins has some probability $p$ of turning up heads when it is flipped. However, this value of $p$ varies from coin to coin. Suppose that the composition of the urn is such that if a coin is selected at random from the urn, then its $p$-value can be regarded as being the value of a random variable that is uniformly distributed over $[0, 1]$. If a coin is selected at random from the urn and flipped twice, compute the probability that (a) the first flip is a head, and (b) both flips are heads.

**63.** In Example 7b let $S$ denote the signal sent and $R$ the signal received.
(a) Compute $E[R]$.
(b) Compute $\text{Var}(R)$.
(c) Is $R$ normally distributed?
(d) Compute $\text{Cov}(R, S)$.

**64.** In the quantizer example, 7c, suppose that $X$ is uniformly distributed over $(0, 1)$. If the discretized regions are determined by $a_0 = 0$, $a_1 = \frac{1}{2}$, $a_2 = 1$, determine the optimal quantizer $Y$ and compute $E[(X - Y)^2]$.

**65.** Let $X$ be a normal random variable with mean $\mu$ and variance $\sigma^2$.
(a) Determine $E[(X - \mu)^3]$
(b) Use (a) to obtain $E[X^3]$
(c) Check your answer in (b) by differentiation of the moment generating function.

**66.** The moment generating function of $X$ is given by $\phi_X(t) = \exp\{2e^t - 2\}$ and that of $Y$ by $\phi_Y(t) = (\frac{1}{4})^{10}(3e^t + 1)^{10}$. If $X$ and $Y$ are independent, what are (a) $P\{X + Y = 2\}$, (b) $P\{XY = 0\}$, and (c) $E[XY]$?

**67.** Let $X$ be the value of the first die and $Y$ the sum of the values when two dice are rolled. Compute the joint moment generating function of $X$ and $Y$.

**68.** The joint density of $X$ and $Y$ is given by

$$f(x, y) = \frac{e^{-x}e^{-y/x}}{x} \qquad 0 < x < \infty, 0 < y < \infty$$

(a) Compute the joint moment generating function of $X$ and $Y$.
(b) Compute the individual moment generating functions.

# 8

# Limit Theorems

## 1 Introduction

The most important theoretical results in probability theory are limit theorems. Of these, the most important are those that are classified either under the heading of "laws of large numbers" or under the heading of "central limit theorems." Usually, theorems are considered to be "laws of large numbers" if they are concerned with stating conditions under which the average of a sequence of random variables converges (in some sense) to the expected average. On the other hand, central limit theorems are concerned with determining conditions under which the sum of a large number of random variables has a probability distribution that is approximately normal.

## 2 Chebyshev's Inequality and the Weak Law of Large Numbers

We start this section by proving a result known as Markov's inequality.

*Proposition 2.1 Markov's Inequality*

*If $X$ is a random variable that takes only nonnegative values, then for any value $a > 0$*

$$P\{X \geq a\} \leq \frac{E[X]}{a}$$

**Proof:** We give a proof for the case where $X$ is continuous with density $f$.

$$\begin{aligned} E[X] &= \int_0^\infty xf(x)\,dx \\ &= \int_0^a xf(x)\,dx + \int_a^\infty xf(x)\,dx \\ &\geq \int_a^\infty xf(x)\,dx \\ &\geq \int_a^\infty af(x)\,dx \\ &= a\int_a^\infty f(x)\,dx \\ &= aP\{X \geq a\} \end{aligned}$$

and the result is proved. (The proof in the general case is exactly as given with $dF(x)$ replacing $f(x)\,dx$ throughout.) ∎

As a corollary, we obtain Proposition 2.2.

---

*Proposition 2.2 Chebyshev's Inequality*

*If $X$ is a random variable with finite mean $\mu$ and variance $\sigma^2$, then for any value $k > 0$*

$$P\{|X - \mu| \geq k\} \leq \frac{\sigma^2}{k^2}$$

---

**Proof:** Since $(X - \mu)^2$ is a nonnegative random variable, we can apply Markov's inequality (with $a = k^2$) to obtain

$$P\{(X - \mu)^2 \geq k^2\} \leq \frac{E[(X - \mu)^2]}{k^2} \tag{2.1}$$

But since $(X - \mu)^2 \geq k^2$ if and only if $|X - \mu| \geq k$, Equation (2.1) is equivalent to

$$P\{|X - \mu| \geq k\} \leq \frac{E[(X - \mu)^2]}{k^2} = \frac{\sigma^2}{k^2}$$

and the proof is complete. ∎

The importance of Markov's and Chebyshev's inequalities is that they enable us to derive bounds on probabilities when only the mean, or both the mean and the variance, of the probability distribution are known. Of course, if the actual distribution were known, then the desired probabilities could be exactly computed and we would not need to resort to bounds.

**Example 2a.** Suppose that it is known that the number of items produced in a factory during a week is a random variable with mean 50.

1. What can be said about the probability that this week's production will exceed 75?
2. If the variance of a week's production is known to equal 25, then what can be said about the probability that this week's production will be between 40 and 60?

***Solution:*** Let $X$ be the number of items that will be produced in a week:

1. By Markov's inequality

$$P\{X > 75\} \leq \frac{E[X]}{75} = \frac{50}{75} = \frac{2}{3}$$

2. By Chebyshev's inequality

$$P\{|X - 50| \geq 10\} \leq \frac{\sigma^2}{10^2} = \frac{1}{4}$$

Hence

$$P\{|X - 50| < 10\} \geq 1 - \tfrac{1}{4} = \tfrac{3}{4}$$

and so the probability that this week's production will be between 40 and 60 is at least .75. ∎

As Chebyshev's inequality is valid for all distributions of the random variable $X$, we cannot expect the bound on the probability to be very close to the actual probability in most cases. For instance, consider Example 2b.

**Example 2b.** If $X$ is uniformly distributed over the interval $(0, 10)$, then, as $E[X] = 5$, $\text{Var}(X) = \frac{25}{3}$, it follows by Chebyshev's inequality that

$$P\{|X - 5| > 4\} \leq \frac{25}{3(16)} \approx .52$$

whereas the exact result is

$$P\{|X - 5| > 4\} = .20$$

Thus, although Chebyshev's inequality is correct, the upper bound that it provides is not particularly close to the actual probability.

Similarly, if $X$ is a normal random variable with mean $\mu$ and variance $\sigma^2$, Chebyshev's inequality states that

$$P\{|X-\mu|>2\sigma\} \le \tfrac{1}{4}$$

whereas the actual probability is given by

$$P\{|X-\mu|>2\sigma\} = P\left\{\left|\frac{X-\mu}{\sigma}\right|>2\right\} = 2[1-\Phi(2)] \approx .0456 \quad ∎$$

Chebyshev's inequality is often used as a theoretical tool in proving results. This is illustrated first by Proposition 2.3 and then, most importantly, by the weak law of large numbers.

---

*Proposition 2.3*

*If* Var $(X) = 0$, *then*

$$P\{X = E[X]\} = 1$$

---

In other words, the only random variables having variances equal to 0 are those that are constant with probability 1.

**Proof:** By Chebyshev's inequality we have, for any $n \ge 1$

$$P\left\{|X-\mu|>\frac{1}{n}\right\} = 0$$

Letting $n \to \infty$ and using the continuity property of probability yields

$$0 = \lim_{n\to\infty} P\left\{|X-\mu|>\frac{1}{n}\right\} = P\left\{\lim_{n\to\infty}\left\{|X-\mu|>\frac{1}{n}\right\}\right\}$$
$$= P\{X \ne \mu\}$$

and the result is established. ∎

---

*Theorem 2.1 The Weak Law of Large Numbers*

*Let $X_1, X_2, \ldots$ be a sequence of independent and identically distributed random variables, each having finite mean $E[X_i] = \mu$. Then, for any $\varepsilon > 0$,*

$$P\left\{\left|\frac{X_1+\cdots+X_n}{n}-\mu\right|>\varepsilon\right\} \to 0 \qquad \text{as } n\to\infty$$

---

**Proof:** We shall prove the result only under the additional assumption that the random variables have a finite variance $\sigma^2$. Now, as

$$E\left[\frac{X_1 + \cdots + X_n}{n}\right] = \mu \quad \text{and} \quad \operatorname{Var}\left(\frac{X_1 + \cdots + X_n}{n}\right) = \frac{\sigma^2}{n}$$

it follows from Chebyshev's inequality that

$$P\left\{\left|\frac{X_1 + \cdots + X_n}{n} - \mu\right| > \varepsilon\right\} \leq \frac{\sigma^2}{n\varepsilon^2}$$

and the result is proved. ∎

The weak law of large numbers was originally proved by Jacob Bernoulli for the special case where the $X_i$ are $0-1$ (that is, Bernoulli) random variables. His statement and proof of this theorem was presented in his book *Ars Conjectandi*, which was published 8 years after his death in 1713 by his nephew Nicholas Bernoulli. It should be noted that as Chebyshev's inequality was not known in his time, Bernoulli had to resort to a quite ingenious proof to establish the result. The general form of the weak law of large numbers presented in Theorem 2.1 was proved by the Russian mathematician Khintchine.

## 3 The Central Limit Theorem

The central limit theorem is one of the most remarkable results in probability theory. Loosely put, it states that the sum of a large number of independent random variables has a distribution that is approximately normal. Hence it not only provides a simple method for computing approximate probabilities for sums of independent random variables, but it also helps explain the remarkable fact that the empirical frequencies of so many natural populations exhibit bell-shaped (that is, normal) curves.

In its simplest form the central limit theorem is as follows.

---

*Theorem 3.1 The Central Limit Theorem*

*Let $X_1, X_2, \ldots$ be a sequence of independent and identically distributed random variables each having mean $\mu$ and variance $\sigma^2$. Then the distribution of*

$$\frac{X_1 + \cdots + X_n - n\mu}{\sigma\sqrt{n}}$$

*tend to the standard normal as $n \to \infty$. That is*

$$P\left\{\frac{X_1 + \cdots + X_n - n\mu}{\sigma\sqrt{n}} \le a\right\} \to \frac{1}{\sqrt{2\pi}} \int_{-\infty}^{a} e^{-x^2/2}\, dx \qquad \text{as } n \to \infty$$

The key to the proof of the central limit theorem is the following lemma, which we state without proof.

### *Lemma 3.1*

*Let $Z_1, Z_2, \ldots$ be a sequence of random variables having distribution functions $F_{Z_n}$ and moment generating functions $\phi_{Z_n}$, $n \ge 1$; and let $Z$ be a random variable having distribution function $F_Z$ and moment generating function $\phi_Z$. If $\phi_{Z_n}(t) \to \phi_Z(t)$ for all t, then $F_{Z_n}(t) \to F_Z(t)$ for all t at which $F_Z(t)$ is continuous.*

If we let $Z$ be a unit normal random variable, then, as $\phi_Z(t) = e^{t^2/2}$, it follows from Lemma 3.1 that if $\phi_{Z_n}(t) \to e^{t^2/2}$ as $n \to \infty$, then $F_{Z_n}(t) \to \Phi(t)$ as $n \to \infty$.

We are now ready to prove the central limit theorem.

**Proof of Central Limit Theorem:** Let us assume at first that $\mu = 0$ and $\sigma^2 = 1$. We shall prove the theorem under the assumption that the moment generating function of the $X_i$, $\phi(t)$, exists and is finite. Now the moment generating function of $X_i/\sqrt{n}$ is given by

$$\phi_{X_i/\sqrt{n}}(t) = E\left[\exp\left\{\frac{tX_i}{\sqrt{n}}\right\}\right]$$
$$= \phi\left(\frac{t}{\sqrt{n}}\right)$$

and thus the moment generating function of $\sum_{i=1}^{n} X_i/\sqrt{n}$ is given by

$$\phi_{\sum_{i=1}^{n} X_i/\sqrt{n}}(t) = \left[\phi\left(\frac{t}{\sqrt{n}}\right)\right]^n$$

Let

$$L(t) = \log \phi(t)$$

and note that

$$L(0) = 0$$

$$\begin{aligned} L'(0) &= \frac{\phi'(0)}{\phi(0)} \\ &= \mu \\ &= 0 \end{aligned}$$

$$\begin{aligned} L''(0) &= \frac{\phi(0)\phi''(0) - [\phi'(0)]^2}{[\phi(0)]^2} \\ &= E[X^2] \\ &= 1 \end{aligned}$$

Now, to prove the theorem we must show that $[\phi(t/\sqrt{n})]^n \to e^{t^2/2}$ as $n \to \infty$, or equivalently, that $nL(t/\sqrt{n}) \to t^2/2$ as $n \to \infty$. To show this, note that

$$\begin{aligned} \lim_{n\to\infty} \frac{L(t/\sqrt{n})}{n^{-1}} &= \lim_{n\to\infty} \frac{-L'(t/\sqrt{n})n^{-3/2}t}{-2n^{-2}} && \text{by L'hospitals rule} \\ &= \lim_{n\to\infty} \left[\frac{L'(t/\sqrt{n})t}{2n^{-1/2}}\right] \\ &= \lim_{n\to\infty} \left[\frac{-L''(t/\sqrt{n})n^{-3/2}t^2}{-2n^{-3/2}}\right] && \text{again by L'hospitals rule} \\ &= \lim_{n\to\infty} \left[L''\left(\frac{t}{\sqrt{n}}\right)\frac{t^2}{2}\right] \\ &= \frac{t^2}{2} \end{aligned}$$

Thus the central limit theorem is proved when $\mu = 0$ and $\sigma^2 = 1$. The result now follows in the general case by considering the standardized random variables $X_i^* = (X_i - \mu)/\sigma$ and applying the above result, since $E[X_i^*] = 0$, $\text{Var}(X_i^*) = 1$. ∎

REMARK. Although Theorem 3.1 only states that, for each $a$,

$$P\left\{\frac{X_1 + \cdots + X_n - n\mu}{\sigma\sqrt{n}} \le a\right\} \to \Phi(a)$$

it can, in fact, be shown that the convergence is uniform in $a$. [We say that $f_n(a) \to f(a)$ uniformly in $a$, if for each $\varepsilon > 0$, there exists an $N$ such that $|f_n(a) - f(a)| < \varepsilon$ for all $a$ whenever $n \ge N$.]

The first version of the central limit theorem was proved by DeMoivre around 1733 for the special case where the $X_i$ are Bernoulli random variables with $p = \frac{1}{2}$. This was subsequently extended by Laplace to the case of arbitrary $p$. (Since a binomial random variable may be regarded as the sum of $n$ independent and identically distributed Bernoulli random variables, this justifies the normal approximation to the binomial that was presented in Section 3.1 of Chapter 5.) Laplace also discovered the more general form of the central limit theorem given in Theorem 3.1. His proof, however, was not completely rigorous and, in fact, cannot easily be made rigorous. A truly rigorous proof of the central limit theorem was first presented by the Russian mathematician Liapounoff in the period 1901–1902.

**Example 3a.** An astronomer is interested in measuring, in light years, the distance from his observatory to a distant star. Although the astronomer has a measuring technique, he knows that, because of changing atmospheric conditions and normal error, each time a measurement is made it will not yield the exact distance but merely an estimate. As a result the astronomer plans to make a series of measurements and then use the average value of these measurements as his estimated value of the actual distance. If the astronomer believes that the values of the measurements are independent and identically distributed random variables having a common mean $d$ (the actual distance) and a common variance of 4 (light years), how many measurements need he make to be reasonably sure that his estimated distance is accurate to within $\pm .5$ light years?

***Solution:*** Suppose the astronomer decides to make $n$ observations. If $X_1, X_2, \ldots, X_n$ are the $n$ measurements, then, from the central limit theorem, it follows that

$$Z_n = \frac{\sum_{i=1}^{n} X_i - nd}{2\sqrt{n}}$$

has approximately a unit normal distribution. Hence

$$P\left\{-.5 \leq \frac{\sum_{i=1}^{n} X_i}{n} - d \leq .5\right\} = P\left\{-.5\frac{\sqrt{n}}{2} \leq Z_n \leq .5\frac{\sqrt{n}}{n}\right\}$$

$$\approx \Phi\left(\frac{\sqrt{n}}{4}\right) - \Phi\left(-\frac{\sqrt{n}}{4}\right) = 2\Phi\left(\frac{\sqrt{n}}{4}\right) - 1$$

Therefore, if the astronomer wanted, for instance, to be 95 percent certain that his estimated value is accurate to within .5 light years, he should make $n^*$ measurements where $n^*$ is such that

$$2\Phi\left(\frac{\sqrt{n^*}}{4}\right) - 1 = .95 \qquad \text{or} \qquad \Phi\left(\frac{\sqrt{n^*}}{4}\right) = .975$$

and thus fromTable 5.1 of Chapter 5

$$\frac{\sqrt{n^*}}{4} = 1.96 \qquad \text{or} \qquad n^* = (7.84)^2 = 61.47$$

As $n^*$ is not integral-valued, he should make 62 observations.

It should, however, be noted that the preceding analysis has been done under the assumption that the normal approximation will be a good approximation when $n = 62$. Although this will usually be the case, in general the question of how large $n$ need be before the approximation is "good" depends on the distribution of the $X_i$. If the astronomer was concerned about this point and wanted to take no chances, he could still solve his problem by using Chebyshev's inequality. Since

$$E\left[\sum_{i=1}^{n} \frac{X_i}{n}\right] = d \qquad \text{Var}\left(\sum_{i=1}^{n} \frac{X_i}{n}\right) = \frac{4}{n}$$

Chebyshev's inequality yields that

$$P\left\{\left|\sum_{i=1}^{n} \frac{X_i}{n} - d\right| > .5\right\} \leq \frac{4}{n(.5)^2} = \frac{16}{n}$$

Hence, if he makes $n = 16/.05 = 320$ observations, he can be 95 percent certain that his estimate will be accurate to within .5 light years. ∎

**Example 3b.** The number of students that enroll in a psychology course is a Poisson random variable with mean 100. The professor in charge of the course has decided that if the number enrolling is 120 or more he will teach the course in two separate sections, whereas if fewer than 120 students enroll he will teach all of the students together in a single section. What is the probability that the professor will have to teach two sections?

***Solution:*** The exact solution $e^{-100} \sum_{i=120}^{\infty} (100)^i/i!$ does not readily yield a numerical answer. However, by recalling that a Poisson random variable with mean 100 is the sum of 100 independent Poisson random variables each with mean 1, we can make use of the central limit theorem to obtain an approximate solution. If $X$ denotes the number of students that enroll in the course, we have

$$P\{X \geq 120\} = P\left(\frac{X - 100}{\sqrt{100}} \geq \frac{120 - 100}{\sqrt{100}}\right\}$$
$$\approx 1 - \Phi(2)$$
$$= .0228$$

where we have used the fact that the variance of a Poisson random variable is equal to its mean. ∎

**Example 3c.** If 10 fair dice are rolled, find the approximate probability that the sum obtained is between 30 and 40.

***Solution:*** Let $X_i$ denote the value of the $i$th die, $i = 1, 2, \ldots, 10$. Since $E(X_i) = \frac{7}{2}$, $\text{Var}(X_i) = E[X_i^2] - (E[X_i])^2 = \frac{35}{12}$, the central limit theorem yields

$$P\left\{30 \leq \sum_{i=1}^{10} X_i \leq 40\right\} = P\left\{\frac{30 - 35}{\sqrt{\frac{350}{12}}} \leq \frac{\sum_{i=1}^{10} X_i - 35}{\sqrt{\frac{350}{12}}} \leq \frac{40 - 35}{\sqrt{\frac{350}{12}}}\right\}$$
$$\approx 2\Phi(\sqrt{6/7}) - 1$$
$$\approx .65$$ ∎

**Example 3d.** Let $X_i$, $i = 1, \ldots, 10$ be independent random variables, each uniformly distributed over $(0, 1)$. Calculate an approximation to $P\left\{\sum_{i=1}^{10} X_i > 6\right\}$.

***Solution:*** Since $E[X_i] = \frac{1}{2}$, $\text{Var}(X_i) = \frac{1}{12}$, we have by the central limit theorem

$$P\left\{\sum_{1}^{10} X_i > 6\right\} = P\left\{\frac{\sum_{1}^{10} X_i - 5}{\sqrt{10\left(\frac{1}{12}\right)}} > \frac{6 - 5}{\sqrt{10\left(\frac{1}{12}\right)}}\right\}$$
$$\approx 1 - \Phi(\sqrt{1.2})$$
$$\approx .16$$

Hence only 16 percent of the time will $\sum_{i=1}^{10} X_i$ be greater than 6. ∎

Central limit theorems also exist when the $X_i$ are independent but not necessarily identically distributed random variables. One version, by no means the most general, is as follows.

---

*Theorem 3.2 Central Limit Theorem for Independent Random Variables*

*Let $X_1, X_2, \ldots$ be a sequence of independent random variables having respective means and variances $\mu_i = E[X_i]$, $\sigma_i^2 = \text{Var}(X_i)$. If (a) the $X_i$ are uniformly bounded, that is, if for some $M$, $P\{|X_i| < M\} = 1$ for all i, and (b) $\sum_{i=1}^{\infty} \sigma_i^2 = \infty$, then*

$$P\left\{\frac{\sum_{i=1}^{n} (X_i - \mu_i)}{\sqrt{\sum_{i=1}^{n} \sigma_i^2}} \leq a\right\} \to \Phi(a) \qquad \text{as } n \to \infty$$

---

## 4 The Strong Law of Large Numbers

The *strong law of large numbers* is probably the best-known result in probability theory. It states that the average of a sequence of independent random variables having a common distribution will, with probability 1, converge to the mean of that distribution.

---

*Theorem 4.1 The Strong Law of Large Numbers*

*Let $X_1, X_2, \ldots$ be a sequence of independent and identically distributed random variables, each having a finite mean $\mu = E[X_i]$. Then, with probability* 1,

$$\frac{X_1 + X_2 + \cdots + X_n}{n} \to \mu \qquad \text{as } n \to \infty$$ [1]

---

As an application of the strong law of large numbers, suppose that a sequence of independent trials of some experiment is performed. Let $E$ be

[1] That is, the strong law of large numbers states that

$$P\left\{\lim_{n\to\infty} (X_1 + \cdots + X_n)/n = \mu\right\} = 1$$

a fixed event of the experiment and denote by $P(E)$ the probability that $E$ occurs on any particular trial. Letting

$$X_i = \begin{cases} 1 & \text{if } E \text{ occurs on the } i\text{th trial} \\ 0 & \text{if } E \text{ does not occur on the } i\text{th trial} \end{cases}$$

we have by the strong law of large numbers that, with probability 1,

$$\frac{X_1 + \cdots + X_n}{n} \to E[X] = P(E) \tag{4.1}$$

Since $X_1 + \cdots + X_n$ represents the number of times that the event $E$ occurs in the first $n$ trials, we may interpret Equation (4.1) as stating that, with probability 1, the limiting proportion of time that the event $E$ occurs is just $P(E)$.

One of the key elements in our proof of the strong law of large numbers will be an inequality due to Kolmogorov, which is itself of independent interest.

---

### *Theorem 4.2 Kolmogorov's Inequality*

*Let $X_1, \ldots, X_n$ be $n$ independent random variables such that $E[X_i] = 0$ and $\text{Var}(X_i) = \sigma_i^2 < \infty$, $i = 1, \ldots, n$. Then, for all $a > 0$,*

$$P\left\{ \max_{i=1,\ldots,n} |X_1 + \cdots + X_i| > a \right\} \le \sum_{i=1}^{n} \frac{\sigma_i^2}{a^2}$$

---

**Proof:** Define the random variable $N$ to be the smallest value of $i$, $i \le n$, such that $(X_1 + \cdots + X_i)^2 > a^2$, and define it to equal $n$ if $(X_1 + \cdots + X_i)^2 \le a^2$ for all $i = 1, \ldots, n$. That is, define

$$\begin{aligned} N &= 1 \quad \text{if } X_1 > a^2 \\ N &= 2 \quad \text{if } X_1^2 \le a^2 \text{ and } (X_1 + X_2)^2 > a^2 \\ &\vdots \\ N &= i \quad \text{if } X_1^2 \le a^2, \ldots, (X_1 + \cdots + X_{i-1})^2 \le a^2, (X_1 + \cdots + X_i)^2 > a^2 \\ &\vdots \\ N &= n \quad \text{if } (X_1 + \cdots + X_i)^2 \le a^2, i = 1, 2, \ldots, n-1 \end{aligned}$$

Now, as the two events

$$\left\{ \max_{i=1,\ldots,n} (X_1 + \cdots + X_i)^2 > a^2 \right\} \quad \text{and} \quad \{(X_1 + \cdots + X_N)^2 > a^2\}$$

are identical, it follows from Markov's inequality that

$$P\left\{\max_{i=1,\ldots,n}(X_1+\cdots+X_i)^2>a^2\right\}=P\{(X_1+\cdots+X_N)^2>a^2\}$$
$$\le\frac{E[(X_1+\cdots+X_N)^2]}{a^2}\tag{4.2}$$

If we could show that

$$E[(X_1+\cdots+X_N)^2]\le E[(X_1+\cdots+X_n)^2]$$

then the result would follow from Equation (4.2) because

$$\begin{aligned}E[(X_1+\cdots+X_n)^2]&=\text{Var}(X_1+\cdots+X_n)\\&=\sum_{i=1}^n\text{Var}(X_i)\\&=\sum_{i=1}^n\sigma_i^2\end{aligned}$$

To prove this inequality, we condition on $N$ and note first that

$$E[(X_1+\cdots+X_n)^2|N=n]=E[(X_1+\cdots+X_N)^2|N=n]$$

and, for $i<n$

$$\begin{aligned}&E[(X_1+\cdots+X_n)^2|N=i]\\&\quad=E[((X_1+\cdots+X_i)+(X_{i+1}+\cdots+X_n))^2|N=i]\\&\quad=E[(X_1+\cdots+X_i)^2|N=i]\\&\qquad+2E[(X_1+\cdots+X_i)(X_{i+1}+\cdots+X_n)|N=i]\\&\qquad+E[(X_{i+1}+\cdots+X_n)^2|N=i]\end{aligned}\tag{4.3}$$

Now, the event $\{N=i\}$ contains information about $X_1,\ldots,X_i$, since it states that $X_1^2\le a^2,\ldots,(X_1+\cdots+X_{i-1})^2\le a^2$, $(X_1+\cdots+X_i)^2>a^2$; but, on the other hand, it says nothing about the values of $X_{i+1},\ldots,X_n$. Hence, as the random variables $X_1,\ldots,X_n$ are all independent, it follows that $X_1+\cdots+X_i$ and $X_{i+1}+\cdots+X_n$ remain conditionally independent given that $N=i$. Therefore,

$$\begin{aligned}&E[(X_1+\cdots+X_i)(X_{i+1}+\cdots+X_n)|N=i]\\&\quad=E[X_1+\cdots+X_i|N=i]E[X_{i+1}+\cdots+X_n|N=i]\\&\quad=E[X_1+\cdots+X_i|N=i]E[X_{i+1}+\cdots+X_n]\\&\quad=0\end{aligned}$$

Thus, from Equation (4.3), we obtain

$$\begin{aligned}E[(X_1+\cdots+X_n)^2|N=i]&\ge E[(X_1+\cdots+X_i)^2|N=i]\\&=E[(X_1+\cdots+X_N)^2|N=i]\end{aligned}$$

Hence we see that for all values of $N$

$$E[(X_1 + \cdots + X_n)^2 | N] \geq E[(X_1 + \cdots + X_N)^2 | N]$$

Taking expectations yields

$$E[(X_1 + \cdots + X_n)^2] \geq E[(X_1 + \cdots + X_N)^2]$$

and the result follows from Equation (4.2). ∎

Kolmogorov's inequality can be regarded as a generalization of Chebyshev's inequality. This is so because if $X$ has mean $\mu$ and variance $\sigma^2$, then by letting $n = 1$ in Kolmogorov's inequality we obtain

$$P\{|X - \mu| > a\} \leq \frac{\sigma^2}{a^2}$$

which, of course, is just Chebyshev's inequality. It should, however, be noted that Kolmogorov's inequality is a much stronger result than Chebyshev's inequality. For if $X_1, \ldots, X_n$ are independent random variables with $E[X_i] = 0$, Var $(X_i) = \sigma_i^2$, then Chebyshev's inequality yields

$$P\{|X_1 + \cdots + X_n| > a\} \leq \sum_{i=1}^{n} \frac{\sigma_i^2}{a^2}$$

whereas Kolmogorov's inequality gives the same bound for the probability of a larger set, namely,

$$\bigcup_{i=1}^{n} \{|X_1 + \cdots + X_i| > a\}$$

Kolmogorov's inequality will now be used as the basis for a proof of the strong law of large numbers in the case where the random variables are assumed independent but not necessarily identically distributed. Before presenting this proof, we shall need the following proposition, known as Kronecker's lemma, which we state without proof.

---

*Proposition 4.1 Kronecker's Lemma*

*If $a_1, a_2, \ldots$ are real numbers such that $\sum_{i=1}^{\infty} a_i/i < \infty$, converges, then*

$$\lim_{n \to \infty} \sum_{i=1}^{n} \frac{a_i}{n} = 0$$

---

---

*Theorem 4.3 Strong Law of Large Numbers for Independent Random Variables*

*Let $X_1, X_2, \ldots$ be independent random variables with $E[X_i] = 0$, $\mathrm{Var}(X_i) = \sigma_i^2 < \infty$. If $\sum_{i=1}^{\infty} \sigma_i^2 / i^2 < \infty$, then with probability 1,*

$$\frac{X_1 + \cdots + X_n}{n} \to 0 \qquad \text{as } n \to \infty$$

---

**Proof:** We will prove the theorem by showing that, with probability 1, $\sum_{i=1}^{\infty} X_i/i$ converges. The result will then follow from Kronecker's Lemma. To begin, for any $a>0$ define the event $E_{j,n}(a)$, $j\leq n$, by

$$E_{j,n}(a) = \left\{ \max_{j\leq k\leq n} \left| \sum_{i=j}^{k} X_i/i \right| > a \right\}$$

and set

$$E_{j,n}(a) = \bigcup_{n=j}^{\infty} E_{j,n}(a) = \left\{ \text{for some } k\geq j, \left| \sum_{i=j}^{k} X_i/i \right| > a \right\}$$

Now by Kolmogorov's Inequality (Theorem 4.2) it follows that

$$P\Big(E_{j,n}(a)\Big) \leq \frac{\sum_{i=j}^{n} \mathrm{Var}(X_i/i)}{a^2} = \frac{1}{a^2} \sum_{i=j}^{n} \sigma_i^2/i^2 \tag{4.4}$$

Also, since the events $E_{j,n}(a)$ are increasing in $n$ to the limiting event $E_j(a)$, it follows by the continuity property of probabilities and Equation (4.4) that

$$\begin{aligned} P\Big(E_j(a)\Big) &= \lim_{n\to\infty} P\Big(E_{j,n}(a)\Big) \\ &\leq \frac{1}{a^2} \sum_{i=j}^{\infty} \sigma_i^2/i^2 \end{aligned} \tag{4.5}$$

Now consider the event

$$D = \bigcup_{m=1}^{\infty} \bigcap_{j=1}^{\infty} E_j(1/m)$$

In words, $D$ is the event that, for some positive value of $m$, for each $j$ there is a larger value $k$ such that $\left| \sum^{k} X_i/i \right| > 1/m$. It is fairly easy to see that the nonoccurence of $D$ implies that $\sum_{i=1}^{\infty} X_i/i$ converges (One way is by noting that

$\sum_{i=1}^{k} X_i/i$ is a Cauchy sequence when $D^c$ occurs). Now

$$P(D) = \left( \bigcup_{m=1}^{\infty} \bigcap_{j=1}^{\infty} E_j\,(1/m) \right)$$

$$= \lim_{m\to\infty} P\left( \bigcap_{j=1}^{\infty} E_j(1/m) \right) \qquad \text{since } \bigcap_{j=1}^{\infty} E_j(1/m) \text{ are increasing in } m$$

$$= \lim_{m\to\infty} \lim_{k\to\infty} P\left( \bigcap_{j=1}^{\infty} E_j(1/m) \right)$$

$$\leq \lim_{m\to\infty} \lim_{k\to\infty} P\left( E_k(1/m) \right) \qquad \text{since } \bigcap_{j=1}^{\mathrm{k}} E_j(1/m) \subset E_k(1/m)$$

$$\leq \lim_{m\to\infty} \lim_{k\to\infty} m^2 \sum_{i=k}^{\infty} \sigma_i^2/i^2 \qquad \text{from (4.5)}$$

$$= 0 \qquad \text{since } \sum_{i=1}^{\infty} \sigma_i^2/i^2 < \infty \text{ implies that } \lim_{k\to\infty} \sum_{i=k}^{\infty} \sigma_i^2/i^2 = 0$$

Thus, $P(D) = 0$, which implies that, with probability 1, $\sum_{i=1}^{\infty} X_i/i$ converges. Hence, by Kronecker's Lemma we obtain that

$$P\left\{ \lim_{n\to\infty} \sum_{i=1}^{n} X_i/n = 0 \right\} = 1$$

and the proof is complete. ∎

If the random variables are assumed to be not only independent, but also identically distributed with mean $\mu$ and finite variance $\sigma^2$, then as $\sum_{i=1}^{\infty} \sigma^2/i^2 < \infty$, it follows that, with probability 1,

$$\lim_{n} \sum_{i=1}^{n} \frac{(X_i - \mu)}{n} = 0$$

or equivalently,

$$\lim_{n} \sum_{i=1}^{n} \frac{X_i}{n} = \mu$$

Hence Theorem 4.3 provides a proof of the strong law in the case of independent and identically distributed (i.i.d.) random variables having finite variances. In fact, it can also be used to prove the strong law for i.i.d. random variables even when the variance is infinite, thus establishing Theorem 4.1.

Many students are initially confused about the difference between the weak and the strong law of large numbers. The weak law of large numbers states that, for any fixed large value of $n$, say $n^*$, $(X_1 + \cdots + X_{n^*})/n^*$ is likely to be near $\mu$. However, it does not say that $(X_1 + \cdots + X_n)/n$ is bound to stay near $\mu$ for all values of $n$ larger than $n^*$. Thus it leaves open the possibility that large values of $|(X_1 + \cdots + X_n)/n - \mu|$ can occur infinitely often (though at infrequent intervals). The strong law shows that this cannot occur. In particular, it shows that, with probability 1, for any positive value $\varepsilon$,

$$\left| \sum_1^n \frac{X_i}{n} - \mu \right|$$

will be greater than $\varepsilon$ only a finite number of times.

The strong law of large numbers was originally proved, in the special case of Bernoulli random variables, by the French mathematician Borel. The general form of the strong law presented in Theorem 4.1 was proved by the Russian mathematician A. N. Kolmogorov.

## 5 Other Inequalities

We are sometimes confronted with situations in which we are interested in obtaining an upper bound for a probability of the form $P\{X - \mu > a\}$, where $a$ is some positive value and when only the mean $\mu = E[X]$ and variance $\sigma^2 = \text{Var}(X)$ of the distribution of $X$ are known. Of course, since $X - \mu > a > 0$ implies that $|X - \mu| > a$, it follows from Chebyshev's inequality that

$$P\{X - \mu > a\} \leq P\{|X - \mu| > a\} \leq \frac{\sigma^2}{a^2} \qquad \text{when } a > 0$$

However, as the following proposition shows, it turns out that we can do better.

---

*Proposition 5.1 One-Sided Chebyshev Inequality*

*If $X$ is a random variable with mean 0 and finite variance $\sigma^2$, then for any $a > 0$*

$$P\{X > a\} \leq \frac{\sigma^2}{\sigma^2 + a^2}$$

---

**Proof:** Since $0 = E[X] = \int_{-\infty}^{\infty} x\, dF(x)$, it follows that

$$\begin{aligned} -a &= \int_{-\infty}^{\infty} (x-a)\, dF(x) \\ &\geq \int_{-\infty}^{a} (x-a)\, dF(x) \end{aligned}$$

Hence

$$\begin{aligned} a &\leq \int_{-\infty}^{a} (a-x)\, dF(x) \\ &= \int_{-\infty}^{\infty} (a-x) I_a(x)\, dF(x) \qquad (5.1) \end{aligned}$$

where

$$I_a(x) = \begin{cases} 1 & \text{if } x \leq a \\ 0 & \text{if } x > a \end{cases}$$

When $a > 0$ we obtain, by squaring both sides of Equation (5.1),

$$a^2 \leq \left( \int_{-\infty}^{\infty} (a-x) I_a(x)\, dF(x) \right)^2$$

or equivalently,

$$a^2 \leq (E[(a-X)I_a(X)])^2 \qquad (5.2)$$

We now make use of a result known as the Cauchy–Schwarz inequality, which states that for any random variables $Y$ and $Z$

$$(E[YZ])^2 \leq E[Y^2]E[Z^2]$$

provided that the right-hand side is finite. (The proof of the Cauchy–Schwarz inequality is provided in Theoretical Exercise 29 of Chapter 7.) Hence, applying the Cauchy–Schwarz inequality to the right-hand side of Equation (5.2), with $Y = (a - X)$ and $Z = I_a(X)$, yields

$$\begin{aligned} a^2 &\leq E[(a-X)^2]E[I_a^2(X)] \\ &= \int_{-\infty}^{\infty} (a-x)^2\, dF(x) \int_{-\infty}^{a} dF(x) \\ &= F(a) \int_{-\infty}^{\infty} (a-x)^2\, dF(x) \\ &= F(a) \left[ \int_{-\infty}^{\infty} a^2\, dF(x) - 2a \int_{-\infty}^{\infty} x\, dF(x) + \int_{-\infty}^{\infty} x^2\, dF(x) \right] \\ &= F(a)(a^2 + \sigma^2) \end{aligned}$$

Hence

$$F(a) \geq \frac{a^2}{a^2 + \sigma^2}$$

or equivalently,

$$P\{X > a\} = 1 - F(a) \leq 1 - \frac{a^2}{a^2 + \sigma^2} = \frac{\sigma^2}{a^2 + \sigma^2}$$

and the result is established. ∎

**Example 5a.** If the number of items produced in a factory during a week is a random variable with mean 100 and variance 400, compute an upper bound on the probability that this week's production will exceed 120.

***Solution:*** It follows from the one-sided Chebyshev inequality that

$$P\{X > 120\} = P\{X - 100 > 20\} \leq \frac{400}{400 + (20)^2} = \frac{1}{2}$$

Hence the probability that this week's production will exceed 120 is at most $\frac{1}{2}$.

If we attempted to obtain a bound by applying Markov's inequality, then we would have obtained

$$P\{X > 120\} \leq \frac{E(X)}{120} = \frac{5}{6}$$

which is a far weaker bound than the previous one. ∎

Suppose now that $X$ has mean $\mu$ and variance $\sigma^2$. As both $X - \mu$ and $\mu - X$ have mean 0 and variance $\sigma^2$, we obtain, from the one-sided Chebyshev inequality, that for $a > 0$,

$$P\{X - \mu > a\} \leq \frac{\sigma^2}{\sigma^2 + a^2}$$

and

$$P\{\mu - X > a\} \leq \frac{\sigma^2}{\sigma^2 + a^2}$$

Thus we have the following corollary.

---

*Corollary 5.1*

*If* $E[X] = \mu$, $\text{Var}(X) = \sigma^2$, *then for* $a > 0$,

$$P\{X > \mu + a\} \leq \frac{\sigma^2}{\sigma^2 + a^2}$$

$$P\{X < \mu - a\} \le \frac{\sigma^2}{\sigma^2 + a^2}$$

---

**Example 5b.** A set of 200 people, consisting of 100 men and 100 women, are randomly divided into 100 pairs of 2 each. Give an upper bound to the probability that less than 30 of these pairs will consist of a man and a woman.

***Solution:*** Number the men, arbitrarily, from 1 to 100 and let for $i = 1, 2, \ldots 100$,

$$X_i = \begin{cases} 1 & \text{if man } i \text{ is paired with a woman} \\ 0 & \text{otherwise} \end{cases}$$

Then $X$, the number of man–woman pairs, can be expressed as

$$X = \sum_{i=1}^{100} X_i$$

As man $i$ is equally likely to be paired with any of the other 199 people, of which 100 are women, we have

$$E[X_i] = P\{X_i = 1\} = \frac{100}{199}$$

Similarly, for $i \neq j$,

$$\begin{aligned} E[X_iX_j] &= P\{X_i = 1, X_j = 1\} \\ &= P\{X_i = 1\}P\{X_j = 1 \mid X_i = 1\} = \frac{100}{199}\frac{99}{197} \end{aligned}$$

where $P\{X_j = 1 \mid X_i = 1\} = 99/197$ since, given that man $i$ is paired with a woman, man $j$ is equally likely to be paired with any of the remaining 197 people, of which 99 are women. Hence we obtain that

$$\begin{aligned} E[X] &= \sum_{i=1}^{100} E[X_i] \\ &= (100)\frac{100}{199} \\ &= 50.25 \end{aligned}$$

$$\begin{aligned} \text{Var}(X) &= \sum_{i=1}^{100} \text{Var}(X_i) + 2\sum_{i<j}\sum \text{Cov}(X_i, X_j) \\ &= 100\frac{100}{199}\frac{99}{199} + 2\binom{100}{2}\left[\frac{100}{199}\frac{99}{197} - \left(\frac{100}{199}\right)^2\right] \\ &= 25.126 \end{aligned}$$

The Chebyshev inequality yields that

$$P\{X < 30\} \leq P\{|X - 50.25| > 20.25\} \leq \frac{25.126}{(20.25)^2} = .061$$

and thus there are fewer than 6 chances in a hundred that fewer than 30 men will be paired with women. However, we can improve on this bound by using the one-sided Chebyshev inequality, which yields that

$$\begin{aligned} P\{X < 30\} &= P\{X < 50.25 - 20.25\} \\ &\leq \frac{25.126}{25.126 + (20.25)^2} \\ &= .058 \end{aligned}$$

∎

The next inequality is one having to do with expectations rather than probabilities. Before stating it, we need the following definition.

---

*Definition*

*A twice differentiable real-valued function $f(x)$ is said to be convex if $f''(x) \geq 0$ for all $x$; similarly, it is said to be concave if $f''(x) \leq 0$.*

---

Some examples of convex functions are $f(x) = x^2, f(x) = e^{ax}$, $f(x) = -x^{1/n}$ for $x \geq 0$. If $f(x)$ is convex, then $g(x) = -f(x)$ is concave, and vice versa.

---

*Proposition 5.2 Jensen's Inequality*

*If $f(x)$ is a convex function, then*

$$E[f(X)] \geq f(E[X])$$

*provided that the expectations exist and are finite.*

---

**Proof:** Expanding $f(x)$ in a Taylor's series expansion about $\mu = E[X]$ yields

$$f(x) = f(\mu) + f'(\mu)(x - \mu) + \frac{f''(\xi)(x - \xi)^2}{2}$$

where $\xi$ is some value between $x$ and $\mu$. Since $f''(\xi) \geq 0$, we obtain

$$f(x) \geq f(\mu) + f'(\mu)(x - \mu)$$

Hence

$$f(X) \geq f(\mu) + f'(\mu)(X - \mu)$$

Taking expectations yields

$$E[f(X)] \geq f(\mu) + f'(\mu)E[X - \mu] = f(\mu)$$

and the inequality is established. ∎

**Example 5c.** An investor is faced with the following choices: she can either invest all of her money in a risky proposition that would lead to a random return $X$ that has mean $m$; or she can put the money into a risk-free venture that will lead to a return of $m$ with probability 1. Suppose her decision will be made on the basis of maximizing the expected value of $u(R)$, where $R$ is her return and $u$ is her "utility" function. By Jensen's inequality it follows that if $u$ is a concave function, then $E[u(X)] \leq u(m)$, and so the risk-free alternative is preferable; whereas if $u$ is convex, then $E[u(X)] \geq u(m)$, and so the risk investment alternative would be preferred. ∎

## *Theoretical Exercises*

**1.** If $X$ has variance $\sigma^2$, then $\sigma$, the positive square root of the variance, is called the *standard deviation.* If $X$ has mean $\mu$ and standard deviation $\sigma$, show that

$$P\{|X - \mu| \geq k\sigma\} \leq \frac{1}{k^2}$$

**2.** If $X$ has mean $\mu$ and standard deviation $\sigma$, the ratio $r \equiv |\mu|/\sigma$ is called the *measurement signal-to-noise ratio* of $X$. The idea being that $X$ can be expressed as $X = \mu + (X - \mu)$ with $\mu$ representing the signal and $X - \mu$ the noise. If we define $|(X - \mu)/\mu| \equiv D$ as the relative deviation of $X$ from its signal (or mean) $\mu$, show that, for $\alpha > 0$,

$$P\{D \leq \alpha\} \geq 1 - \frac{1}{r^2\alpha^2}$$

**3.** Compute the measurement signal-to-noise ratio—that is, $|\mu|/\sigma$ where $\mu = E[X]$, $\sigma^2 = \text{Var}(X)$—of the following random variables:
(a) Poisson with mean $\lambda$;
(b) binomial with parameters $n$ and $p$;
(c) geometric with mean $1/p$;
(d) uniform over $(a, b)$;
(e) exponential with mean $1/\lambda$;
(f) normal with parameters $\mu, \sigma^2$.

**4.** Let $Z_n$, $n \geq 1$ be a sequence of random variables and $c$ a constant such that for each $\varepsilon > 0$, $P\{|Z_n - c| > \varepsilon\} \to 0$ as $n \to \infty$. Show that for any bounded continuous function $g$,

$$E[g(Z_n)] \to g(c) \text{ as } n \to \infty$$

**5.** Let $f(x)$ be a continuous function defined for $0 \le x \le 1$. Consider the functions

$$B_n(x) = \sum_{k=0}^{n} f\left(\frac{k}{n}\right)\binom{n}{k} x^k (1-x)^{n-k}$$

(called Bernstein polynomials) and prove that

$$\lim_{n\to\infty} B_n(x) = f(x)$$

HINT: Let $X_1, X_2, \ldots$ be independent Bernoulli random variables with mean $x$. Show and then use the fact (by making use of the result of Theoretical Exercise 4) that

$$B_n(x) = E\left[f\left(\frac{X_1 + \cdots + X_n}{n}\right)\right]$$

As it can be shown that the convergence of $B_n(x)$ to $f(x)$ is uniform in $x$, the above provides a probabilistic proof to the famous Weierstrass theorem of analysis that states that any continuous function on a closed interval can be approximated arbitrarily closely by a polynomial.

**6.** Let $X_1, X_2, \ldots,$ be a sequence of independent random variables. Suppose that $E[X_i] = 0$ and $\text{Var}(X_i) = \sigma_i^2$, and suppose that

$$\lim_{n\to\infty} \sum_{i=1}^{n} \frac{\sigma_i^2}{n^2} = 0$$

Prove that for any $\varepsilon > 0$,

$$P\left\{\left|\frac{X_1 + \cdots + X_n}{n}\right| > \varepsilon\right\} \to 0 \qquad \text{as } n \to \infty$$

Use this to prove that if $Y_i$, $i \ge 1$, are independent Bernoulli random variables, then, for any $\varepsilon > 0$,

$$P\left\{\left|\frac{Y_1 + \cdots + Y_n}{n} - P(n)\right| \le \varepsilon\right\} \to 1 \qquad \text{as } n \to \infty$$

where $E[Y_i] = P_i$, and $P(n) = \sum_{i=1}^{n} P_i/n$.

**7.** Let $X_1, \ldots, X_n$ be (possibly dependent) random variables such that $E[X_j] = \mu_j$, $\text{Var}(X_j) = \sigma_j^2$, $j = 1, \ldots, n$. Prove that

$$P\left\{\left|\frac{X_j - \mu_j}{\sigma_j}\right| \le \sqrt{n}\, t \text{ for all } j = 1, \ldots, n\right\} \ge 1 - \frac{1}{t^2}$$

for all $t > 0$.

**8.** (a) Let $X$ be a discrete random variable, whose possible values are $1, 2, \ldots$. If $P\{X = k\}$ is nonincreasing in $k = 1, 2, \ldots$, prove that

$$P\{X = k\} \leq 2\frac{E[X]}{k^2}$$

(b) Let $X$ be a nonnegative continuous random variable having a nonincreasing density function. Show that

$$f(x) \leq \frac{2E[X]}{x^2} \qquad \text{for all } x > 0$$

**9.** Suppose that a fair die is rolled 100 times. Let $X_i$ be the value obtained on the $i$th roll. Compute an approximation for

$$P\left\{\prod_{1}^{100} X_i \leq a^{100}\right\} \qquad 1 < a < 6$$

**10.** Explain why a gamma random variable with parameters $(t, \lambda)$ has an approximately normal distribution when $t$ is large.

**11.** Prove that the strong law of large numbers remains valid when $E(X_i) = +\infty$.
HINT: For given $M$, define

$$X_i^M = \begin{cases} X_i & \text{if } X_i \leq M \\ M & \text{if } X_i > M \end{cases}$$

Use (a) the strong law of large numbers on the sequence $X_i^M, i \geq 1$; (b) the fact that $X_i^M \leq X_i$; and (c) let $M \to \infty$.

**12.** Suppose a fair coin is tossed 1000 times. If the first 100 tosses all result in heads, what proportion of heads would you expect on the final 900 tosses? Comment on the statement that "the strong law of large numbers swamps but does not compensate."

**13.** If $E[X] < 0$ and $\theta \neq 0$ is such that $E[e^{\theta X}] = 1$, show that $\theta > 0$.

## Problems

**1.** Suppose that $X$ is a random variable with mean and variance both equal to 20. What can be said about $P\{0 \leq X \leq 40\}$?

**2.** From past experience a professor knows that the test score of a student taking his final examination is a random variable with mean 75.
(a) Give an upper bound to the probability that a student's test score will exceed 85.
Suppose in addition the professor knows that the variance of a student's test score is equal to 25.

(b) What can be said about the probability that a student will score between 65 and 85?

(c) How many students would have to take the examination so as to ensure, with probability at least .9, that the class average would be within 5 of 75. Do not use the central limit theorem.

**3.** Use the central limit theorem to solve part (c) of Problem 2.

**4.** Let $X_1, \ldots, X_{20}$ be independent Poisson random variables with mean 1.

(a) Use the Markov inequality to obtain a bound on $P\left\{\sum_1^{20} X_i > 15\right\}$.

(b) Use the central limit theorem to approximate $P\left\{\sum_1^{20} X_i > 15\right\}$.

**5.** Fifty numbers are rounded off to the nearest integer and then summed. If the individual round-off-errors are uniformly distributed over $(-.5, .5)$ what is the probability that the resultant sum differs from the exact sum by more than 3?

**6.** A die is continually rolled until the total sum of all rolls exceeds 300. What is the probability that at least 80 rolls are necessary?

**7.** One has 100 lightbulbs whose lifetimes are independent exponentials with mean 5 hours. If the bulbs are used one at a time, with a failed bulb being immediately replaced by a new one, what is the probability that there is still a working bulb after 525 hours?

**8.** In Problem 7 suppose it takes a random time, uniformly distributed over $(0,.5)$ to replace a failed bulb. What is the probability that all bulbs have failed by time 550?

**9.** If $X$ is a gamma random variable with parameters $(n,1)$ how large need $n$ be so that

$$P\{|X/n - 1| > .01\} < .01$$

**10.** Civil engineers believe that $W$, the amount of weight (in units of 1000 pounds) that a certain span of a bridge can withstand without structural damage resulting, is normally distributed with mean 400 and standard deviation 40. Suppose that the weight (again, in units of 1000 pounds) of a car is a random variable with mean 3 and standard deviation .3. How many cars would have to be on the bridge span for the probability of structural damage to exceed .1?

**11.** Many people believe that the daily change of price of a company's stock on the stock market is a random variable with mean 0 and variance $\sigma^2$. That is, if $Y_n$ represents the price of the stock as the $n$th day, then

$$Y_n = Y_{n-1} + X_n \qquad n \geq 1$$

where $X_1, X_2, \ldots$ are independent and identically distributed random variables with mean 0 and variance $\sigma^2$. Suppose that the stock's price today is 100. If $\sigma^2 = 1$, what can you say about the probability that, in the next 10 days, the stock's price will always remain between 95 and 105?

**12.** We have 100 components that we will put in use in a sequential fashion. That is, component 1 is initially put in use, and upon failure it is replaced by component 2, which is itself replaced upon failure by component 3, and so on. If the lifetime of component $i$ is exponentially distributed with mean $10 + i/10$, $i = 1, \ldots, 100$, estimate the probability that the total life of all components will exceed 1200. Now repeat when the life distribution of component $i$ is uniformly distributed over $(0, 20 + i/5)$, $i = 1, \ldots, 100$.

**13.** Redo Example 5b under the assumption that the number of man–woman pairs is (approximately) normally distributed. Does this seem like a reasonable supposition?

**14.** Repeat part (a) of Problem 2 when it is known that the variance of a student's test score is equal to 25.

**15.** A lake contains 4 distinct types of fish. Suppose that each fish caught is equally likely to be any one of these types. Let $Y$ denote the number of fish that need be caught in order to obtain at least one of each type.
(a) Give an interval $(a, b)$ such that $P\{a \leq Y \leq b\} \geq .90$.
(b) Using the one-sided Chebyshev inequality, how many fish need we plan on catching so as to be at least 90 percent certain of obtaining at least one of each type?

**16.** If $X$ is a nonnegative random variable with mean 25, what can be said about the following?

(a) $E[X^3]$;
(b) $E[\sqrt{X}]$;
(c) $E[\log X]$;
(d) $E[e^{-X}]$?

**17.** Let $X$ be a nonnegative random variable. Prove that

$$E[X] \leq (E[X^2])^{1/2} \leq (E[X^3])^{1/3} \leq \cdots$$

**18.** Would the results of Example 5c change if the investor were allowed to divide her money and invest the fraction $\alpha, 0 < \alpha < 1$ in the risky proposition and invest the remainder in the risk-free vecture? Her return for such a split investment would be $R = \alpha X + (1 - \alpha)m$.

# 9

# Additional Topics in Probability

## 1. The Poisson Process

Before defining a Poisson process, recall that a function $f$ is said to be $o(h)$ if $\lim_{h\to 0} f(h)/h = 0$. That is, $f$ is $o(h)$ if, for small values of $h, f(h)$ is small even in relation to $h$.

Suppose now that "events" are occurring at random time points and let $N(t)$ denote the number of events that occur in the time interval $[0, t]$. The stochastic process $\{N(t), t \geq 0\}$ is said to be a *Poisson process having rate* $\lambda, \lambda > 0$ if

(i) $N(0) = 0$.
(ii) The number of events that occur in disjoint time intervals are independent.
(iii) The distribution of the number of events that occur in a given interval depends only on the length of that interval and not on its location.
(iv) $P\{N(h) = 1\} = \lambda h + o(h)$.
(v) $P\{N(h) \geq 2\} = o(h)$.

Thus Condition (i) states that the process begins at time 0. Condition (ii), the *independent increment* assumption, states for instance, that the number of events by time $t$ [that is, $N(t)$] is independent of the number of events that occur between $t$ and $t + s$ [that is, $N(t + s) - N(t)$]. Condition (iii), the *stationary increment* assumption, states that the probability distribution of $N(t + s) - N(t)$ is the same for all values of $t$.

In Chapter 4 we presented an argument, based on the Poisson distribution being a limiting version of the binomial distribution, that the above conditions imply that $N(t)$ has a Poisson distribution with mean $\lambda t$. We will now obtain this result by a different method.

---

### *Lemma 1.1*

*For a Poisson process with rate* $\lambda$

$$P\{N(t) = 0\} = e^{-\lambda t}$$

---

**Proof:** Let $P_0(t) = P\{N(t) = 0\}$. We derive a differential equation for $P_0(t)$ in the following manner:

$$\begin{aligned} P_0(t+h) &= P\{N(t+h) = 0\} \\ &= P\{N(t) = 0, N(t+h) - N(t) = 0\} \\ &= P\{N(t) = 0\}P\{N(t+h) - N(t) = 0\} \\ &= P_0(t)[1 - \lambda h + o(h)] \end{aligned}$$

where the final two equations follow from Assumption (ii) plus the fact that Assumptions (iii) and (iv) imply that $P\{N(h) = 0\} = 1 - \lambda h + o(h)$. Hence

$$\frac{P_0(t+h) - P_0(t)}{h} = -\lambda P_0(t) + \frac{o(h)}{h}$$

Now, letting $h \to 0$, we obtain

$$P_0'(t) = -\lambda P_0(t)$$

or equivalently,

$$\frac{P_0'(t)}{P_0(t)} = -\lambda$$

which implies, by integration, that

$$\log P_0(t) = -\lambda t + c$$

or

$$P_0(t) = Ke^{-\lambda t}$$

Since $P_0(0) = P\{N(0) = 0\} = 1$, we arrive at

$$P_0(t) = e^{-\lambda t}$$ ∎

For a Poisson process, let us denote by $T_1$ the time of the first event. Further, for $n > 1$, let $T_n$ denote the elapsed time between the $(n-1)$st and the $n$th event. The sequence $\{T_n, n = 1, 2, \ldots\}$ is called the *sequence of interarrival times*. For instance, if $T_1 = 5$ and $T_2 = 10$, then the first event of the Poisson process would have occurred at time 5 and the second at time 15.

We shall now determine the distribution of the $T_n$. To do so, we first note that the event $\{T_1 > t\}$ takes place if and only if no events of the Poisson process occur in the interval $[0, t]$, and thus,

$$P\{T_1 > t\} = P\{N(t) = 0\} = e^{-\lambda t}$$

Hence $T_1$ has an exponential distribution with mean $1/\lambda$. Now,

$$P\{T_2 > t\} = E[P\{T_2 > t | T_1\}]$$

However,

$$\begin{aligned} P\{T_2 > t | T_1 = s\} &= P\{0 \text{ events in } (s, s+t] | T_1 = s\} \\ &= P\{0 \text{ events in } (s, s+t]\} \\ &= e^{-\lambda t} \end{aligned}$$

where the last two equations followed from independent and stationary increments. Therefore, from the above we conclude that $T_2$ is also an exponential random variable with mean $1/\lambda$, and furthermore, that $T_2$ is independent of $T_1$. Repeating the same argument yields Proposition 1.1.

---

*Proposition 1.1*

$T_1, T_2, \ldots$ are independent exponential random variables each with mean $1/\lambda$.

---

Another quantity of interest is $S_n$, the arrival time of the $n$th event, also called the *waiting time* until the $n$th event. It is easily seen that

$$S_n = \sum_{i=1}^{n} T_i \qquad n \geq 1$$

and hence from Proposition 1.1 and the results of Section 2 of Chapter 5, it follows that $S_n$ has a gamma distribution with parameters $n$ and $\lambda$. That is, the probability density of $S_n$ is given by

$$f_{S_n}(x) = \lambda e^{-\lambda x} \frac{(\lambda x)^{n-1}}{(n-1)!} \qquad x \geq 0$$

We are now ready to prove that $N(t)$ is a Poisson random variable with mean $\lambda t$.

*Theorem 1.1*

For a Poisson process with rate $\lambda$,

$$P\{N(t) = n\} = \frac{e^{-\lambda t}(\lambda t)^n}{n!}$$

**Proof:** Note that the $n$th event of the Poisson process will occur before or at time $t$ if and only if the number of events that occur by $t$ is at least $n$. That is,

$$N(t) \geq n \Leftrightarrow S_n \leq t$$

and so

$$\begin{aligned} P\{N(t) = n\} &= P\{N(t) \geq n\} - P\{N(t) \geq n + 1\} \\ &= P\{S_n \leq t\} - P\{S_{n+1} \leq t\} \\ &= \int_0^t \lambda e^{-\lambda x} \frac{(\lambda x)^{n-1}}{(n-1)!}\,dx - \int_0^t \lambda e^{-\lambda x} \frac{(\lambda x)^n}{n!}\,dx \end{aligned}$$

But the integration by parts formula $\int u\,dv = uv - \int v\,du$ yields, with $u = e^{-\lambda x}$, $dv = \lambda[(\lambda x)^{n-1}/(n-1)!]\,dx$,

$$\int_0^t \lambda e^{-\lambda x} \frac{(\lambda x)^{n-1}}{(n-1)!}\,dx = e^{-\lambda t}\frac{(\lambda t)^n}{n!} + \int_0^t \lambda e^{-\lambda x} \frac{(\lambda x)^n}{n!}\,dx$$

which completes the proof.

## 2 Markov Chains

Consider a sequence of random variables $X_0, X_1, \ldots$, and suppose that the set of possible values of these random variables is $\{0, 1, \ldots, M\}$. It will be helpful to interpret $X_n$ as being the state of some system at time $n$, and, in accordance with this interpretation, we say that the system is in state $i$ at time $n$ if $X_n = i$. The sequence of random variables is said to form a *Markov chain* if each time the system is in state $i$ there is some fixed probability—call it $P_{ij}$—that it will next be in state $j$. That is, for all $i_0, \ldots, i_{n-1}, i, j$,

$$P\{X_{n+1} = j \mid X_n = i, X_{n-1} = i_{n-1}, \ldots, X_1 = i_1, X_0 = i_0\} = P_{ij}$$

The values $P_{ij}$, $0 \le i \le M$, $0 \le j \le N$, are called the transition probabilities of the Markov chain and they satisfy (why?)

$$P_{ij} \geq 0 \qquad \sum_{j=0}^{M} P_{ij} = 1 \qquad i = 0, 1, \ldots, M$$

It is convenient to arrange the transition probabilities $P_{ij}$ in a square array as follows:

$$\begin{Vmatrix} P_{00} & P_{01} \cdots P_{0M} \\ P_{10} & P_{11} \cdots P_{1M} \\ \vdots & \\ P_{M0} & P_{M1} \cdots P_{MM} \end{Vmatrix}$$

Such an array is called a *matrix.*

Knowledge of the transition probability matrix and the distribution of $X_0$ enables us, in theory, to compute all probabilities of interest. For instance, the joint probability mass function of $X_0, \ldots, X_n$ is given by

$$\begin{aligned} P\{X_n &= i_n, X_{n-1} = i_{n-1}, \ldots, X_1 = i_1, X_0 = i_0\} \\ &= P\{X_n = i_n | X_{n-1} = i_{n-1}, \ldots, X_0 = i_0\} P\{X_{n-1} = i_{n-1}, \ldots, X_0 = i_0\} \\ &= P_{i_{n-1}, i_n} P\{X_{n-1} = i_{n-1}, \ldots, X_0 = i_0\} \end{aligned}$$

and continual repetition of this argument yields that the above is equal to

$$= P_{i_{n-1}, i_n} P_{i_{n-2}, i_{n-1}} \cdots P_{i_1, i_2} P_{i_0, i_1} P\{X_0 = i_0\}$$

**Example 2a.** Suppose that whether or not it rains tomorrow depends on previous weather conditions only through whether or not it is raining today. Suppose further that if it is raining today, then it will rain tomorrow with probability $\alpha$, and, if it is not raining today, then it will rain tomorrow with probability $\beta$.

If we say that the system is in state 0 when it rains and state 1 when it does not, then the above is a two-state Markov chain having transition probability matrix

$$\begin{Vmatrix} \alpha & 1 - \alpha \\ \beta & 1 - \beta \end{Vmatrix}$$

That is, $P_{00} = \alpha = 1 - P_{01}$, $P_{10} = \beta = 1 - P_{11}$. ∎

**Example 2b.** Consider a gambler who, at each play of the game, either wins 1 unit with probability $p$ or loses 1 unit with probability $1 - p$. If we suppose that the gambler will quit playing when his fortune hits either 0 or $M$, then the gambler's sequence of fortunes is a Markov chain having transition probabilities

$$\begin{aligned} P_{i,i+1} = p = 1 - P_{i,i-1} \qquad & i = 1, \ldots, M - 1 \\ P_{00} = P_{MM} = 1 \qquad & \end{aligned}$$ ∎

**Example 2c.** The physicists P. and T. Ehrenfest considered a conceptual model for the movement of molecules in which $M$ molecules are distributed among 2 urns. At each time point one of the molecules is chosen at random, and is removed from its urn and placed in the other one. If we let $X_n$ denote the number of molecules in the first urn immediately after the $n$th exchange, then $\{X_0, X_1, \ldots\}$ is a Markov chain with transition probabilities

$$P_{i,i+1} = \frac{M-i}{M} \qquad 0 \le i \le M$$

$$P_{i,i-1} = \frac{i}{M} \qquad 0 \le i \le M$$

$$P_{ij} = 0 \qquad \text{if } |j-i| > 1$$

∎

Thus, for a Markov chain, $P_{ij}$ represents the probability that a system in state $i$ will enter state $j$ at the next transition. We can also define the two-stage transition probability, $P_{ij}^{(2)}$, that a system, presently in state $i$, will be in state $j$ after two additional transitions. That is,

$$P_{ij}^{(2)} = P\{X_{m+2} = j \mid X_m = i\}$$

The $P_{ij}^{(2)}$ can be computed from the $P_{ij}$ as follows:

$$\begin{aligned}
P_{ij}^{(2)} &= P\{X_2 = j \mid X_0 = i\} \\
&= \sum_{k=0}^{M} P\{X_2 = j, X_1 = k \mid X_0 = i\} \\
&= \sum_{k=0}^{M} P\{X_2 = j \mid X_1 = k, X_0 = i\} P\{X_1 = k \mid X_0 = i\} \\
&= \sum_{k=0}^{M} P_{kj} P_{ik}
\end{aligned}$$

In general, we define the $n$-stage transition probabilities, denoted as $P_{ij}^{(n)}$, by

$$P_{ij}^{(n)} = P\{X_{n+m} = j \mid X_m = i\}$$

Proposition 2.1, known as the Chapman–Kolmogorov equations, shows how the $P_{ij}^{(n)}$ can be computed.

---

*Proposition 2.1 The Chapman–Kolmogorov Equations*

$$P_{ij}^{(n)} = \sum_{k=0}^{M} P_{ik}^{(r)} P_{kj}^{(n-r)} \qquad \textit{for all } 0 < r < n$$

---

**Proof**

$$\begin{aligned}
P_{ij}^{(n)} &= P\{X_n = j \mid X_0 = i\} \\
&= \sum_k P\{X_n = j, X_r = k \mid X_0 = i\} \\
&= \sum_k P\{X_n = j \mid X_r = k, X_0 = i\} P\{X_r = k \mid X_0 = i\} \\
&= \sum_k P_{kj}^{(n-r)} P_{ik}^{(r)}
\end{aligned}$$

∎

**Example 2d.** A Random Walk. An example of a Markov chain having a countably infinite state space is the so-called random walk, which tracks a particle as it moves along a one-dimensional axis. Suppose that at each point in time the particle will either move one step to the right or one step to the left with respective probabilities $p$ and $1 - p$. That is, suppose the particle's path follows a Markov chain with transition probabilities

$$P_{i,i+1} = p = 1 - P_{i,i-1} \qquad i = 0, \pm 1, \ldots$$

If the particle is at state $i$, then the probability it will be at state $j$ after $n$ transitions is the probability that $(n - i + j)/2$ of these steps are to the right and $n - [(n - i + j)/2] = (n + i - j)/2$ are to the left. As each step will be to the right, independently of the other steps, with probability $p$, it follows that the above is just the binomial probability

$$P_{ij}^{n} = \binom{n}{\dfrac{n - i + j}{2}} p^{(n-i+j)/2}(1 - p)^{(n+i-j)/2}$$

where $\binom{n}{x}$ is taken to equal 0 when $x$ is not a nonnegative integer less than or equal to $n$. The above can be rewritten as

$$P_{i,i+2k}^{2n} = \binom{2n}{n + k} p^{n+k}(1 - p)^{n-k} \qquad k = 0, \pm 1, \ldots, \pm n$$

$$P_{i,i+2k-1}^{2n+1} = \binom{2n + 1}{n + k + 1} p^{n+k+1}(1 - p)^{n-k}$$

$$k = 0, \pm 1, \ldots, \pm n, -(n + 1)$$

∎

Although the $P_{ij}^{(n)}$ denote conditional probabilities, we can, by conditioning on the initial state, use them to derive expressions for unconditional probabilities. For instance,

$$\begin{aligned}
P\{X_n = j\} &= \sum_i P\{X_n = j \mid X_0 = i\} P\{X_0 = i\} \\
&= \sum_i P_{ij}^{(n)} P\{X_0 = i\}
\end{aligned}$$

For a large number of Markov chains it turns out that $P_{ij}^{(n)}$ converges, as $n \to \infty$, to a value $\Pi_j$ that depends only on $j$. That is, for large values of $n$, the probability of being in state $j$ after $n$ transitions is approximately equal to $\Pi_j$ no matter what the initial state was. It can be shown that a sufficient condition for a Markov chain to possess this property is that for some $n > 0$,

$$P_{ij}^{(n)} > 0 \qquad \text{for all } i, j = 0, 1, \ldots, M \tag{2.1}$$

Markov chains that satisfy Equation (2.1) are said to be *ergodic*. Since Proposition 2.1 yields

$$P_{ij}^{(n+1)} = \sum_{k=0}^{M} P_{ik}^{(n)} P_{kj}$$

it follows, by letting $n \to \infty$, that for egodic chains

$$\Pi_j = \sum_{k=0}^{M} \Pi_k P_{kj} \tag{2.2}$$

Furthermore, since $1 = \sum_{j=0}^{M} P_{ij}^{(n)}$, we also obtain, by letting $n \to \infty$,

$$\sum_{j=0}^{M} \Pi_j = 1 \tag{2.3}$$

In fact, it can be shown that the $\Pi_j$, $0 \le j \le M$, are the unique nonnegative solutions of Equations (2.2) and (2.3). All this is summed up in Theorem 2.1, which we state without proof.

---

*Theorem 2.1*

*For an ergodic Markov chain*

$$\Pi_j = \lim_{n \to \infty} P_{ij}^{(n)}$$

*exists, and the $\Pi_j$, $0 \le j \le M$ are the unique nonnegative solutions of*

$$\Pi_j = \sum_{k=0}^{M} \Pi_k P_{kj}$$

$$\sum_{j=0}^{M} \Pi_j = 1$$

---

**Example 2e.** Consider Example 2a, in which we assume that if it rains today, then it will rain tomorrow with probability $\alpha$; and, if it does not rain today, then it will rain tomorrow with probability $\beta$. From Theorem

2.1 it follows that the limiting probabilities of rain and of no rain, $\Pi_0$ and $\Pi_1$, are given by

$$\begin{aligned}\Pi_0 &= \alpha\Pi_0 + \beta\Pi_1\\ \Pi_1 &= (1-\alpha)\Pi_0 + (1-\beta)\Pi_1\\ \Pi_0 + \Pi_1 &= 1\end{aligned}$$

which yields

$$\Pi_0 = \frac{\beta}{1+\beta-\alpha} \qquad \Pi_1 = \frac{1-\alpha}{1+\beta-\alpha}$$

For instance, if $\alpha = .6$, $\beta = .3$, then the limiting probability of rain on the $n$th day is $\Pi_0 = \frac{3}{7}$. ∎

The quantity $\Pi_j$ is also equal to the long run proportion of time that the Markov chain is in state $j$, $j = 0, \ldots, M$. To intuitively see why this might be so, let $P_j$ denote the long run proportion of time the chain is in state $j$. (It can be proven, using the strong law of large numbers, that for an ergodic chain such long run proportions exist and are constants). Now, since the proportion of time the chain is in state $k$ is $P_k$ and since, when in state $k$, the chain goes to state $j$ with probability $P_{kj}$, it follows that the proportion of time the Markov chain is entering state $j$ from state $k$ is equal to $P_k P_{kj}$. Summing over all $k$ shows that $P_j$, the proportion of time the Markov chain is entering state $j$, satisfies

$$P_j = \sum_k P_k P_{kj}$$

Since clearly it is also true that

$$\sum_j P_j = 1$$

it thus follows, since by Theorem 2.1 the $\Pi_j$, $j = 0, \ldots, M$ are the unique solution of the above, that $P_j = \Pi_j$, $j = 0, \ldots, M$.

**Example 2f.** Suppose, in Example 2c, we are interested in the proportion of time there are $j$ molecules in urn 1, $j = 0, \ldots, M$. By Theorem 2.1 these quantities will be the unique solution of

$$\begin{aligned}\Pi_0 &= \Pi_1 \times \frac{1}{M}\\ \Pi_j &= \Pi_{j-1} \times \frac{M-j+1}{M} + \Pi_{j+1} \times \frac{j+1}{M}, \qquad j = 1, \ldots, M\\ \Pi_M &= \Pi_{M-1} \times \frac{1}{M}\\ \sum_{j=0}^{M} \Pi_j &= 1\end{aligned}$$

However, as it is easily checked that

$$\Pi_j = \binom{M}{j} \left(\tfrac{1}{2}\right)^M, \qquad j = 0, \ldots, M$$

satisfy the above equations it follows that these are the long run proportions of time that the Markov chain is in each of the states. (See Problem 11 for an explanation of how one might have guessed at the above solution). ∎

## 3 Surprise, Uncertainty, and Entropy

Consider an event $E$ that can occur when an experiment is performed. How surprised would we be to hear that $E$ does, in fact, occur? It seems reasonable to suppose that the amount of surprise engendered by the information that $E$ has occurred should depend on the probability of $E$. For instance, if the experiment consists of rolling a pair of dice, then we would not be too surprised to hear that $E$ has occurred when $E$ represents the event that the sum of the dice is even (and thus has probability $\frac{1}{2}$), whereas we would certaintly be more surprised to hear that $E$ has occurred when $E$ is the event that the sum of the dice is 12 (and thus has probability $\frac{1}{36}$).

In this section we shall attempt to quantify the concept of surprise. To begin, let us agree to suppose that the "surprise" one feels upon learning that an event $E$ has occurred depends only on the probability of $E$; and let us denote by $S(p)$ the surprise evoked by the occurrence of an event having probability $p$. We shall attempt to determine the functional form of $S(p)$ by first agreeing on a set of reasonable conditions that $S(p)$ should satisfy and then proving that these axioms require that $S(p)$ has a specified form. We shall assume throughout that $S(p)$ is defined for all $0 < p \leq 1$ but is not defined for events having $p = 0$.

Our first condition is just a statement of the intuitive fact that there is no surprise in hearing that an event sure to occur has indeed occurred.

---

*Axiom 1*

$$S(1) = 0$$

---

Our second condition states that the more unlikely an event is to occur, the greater is the surprise evoked by its occurrence.

---

### Axiom 2

*$S(p)$ is a strictly decreasing function of $p$, that is, if $p < q$, then $S(p) > S(q)$.*

---

The third condition is a mathematical statement of the fact that we would intuitively expect a small change in $p$ to correspond to a small change in $S(p)$.

---

### Axiom 3

*$S(p)$ is a continuous function of $p$.*

---

To motivate the final condition, consider two independent events $E$ and $F$, having respective probabilities $P(E) = p$ and $P(F) = q$. Since $P(EF) = pq$, the surprise evoked by the information that both $E$ and $F$ have occurred is $S(pq)$. Now, suppose that we are first told that $E$ has occurred and then, afterwards, that $F$ has also occurred. As $S(p)$ is the surprise evoked by the occurrence of $E$, it follows that $S(pq) - S(p)$ represents the additional surprise evoked when we are informed that $F$ has also occurred. However, as $F$ is independent of $E$, the knowledge that $E$ occurred does not change the probability of $F$, and hence the additional surprise should just be $S(q)$. This reasoning suggests the final condition.

---

### Axiom 4

$$S(pq) = S(p) + S(q) \qquad 0 < p \leq 1, 0 < q \leq 1$$

---

We are now ready for Theoren 3.1, which yields the structure of $S(p)$.

---

*Theorem 3.1*

*If $S(\cdot)$ satisfies Axioms* 1 *through* 4, *then*

$$S(p) = -C \log_2 p$$

*where $C$ is an arbitrary positive integer.*

---

**Proof:** It follows from Axiom 4 that

$$S(p^2) = S(p) + S(p) = 2S(p)$$

and by induction

$$S(p^m) = mS(p) \tag{3.1}$$

Also, since for any integral $n$, $S(p) = S(p^{1/n} \cdots p^{1/n}) = nS(p^{1/n})$, it follows that

$$S(p^{1/n}) = \frac{1}{n} S(p) \tag{3.2}$$

Thus, from Equations (3.1) and (3.2), we obtain

$$\begin{aligned} S(p^{m/n}) &= mS(p^{1/n}) \\ &= \frac{m}{n} S(p) \end{aligned}$$

which is equivalent to

$$S(p^x) = xS(p) \tag{3.3}$$

whenever $x$ is a positive rational number. But this implies by the continuity of $S$ (Axiom 3) that Equation (3.3) is valid for all nonnegative $x$. (Reason this out.)

Now, for any $p, 0 < p \leq 1$, let $x = -\log_2 p$. Then $p = (\frac{1}{2})^x$ and from Equation (3.3)

$$S(p) = S((\tfrac{1}{2})^x) = xS(\tfrac{1}{2}) = -C \log_2 p$$

where $C = S(\frac{1}{2}) > S(1) = 0$ by Axioms 2 and 1. ∎

It is usual, and we shall do so, to let $C$ equal 1. In this case the surprise is said to be expressed in units of *bits* (short for binary digits).

Consider now a random variable $X$, which must take on one of the values $x_1, \ldots, x_n$ with respective probabilities $p_1, \ldots, p_n$. As $-\log p_i$ represents the surprise evoked if $X$ takes on the value $x_i$,[1] it follows that the expected amount of surprise we shall receive upon learning the value of $X$ is given by

$$H(X) = -\sum_{i=1}^{n} p_i \log p_i$$

The quantity $H(X)$ is known in information theory as the *entropy* of the random variable $X$. (In case one of the $p_i = 0$, we take $0 \log 0$ to equal 0.) It can be shown (and we leave it as an exercise) that $H(X)$ is maximized when all of the $p_i$ are equal. (Is this intuitive?)

As $H(X)$ represents the average amount of surprise one receives upon learning the value of $X$, it can also be interpreted as representing the amount of *uncertainty* that exists as to the value of $X$. In fact, in information theory, $H(X)$ is interpreted as the average amount of *information* received when the value of $X$ is observed. Thus the average surprise evoked by $X$, the uncertainty of $X$, or the average amount of information yielded by $X$, all represent the same concept viewed from three slightly different points of view.

Consider now two random variables $X$ and $Y$, which take on respective values $x_1, \ldots, x_n$ and $y_1, \ldots, y_m$ with joint mass function

$$p(x_i, y_j) = P\{X = x_i, Y = y_j\}$$

It follows that the uncertainty as to the value of the random vector $(X, Y)$, denoted by $H(X, Y)$, is given by

$$H(X, Y) = -\sum_i \sum_j p(x_i, y_j) \log p(x_i, y_j)$$

Suppose now that $Y$ is observed to equal $y_j$. In this situation the amount of uncertainty remaining in $X$ is given by

$$H_{Y=y_j}(X) = -\sum_i p(x_i \mid y_j) \log p(x_i \mid y_j)$$

where

$$p(x_i \mid y_j) = P\{X = x_i \mid Y = y_j\}$$

Hence the average amount of uncertainty that will remain in $X$ after $Y$ is observed is given by

$$H_Y(X) = \sum_j H_{Y=y_j}(X) p_Y(y_j)$$

where

$$p_Y(y_j) = P\{Y = y_j\}$$

[1] For the remainder of this chapter we shall write $\log x$ for $\log_2 x$. Also, we shall use $\ln x$ for $\log_e x$.

Proposition 3.1 relates $H(X, Y)$ to $H(Y)$ and $H_Y(X)$. It states that the uncertainty as to the value of $X$ and $Y$ is equal to the uncertainty of $Y$ plus the average uncertainty remaining in $X$ when $Y$ is to be observed.

---

*Proposition 3.1*

$$H(X, Y) = H(Y) + H_Y(X)$$

---

**Proof:** Using the identity $p(x_i, y_j) = p_Y(y_j)p(x_i \mid y_j)$ yields

$$\begin{aligned} H(X, Y) &= -\sum_i \sum_j p(x_i, y_j) \log p(x_i, y_j) \\ &= -\sum_i \sum_j p_Y(y_j)p(x_i \mid y_j)[\log p_Y(y_j) + \log p(x_i \mid y_j)] \\ &= -\sum_j p_Y(y_j) \log p_Y(y_j) \sum_i p(x_i \mid y_j) \\ &\quad -\sum_j p_Y(y_j) \sum_i p(x_i \mid y_j) \log p(x_i \mid y_j) \\ &= H(Y) + H_Y(X) \end{aligned}$$

∎

It is a fundamental result in information theory that the amount of uncertainty in a random variable $X$ will, on the average, decrease when a second random variable $Y$ is observed. Before proving this, we need the following lemma, whose proof is left as an exercise.

---

*Lemma 3.1*

$$\ln x \leq x - 1 \qquad x > 0$$

*with equality only at $x = 1$.*

---

---

*Theorem 3.2*

$$H_Y(X) \leq H(X)$$

*with equality if and only if $X$ and $Y$ are independent.*

---

**Proof**

$$
\begin{aligned}
H_Y(X) - H(X) &= -\sum_i \sum_j p(x_i \mid y_j) \log [p(x_i \mid y_j)] p(y_j) \\
&\quad + \sum_i \sum_j p(x_i, y_j) \log p(x_i) \\
&= \sum_i \sum_j p(x_i, y_j) \log \left[ \frac{p(x_i)}{p(x_i \mid y_j)} \right] \\
&\le \log e \sum_i \sum_j p(x_i, y_j) \left[ \frac{p(x_i)}{p(x_i \mid y_j)} - 1 \right] \qquad \text{by Lemma 3.1} \\
&= \log e \left[ \sum_i \sum_j p(x_i) p(y_j) - \sum_i \sum_j p(x_i, y_j) \right] \\
&= \log e [1 - 1] \\
&= 0
\end{aligned}
$$

∎

## 4 Coding Theory and Entropy

Suppose that the value of a discrete random vector $X$ is to be observed at location $A$ and then transmitted to location $B$ via a communication network that consists of two signals, 0 and 1. In order to do this, it is first necessary to encode each possible value of $X$ in terms of a sequence of 0's and 1's. To avoid any ambiguity, it is usually required that no encoded sequence can be obtained from a shorter encoded sequence by adding more terms to the shorter.

For instance, if $X$ can take on four possible values $x_1, x_2, x_3, x_4$, then one possible coding would be

$$
\begin{aligned}
x_1 &\leftrightarrow 00 \\
x_2 &\leftrightarrow 01 \\
x_3 &\leftrightarrow 10 \\
x_4 &\leftrightarrow 11
\end{aligned}
\tag{4.1}
$$

That is, if $X = x_1$, then the message 00 is sent to location $B$, whereas if $X = x_2$, then 01 is sent to $B$, and so on. A second possible coding is

$$
\begin{aligned}
x_1 &\leftrightarrow 0 \\
x_2 &\leftrightarrow 10 \\
x_3 &\leftrightarrow 110 \\
x_4 &\leftrightarrow 111
\end{aligned}
\tag{4.2}
$$

However, a coding such as

$$\begin{aligned} x_1 &\leftrightarrow 0 \\ x_2 &\leftrightarrow 1 \\ x_3 &\leftrightarrow 00 \\ x_4 &\leftrightarrow 01 \end{aligned}$$

is not allowed because the coded sequences for $x_3$ and $x_4$ are both extensions of the one for $x_1$.

One of the objectives in devising a code is to minimize the expected number of bits (that is, binary digits) that need be sent from location $A$ to location $B$. For example, if

$$\begin{aligned} P\{X = x_1\} &= \tfrac{1}{2} \\ P\{X = x_2\} &= \tfrac{1}{4} \\ P\{X = x_3\} &= \tfrac{1}{8} \\ P\{X = x_4\} &= \tfrac{1}{8} \end{aligned}$$

then the code given by Equation (4.2) would expect to send $\frac{1}{2}(1) + \frac{1}{4}(2) + \frac{1}{8}(3) + \frac{1}{8}(3) = 1.75$ bits; whereas the code given by Equation (4.1) would expect to send 2 bits. Hence, for the above set of probabilities, the encoding in Equation (4.2) is more efficient than that in Equation (4.1).

The above raises the following question: For a given random vector $X$, what is the maximum efficiency achievable by an encoding scheme? The answer is that for any coding, the average number of bits that will be sent is at least as large as the entropy of $X$. In order to prove this result, known in information theory as the noiseless coding theorem, we shall need Lemma 4.1.

---

*Lemma 4.1*

*Let $X$ take on the possible values $x_1, \ldots, x_N$. Then, in order for it to be possible to encode the values of $X$ in binary sequences (none of which is an extension of another) of respective lengths $n_1, \ldots, n_N$, it is necessary and sufficient that*

$$\sum_{i=1}^{N} \left(\tfrac{1}{2}\right)^{n_i} \leq 1$$

---

**Proof:** For a fixed set of $N$ positive integers $n_1, \ldots, n_N$, let $w_j$ denote the number of the $n_i$ that are equal to $j$, $j = 1, \ldots$. In order for there to be a coding that assigns $n_i$ bits to the value $x_i$, $i = 1, \ldots, N$, it is clearly necessary that $w_1 \leq 2$. Furthermore, as no binary sequence is allowed to be an extension of any other, we must have that $w_2 \leq 2^2 - 2w_1$. (This follows because $2^2$ is the number of binary sequences of length 2, whereas $2w_1$ is the number of sequences that are extensions of the $w_1$ binary sequence of length 1.) In general, the same reasoning shows that we must have

$$w_n \leq 2^n - w_1 2^{n-1} - w_2 2^{n-2} - \cdots - w_{n-1} 2 \tag{4.3}$$

for $n = 1, \ldots$. In fact, a little thought should convince the reader that these conditions are not only necessary but are also sufficient for a code to exist that assigns $n_i$ bits to $x_i$, $i = 1, \ldots, N$.

Rewriting inequality (4.3) as

$$w_n + w_{n-1} 2 + w_{n-2} 2^2 + \cdots + w_1 2^{n-1} \leq 2^n \qquad n = 1, \ldots$$

and dividing by $2^n$ yields that the necessary and sufficient conditions are

$$\sum_{j=1}^{n} w_j \left(\tfrac{1}{2}\right)^j \leq 1 \qquad \text{for all } n \tag{4.4}$$

However, as $\sum_{j=1}^{n} w_j \left(\tfrac{1}{2}\right)^j$ is increasing in $n$, it follows that Equation (4.4) will be true if and only if

$$\sum_{j=1}^{\infty} w_j \left(\tfrac{1}{2}\right)^j \leq 1$$

The result is now established, since by the definition of $w_j$ as the number of $n_i$ that equal $j$, it follows that

$$\sum_{j=1}^{\infty} w_j \left(\tfrac{1}{2}\right)^j = \sum_{i=1}^{N} \left(\tfrac{1}{2}\right)^{n_i} \qquad \blacksquare$$

We are now ready to prove Theorem 4.1.

---

### *Theorem 4.1 The Noiseless Coding Theorem*

*Let $X$ take on the values $x_1, \ldots, x_N$ with respective probabilities $p(x_1), \ldots, p(x_N)$. Then, for any coding of $X$ that assigns $n_i$ bits to $x_i$,*

$$\sum_{i=1}^{N} n_i p(x_i) \geq H(X) = -\sum_{i=1}^{N} p(x_i) \log p(x_i)$$

---

**Proof:** Let $P_i = p(x_i)$, $q_i = 2^{-n_i} \Big/ \sum_{j=1}^{N} 2^{-n_j}$, $i = 1, \ldots, N$. Then

$$
\begin{aligned}
-\sum_{i=1}^{N} P_i \log\left(\frac{P_i}{q_i}\right) &= -\log e \sum_{i=1}^{N} P_i \ln\left(\frac{P_i}{q_i}\right) \\
&= \log e \sum_{i=1}^{N} P_i \ln\left(\frac{q_i}{P_i}\right) \\
&\leq \log e \sum_{i=1}^{N} P_i\left(\frac{q_i}{P_i} - 1\right) \qquad \text{by Lemma 3.1} \\
&= 0 \qquad \text{since } \sum_{i=1}^{N} P_i = \sum_{i=1}^{N} q_i = 1
\end{aligned}
$$

Hence

$$
\begin{aligned}
-\sum_{i=1}^{N} P_i \log P_i &\leq -\sum_{i=1}^{N} P_i \log q_i \\
&= \sum_{i=1}^{N} n_i P_i + \log\left(\sum_{j=1}^{N} 2^{-n_j}\right) \\
&\leq \sum_{i=1}^{N} n_i P_i \qquad \text{by Lemma 4.1}
\end{aligned}
$$

∎

**Example 4a.** Consider a random variable $X$ with probability mass function

$$p(x_1) = \tfrac{1}{2} \qquad p(x_2) = \tfrac{1}{4} \qquad p(x_3) = p(x_4) = \tfrac{1}{8}$$

Since

$$
\begin{aligned}
H(X) &= -[\tfrac{1}{2}\log\tfrac{1}{2} + \tfrac{1}{4}\log\tfrac{1}{4} + \tfrac{1}{4}\log\tfrac{1}{8}] \\
&= \tfrac{1}{2} + \tfrac{2}{4} + \tfrac{3}{4} \\
&= 1.75
\end{aligned}
$$

it follows, from Theorem 4.1, that there is no more efficient coding scheme than

$$
\begin{aligned}
x_1 &\leftrightarrow 0 \\
x_2 &\leftrightarrow 10 \\
x_3 &\leftrightarrow 110 \\
x_4 &\leftrightarrow 111
\end{aligned}
$$

∎

For most random vectors there does not exist a coding for which the average number of bits sent attains the lower bound $H(X)$. However, it is always possible to devise a code such that the average number of bits is within 1 of $H(X)$. To prove this, define $n_i$ to be the integer satisfying

$$-\log p(x_i) \leq n_i \leq -\log p(x_i) + 1$$

Now,

$$\sum_{i=1}^{N} 2^{-n_i} \leq \sum_{i=1}^{N} 2^{\log p(x_i)} = \sum_{i=1}^{N} p(x_i) = 1$$

and so by Lemma 4.1 we can associate sequences of bits, having lengths $n_i$ to the $x_i$, $i = 1, \ldots, N$. The average length of such a sequence,

$$L = \sum_{i=1}^{N} n_i p(x_i)$$

satisfies

$$-\sum_{i=1}^{N} p(x_i) \log p(x_i) \leq L \leq -\sum_{i=1}^{N} p(x_i) \log p(x_i) + 1$$

or

$$H(X) \leq L \leq H(X) + 1$$

**Example 4b.** Suppose that 10 independent tosses of a coin, having probability $p$ of coming up heads, are made at location $A$ and the result is to be transmitted to location $B$. The outcome of this experiment is a random vector $X = (X_1, \ldots, X_{10})$, where $X_i$ is 1 or 0 according to whether or not the outcome of the $i$th toss is heads. By the results of this section it follows that $L$, the average number of bits transmitted by any code satisfies

$$H(X) \leq L$$

with

$$L \leq H(X) + 1$$

for at least one code. Now, since the $X_i$ are independent, it follows from Proposition 3.1 and Theorem 3.2 that

$$\begin{aligned} H(X) = H(X_1, \ldots, X_n) &= \sum_{i=1}^{n} H(X_i) \\ &= -10[p \log p + (1-p) \log (1-p)] \end{aligned}$$

If $p = \frac{1}{2}$, then $H(X) = 10$, and it follows that we can do no better than just encoding $X$ by its actual value. That is, for example, if the first 5 tosses come up heads and the last 5 tails, then the message 1111100000 is transmitted to location $B$.

However, if $p \neq \frac{1}{2}$, we can often do better by using a different coding scheme. For instance, if $p = \frac{1}{4}$, then

$$H(X) = -10(\tfrac{1}{4} \log \tfrac{1}{4} + \tfrac{3}{4} \log \tfrac{3}{4}) = 8.11$$

and thus there is an encoding for which the average length of the encoded message is no greater than 9.11.

One simple coding that is more efficient, in this case, than the identity code is to break up $(X_1, \ldots, X_{10})$ into 5 pairs of 2 random variables each and then code each of the pairs as follows:

$$\begin{aligned}
X_i = 0, X_{i+1} &= 0 \leftrightarrow 0 \\
X_i = 0, X_{i+1} &= 1 \leftrightarrow 10 \\
X_i = 1, X_{i+1} &= 0 \leftrightarrow 110 \\
X_i = 1, X_{i+1} &= 1 \leftrightarrow 111
\end{aligned}$$

for $i = 1, 3, 5, 7, 9$. The total message then transmitted is the successive encodings of the above pairs.

For instance, if the outcome *TTTHHTTTTH* is observed, then the message 010110010 is sent. The average number of bits needed to transmit the message using this code is

$$\begin{aligned}
5[1(\tfrac{3}{4})^2 + 2(\tfrac{1}{4})(\tfrac{3}{4}) + 3(\tfrac{1}{4})(\tfrac{3}{4}) + 3(\tfrac{1}{4})^2] &= \tfrac{135}{16} \\
&= 8.44
\end{aligned}$$

∎

Up to this point we have assumed that the message sent at location $A$ is received, without error, at location $B$. However, there are always certain errors that can occur because of random disturbances along the communications channel. Such random disturbances might lead, for example, to the message 00101101, sent at $A$, being received at $B$ in the form 01101101.

Let us suppose that a bit transmitted at location $A$ will be correctly received at location $B$ with probability $p$, independently from bit to bit. Such a communications system is called a binary symmetric channel. Suppose further that $p = .8$ and we want to transmit a message, consisting of a large number of bits, from $A$ to $B$. Thus direct transmission of the message will result in an error probability of .20, for each bit, which is quite high. One way to reduce this probability of bit error would be to transmit each bit 3 times and then decode by majority rule. That is, we could use the following scheme:

| *Encode* | *Decode* |
|---|---|
| $0 \rightarrow 000$ | 000, 001, 010, 100 $\rightarrow 0$ |
| $1 \rightarrow 111$ | 111, 110, 101, 011 $\rightarrow 1$ |

Note that if no more than one error occurs in transmission, then the bit will be correctly decoded. Hence the probability of bit error is reduced to

$$(.2)^3 + 3(.2)^2(.8) = .104$$

a considerable improvement. In fact, it is clear that we can make the probability of bit error as small as we want by repeating the bit many times and then decoding by majority rule. For instance, the following scheme:

| *Encode* | *Decode* |
|---|---|
| 0 → string of 17 0's | By majority rule |
| 1 → string or 17 1's | |

will reduce the probability of bit error to below .01.

The problem with the above type of encoding scheme is that although it decreases the probability of bit error, it does so at the cost of also decreasing the effective rate of bits sent per signal. (See Table 9.1.)

**Table 9.1 Repetition of Bits Encoding Scheme**

| *Probability of error (per bit)* | *Rate (bits transmitted per signal)* |
|---|---|
| .20 | 1 |
| .10 | .33 $(=\frac{1}{3})$ |
| .01 | .06 $(=\frac{1}{17})$ |

In fact, at this point it may appear inevitable to the reader that decreasing the probability of bit error to 0 always results in also decreasing the effective rate at which bits are transmitted per signal to 0. However, it is a remarkable result of information theory, known as the noisy coding theorem and due to Claude Shannon, that this is not the case. We now state this result as Theorem 4.2.

---

*Theorem 4.2 The Noisy Coding Theorem*

*There is a number C such that for any value R which is less than C, and any $\varepsilon > 0$, there exists a coding-decoding scheme that transmits at an average rate of R bits sent per signal and with an error (per bit) probability of less than $\varepsilon$. The largest such value of C, call it $C^*$,[2] is called the channel capacity, and for the binary symmetric channel,*

$$C^* = 1 + p \log p + (1 - p) \log (1 - p)$$

---

## Theoretical Exercises and Problems

**1.** Customers arrive at a bank at a Poisson rate $\lambda$. Suppose two customers arrived during the first hour. What is the probability that
(a) both arrived during the first 20 minutes;
(b) at least one arrived during the first 20 minutes?

**2.** Cars cross a certain point in the highway in accordance with a Poisson process with rate $\lambda = 3$ per minute. If Al blindly runs across the highway, then what is the probability that he will be uninjured if the amount of time that it takes him to cross the road is $s$ seconds? (Assume that if he is on the highway when a car passes by, then he will be injured.) Do it for $s = 2, 5, 10, 20$.

**3.** Suppose in Problem 2, that Al is agile enough to escape from a single car but if he encounters two or more cars while attempting to cross the road, then he is injured. What is the probability that he will be unhurt if it takes him $s$ seconds to cross? Do it for $s = 5, 10, 20, 30$.

**4.** Suppose that 3 white and 3 black balls are distributed in two urns in such a way that each contains 3 balls. We say that the system is in state $i$ if the first urn contains $i$ white balls, $i = 0, 1, 2, 3$. At each stage 1 ball is drawn from each urn and the ball drawn from the first urn is placed in the second, and conversely with the ball from the second urn. Let $X_n$ denote the state of the system after the $n$th stage, and compute the transition probabilities of the Markov chain $\{X_n, n \geq 0\}$.

**5.** Consider Example 2a. If there is a 50 : 50 chance of rain today, compute the probability that it will rain 3 days from now if $\alpha = .7$, $\beta = .3$.

**6.** Compute the limiting probabilities for the model of Problem 4.

[2] For an entropy interpretation of $C^*$ see Problem 18.

**7.** A transition probability matrix is said to be doubly stochastic if

$$\sum_{i=0}^{M} P_{ij} = 1$$

for all states $j = 0, 1, \ldots, M$. If such a Markov chain is ergodic, show that $\Pi_j = 1/(M+1)$, $j = 0, 1, \ldots, M$.

**8.** On any given day Rebecca is either cheerful (c), so-so (s), or gloomy (g). If she is cheerful today then she will be c, s, or g tomorrow with respective probabilities .7, .2, .1. If she is so-so today then she will be c, s, or g tomorrow with respective probabilities .4, .3, .3. If she is gloomy today then she will be c, s, or g tomorrow with probabilities .2, .4, .4. What proportion of time is Rebecca cheerful?

**9.** Suppose that whether or not it rains tomorrow depends on past weather conditions only through the last 2 days. Specifically, suppose that if it has rained yesterday and today then it will rain tomorrow with probability .8; if it rained yesterday but not today then it will rain tomorrow with probability .3; if it rained today but not yesterday then it will rain tomorrow with probability .4; and if it has not rained either yesterday or today then it will rain tomorrow with probability .2. What proportion of days does it rain?

**10.** A certain individual goes for a run each morning. When he leaves his house for his run he is equally likely to go out either the front or the back door; and similarly when he returns he is equally likely to go to either the front or back door. The runner owns 5 pairs of running shoes which he takes off after the run at whichever door he happens to be at. If there are no shoes at the door from which he leaves to go running he runs barefooted. We are interested in determining the proportion of time that he runs barefooted.

(a) Set this up as a Markov chain. Give the states and the transition probabilities.

(b) Determine the proportion of days that he runs barefooted.

**11.** This problem refers to Example 2f.

(a) Verify that the proposed value of $\Pi_j$ satisfy the necessary equations.

(b) For any given molecule what do you think is the (limiting) probability that it is in urn 1?

(c) Do you think that the events that molecule $j$ is in urn 1 at a very large time, $j \geq 1$, would be (in the limit) independent?

(d) Explain why the limiting probabilities are as given.

**12.** Determine the entropy of the sum obtained when a pair of the fair dice are rolled.

**13.** If $X$ can take on any of $n$ possible values with respective probabilities $P_1, \ldots, P_n$, prove that $H(X)$ is maximized when $P_i = 1/n$, $i = 1, \ldots, n$. What is $H(X)$ equal to in this case?

**14.** A pair of fair dice are rolled. Let

$$X = \begin{cases} 1 & \text{if the sum of the dice is 6} \\ 0 & \text{otherwise} \end{cases}$$

and let $Y$ equal the value of the first die. Compute (a) $H(Y)$, (b) $H_Y(X)$, and (c) $H(X, Y)$.

**15.** A coin having probability $p = \frac{2}{3}$ of coming up heads is flipped 6 times. Compute the entropy of the outcome of this experiment.

**16.** A random variable can take on any of $n$ possible values $x_1, \ldots, x_n$ with respective probabilities $p(x_i)$, $i = 1, \ldots, n$. We shall attempt to determine the value of $X$ by asking a series of questions, each of which can be answered by "yes" or "no." For instance, we may ask "Is $X = x_1$?" or "Is $X$ equal to either $x_1$ or $x_2$ or $x_3$?", and so on. What can you say about the average number of such questions that you will need to ask in order to determine the value of $X$?

**17.** For any discrete random variable $X$ and function $f$ show that

$$H(f(X)) \leq H(X)$$

**18.** In transmitting a bit from location $A$ to location $B$, if we let $X$ denote the value of the bit sent at location $A$ and $Y$ the value received at location $B$, then $H(X) - H_Y(X)$ is called the rate of transmission of information from $A$ to $B$. The maximal rate of transmission, as a function of $P\{X = 1\} = 1 - P\{X = 0\}$, is called the channel capacity. Show that for a binary symmetric channel with $P\{Y = 1 \mid X = 1\} = P\{Y = 0 \mid X = 0\} = p$, the channel capacity is attained by the rate of transmission of information when $P\{X = 1\} = \frac{1}{2}$ and its value is $1 + p \log p + (1 - p) \log (1 - p)$.

## References

### Sections 1 and 2

Kemeny, J., L. Snell, and A. Knapp. *Denumerable Markov Chains.* New York: D. Van Nostrand Company, 1966.
Parzen, E. *Stochastic Processes.* San Francisco: Holden-Day, Inc., 1962.
Ross, S. M. *Introduction to Probability Models,* 3rd ed. New York: Academic Press, Inc., 1984.
Ross, S. M. *Stochastic Processes.* New York: John Wiley & Sons, Inc., 1983.

### Sections 3 and 4

Abramson, N. *Information Theory and Coding.* New York: McGraw-Hill Book Company, 1963.
McEliece, R. *Theory of Information and Coding.* Reading, Mass.: Addison-Wesley Publishing Co., Inc., 1977.
Peterson, W., and E. Weldon. *Error Correcting Codes,* 2nd ed. Cambridge, Mass.: The M.I.T. Press, 1972.

# 10

# Simulation

## 1 Introduction

How can we determine the probability of our winning a game of solitaire? (By solitaire we mean any one of the standard solitaire games played with an ordinary deck of 52 playing cards and with some fixed playing strategy.) One possible approach is to start with the reasonable hypothesis that all (52)! possible arrangements of the deck of cards are equally likely to occur and then attempt to determine how many of these lead to a win. Unfortunately, there does not appear to be any systematic method for determining the number of arrangements that lead to a win and, as (52)! is a rather large number and the only way to determine whether or not a particular arrangement leads to a win seems to be by playing the game out, it can be seen that this approach will not work.

In fact, it might appear that the determination of the win probability for solitaire is mathematically intractable. However, all is not lost, for probability falls not only within the realm of mathematics, but also within the realm of applied science; and, as in all applied sciences, experimentation is a valuable technique. For our solitaire example, experimentation takes the form of playing a large number of such games, or, better yet, programming a computer to do so. After playing, say $n$ games, if we let

$$X_i = \begin{cases} 1 & \text{if the } i\text{th game results in a win} \\ 0 & \text{otherwise} \end{cases}$$

then $X_i$, $i = 1, \ldots, n$ will be independent Bernoulli random variables for which

$$E[X_i] = P\{\text{win at solitaire}\}$$

Hence, by the strong law of large numbers, we know that

$$\sum_{i=1}^{n} \frac{X_i}{n} = \frac{\text{number of games won}}{\text{number of games played}}$$

will, with probability 1, converge to $P\{\text{win at solitaire}\}$. That is, by playing a large number of games we can use the proportion of won games as an estimate of the probability of winning. This method of empirically determining probabilities by means of experimentation is known as *simulation*.

In order to use a computer to initiate a simulation study, we must be able to generate the value of a uniform (0, 1) random variable; such variates are called random numbers. To generate such numbers most computers have a built-in subroutine, called a random number generator, whose output is a sequence of pseudo random numbers. This is a sequence of numbers that is, for all practical purposes, indistinguishable from a sample from the uniform (0, 1) distribution. Most random number generators start with an initial value $X_0$, called the seed, and then recursively compute values by specifying positive integers $a$, $c$, and $m$, and then letting

$$X_{n+1} = (aX_n + c) \text{ modulo } m \qquad n \geq 0$$

where the foregoing means that $aX_n + c$ is divided by $m$ and the remainder is taken as the value of $X_{n+1}$. Thus each $X_n$ is either $0, 1, \ldots, m - 1$ and the quantity $X_n/m$ is taken as an approximation to a uniform (0, 1) random variable. It can be shown that, subject to suitable choices for $a$, $c$, and $m$, the foregoing gives rise to a sequence of numbers that look as if they were generated from independent uniform (0, 1) random variables.

As our starting point in simulation, we shall suppose that we can simulate from the uniform (0, 1) distribution and we shall use the term "random numbers" to mean independent random variables from this distribution.

In the solitaire example we would need to program a computer to play out the game starting with a given ordering of the cards. However, since the initial ordering is supposed to be equally likely to be any of the (52)! possible permutations, it is also necessary to be able to generate a random permutation. Using only random numbers, the following algorithm shows how this can be accomplished. The algorithm begins by randomly choosing one of the elements and then putting it in position $n$; it then randomly chooses among the remaining elements and puts the choice in position $n - 1$; and so on. It efficiently makes a random choice among the remaining elements by keeping these elements in an ordered list and then randomly choosing a position on that list.

**Example 1a. Generating a Random Permutation.** Suppose we are interested in generating a permutation of the integers $1, 2, \ldots, n$ that is such that all $n!$ possible orderings are equally likely. Starting with any initial permutation we will accomplish this after $n - 1$ steps where at each step we will interchange the positions of two of the numbers of the permutation. Throughout, we will keep track of the permutation by letting $X(i)$, $i = 1, \ldots, n$ denote the number currently in position $i$. The algorithm operates as follows:

1. Consider any arbitrary permutation and let $X(i)$ denote the element in position $i$, $i = 1, \ldots, n$. (For instance, we could take $X(i) = i$, $i = 1, \ldots, n$).
2. Generate a random variable $N_n$ that is equally likely to equal any of the values $1, 2, \ldots, n$.
3. Interchange the values of $X(N_n)$ and $X(n)$. The value of $X(n)$ will now remain fixed. [For instance, suppose $n = 4$ and initially $X(i) = i$, $i = 1, 2, 3, 4$. If $N_4 = 3$, then the new permutation is $X(1) = 1$, $X(2) = 2$, $X(3) = 4$, $X(4) = 3$, and element 3 will remain in position 4 throughout.]
4. Generate a random variable $N_{n-1}$ that is equally likely to be either $1, 2, \ldots, n - 1$.
5. Interchange the values of $X(N_{n-1})$ and $X(n - 1)$. [If $N_3 = 1$, then the new permutation is $X(1) = 4$, $X(2) = 2$, $X(3) = 1$, $X(4) = 3$.]
6. Generate $N_{n-2}$, which is equally likely to be either $1, 2, \ldots, n - 2$.
7. Interchange the values of $X(N_{n-2})$ and $X(2)$. [If $N_2 = 1$, then the new permutation is $X(1) = 2$, $X(2) = 4$, $X(3) = 1$, $X(4) = 3$ and this is the final permutation.]
8. Generate $N_{n-3}$, and so on. The algorithm continues until $N_2$ is generated and after the next interchange the resulting permutation is the final one.

To implement this algorithm, it is necessary to be able to generate a random variable that is equally likely to be any of the values $1, 2, \ldots, k$. To accomplish this, let $U$ denote a random number—that is, $U$ is uniformly distributed on (0, 1), and note that $kU$ is uniform on $(0, k)$. Hence,

$$P\{i - 1 < kU < i\} = \frac{1}{k} \qquad i = 1, \ldots, k$$

and so if we take $N_k = [kU] + 1$, where $[x]$ is the integer part of $x$ (that is, it is the largest integer less than or equal to $x$), then $N_k$ will have the desired distribution.

The algorithm can now be succinctly written as follows:

*Step 1:* Let $X(1), \ldots, X(n)$ be any permutation of $1, 2, \ldots, n$. (For instance, we can set $X(i) = i$, $i = 1, \ldots, n$.)
*Step 2:* Let $I = n$.
*Step 3:* Generate a random number $U$ and set $N = [IU] + 1$.
*Step 4:* Interchange the values of $X(N)$ and $X(I)$.
*Step 5:* Reduce the value of $I$ by 1 and if $I > 1$ go to Step 3.
*Step 6:* $X(1), \ldots, X(n)$ is the desired random generated permutation. ∎

The foregoing algorithm for generating a random permutation is extremely useful. For instance, suppose that a statistician is developing an experiment to compare the effects of $m$ different treatments on a set of $n$ subjects. He decides to split the subjects into $m$ different groups of respective sizes $n_1, n_2, \ldots, n_m$ where $\sum_{i=1}^{m} n_i = n$ with the members of the $i$th group to receive treatment $i$. To eliminate any bias in the assignment of subjects to treatments (for instance, it would cloud the meaning of the experimental results if it turned out that all the "best" subjects had been put in the same group), it is imperative that the assignment of a subject to a given group be done "at random." How is this to be accomplished?[1]

A simple and efficient procedure is to arbitrarily number the subjects 1 through $n$ and then generate a random permutation $X(1), \ldots, X(n)$ of 1, 2, . . . , $n$. Now assign subjects $X(1), X(2), \ldots, X(n_1)$ to be in group 1, $X(n_1 + 1), \ldots, X(n_1 + n_2)$ to be in group 2, and in general group $j$ is to consist of subjects numbered $X(n_1 + n_2 + \cdots + n_{j-1} + k)$, $k = 1, \ldots, n_j$.

## 2 General Techniques for Simulating Continuous Random Variables

In this section we present two general methods for using random numbers to simulate continuous random variables.

### 2.1 *The Inverse Transformation Method*

A general method for simulating a random variable having a continuous distribution—called the *Inverse Transformation Method*—is based on the following proposition.

---

*Proposition 2.1*

*Let $U$ be a uniform (0, 1) random variable. For any continuous distribution function $F$ if we define the random variable $Y$ by*

$$Y = F^{-1}(U)$$

*then the random variable $Y$ has distribution function $F$. [$F^{-1}(x)$ is defined to equal that value $y$ for which $F(y) = x$.]*

---

[1] When $m = 2$, another technique for randomly dividing the subjects was presented in Example 2f of Chapter 6. The preceding procedure is faster but requires more space than the one of Example 2f.

**Proof**

$$F_Y(a) = P\{Y \leq a\} \tag{2.1}$$
$$= P\{F^{-1}(U) \leq a\}$$

Now, since $F(x)$ is a monotone function, it follows that $F^{-1}(U) \leq a$ if and only if $U \leq F(a)$. Hence, from Equation (2.1), we see that

$$F_Y(a) = P\{U \leq F(a)\}$$
$$= F(a)$$ ∎

It follows from Proposition 2.1 that we can simulate a random variable $X$ having a continuous distribution function $F$ by generating a random number $U$ and then setting $X = F^{-1}(U)$.

**Example 2a. Simulating an Exponential Random Variable.** If $F(x) = 1 - e^{-x}$, then $F^{-1}(u)$ is that value of $x$ such that

$$1 - e^{-x} = u$$

or

$$x = -\log(1 - u)$$

Hence, if $U$ is a uniform (0, 1) variable, then

$$F^{-1}(U) = -\log(1 - U)$$

is exponentially distributed with mean 1. Since $1 - U$ is also uniformly distributed on (0, 1), it follows that $-\log U$ is exponential with mean 1. Since $cX$ is exponential with mean $c$ when $X$ is exponential with mean 1, it follows that $-c \log U$ is exponential with mean $c$. ∎

The results of Example 2a can also be utilized to simulate a gamma random variable.

**Example 2b. Simulating a Gamma $(n, \lambda)$ Random Variable.** To simulate from a gamma distribution with parameters $(n, \lambda)$, when $n$ is an integer, we use the fact that the sum of $n$ independent exponential random variables each having rate $\lambda$ has this distribution. Hence, if $U_1, \ldots, U_n$ are independent uniform (0, 1) random variables,

$$X = -\sum_{i=1}^{n} \frac{1}{\lambda} \log U_i = -\frac{1}{\lambda} \log\left(\prod_{i=1}^{n} U_i\right)$$

has the desired distribution. ∎

## 2.2 *The Rejection Method*

Suppose that we have a method for simulating a random variable having density function $g(x)$. We can use this as the basis for simulating from the continuous distribution having density $f(x)$ by simulating $Y$ from $g$ and then accepting this simulated value with a probability proportional to $f(Y)/g(Y)$.

Specifically, let $c$ be a constant such that

$$\frac{f(y)}{g(y)} \leq c \qquad \text{for all } y$$

We then have the following technique for simulating a random variable having density $f$.

*Rejection Method*

*Step 1:* Simulate $Y$ having density $g$ and simulate a random number $U$.

*Step 2:* If $U \leq f(Y)/cg(Y)$, set $X = Y$. Otherwise return to Step 1.

The rejection method is pictorially expressed in Figure 10.1.

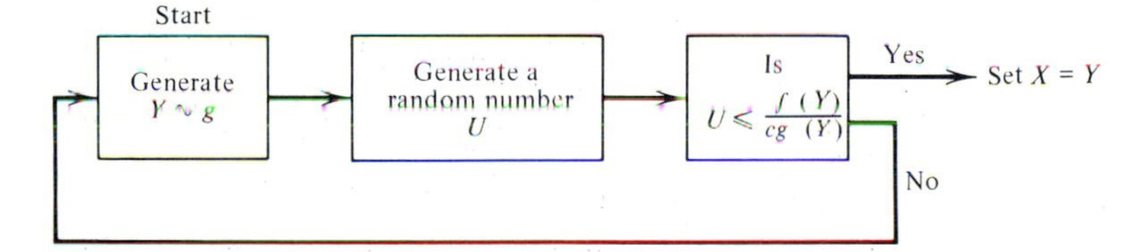


**Figure 10.1** The rejection method for simulating a random variable $X$ having density function $f$

We now prove that the rejection method works.

---

## Proposition 2.2

*The random variable X generated by the Rejection Method has density function $f$.*

---

**Proof:** Let $X$ be the value obtained and let $N$ denote the number of necessary iterations. Then

$$\begin{aligned} P\{X \leq x\} &= P\{Y_N \leq x\} \\ &= P\left\{Y \leq x \middle| U \leq \frac{f(Y)}{cg(Y)}\right\} \\ &= \frac{P\left\{Y \leq x, U \leq \dfrac{f(Y)}{cg(Y)}\right\}}{K} \end{aligned}$$

where $K = P\{U \leq f(Y)/cg(Y)\}$. Now the joint density function of $Y$ and $U$ is, by independence,

$$f(y, u) = g(y) \qquad 0 < u < 1$$

and so using the foregoing we have

$$\begin{aligned} P\{X \leq x\} &= \frac{1}{K} \iint\limits_{\substack{y \leq x \\ 0 \leq u \leq f(y)/cg(y)}} g(y)\, du\, dy \\ &= \frac{1}{K} \int_0^x \int_0^{f(y)/cg(y)} du\, g(y)\, dy \qquad (2.2) \\ &= \frac{1}{cK} \int_0^x f(y)\, dy \end{aligned}$$

Letting $x$ approach $\infty$ and using the fact that $f$ is a density gives

$$1 = \frac{1}{cK} \int_0^\infty f(y)\, dy = \frac{1}{cK}$$

Hence, from Equation (2.2) we obtain that

$$P\{X \leq x\} = \int_0^x f(y)\, dy$$

which completes the proof. ∎

REMARKS: (a) It should be noted that the way in which we "accept the value $Y$ with probability $f(Y)/cg(Y)$" is by generating a random number $U$ and then accepting $Y$ if $U \leq f(Y)/cg(Y)$.

(b) Since each iteration will, independently, result in an accepted value with probability $P\{U \leq f(Y)/cg(Y)\} = K = 1/c$, it follows that the number of iterations has a geometric distribution with mean $c$.

**Example 2c. Simulating a Normal Random Variable.** To simulate a unit normal random variable $Z$ (that is, one with mean 0 and variance 1), note first that the absolute value of $Z$ has probability density function

$$f(x) = \frac{2}{\sqrt{2\pi}} e^{-x^2/2} \qquad 0 < x < \infty \qquad (2.3)$$

We will start by simulating from the preceding density function by using the rejection method with $g$ being the exponential density function with mean 1—that is

$$g(x) = e^{-x} \qquad 0 < x < \infty$$

Now, note that

$$\begin{aligned}\frac{f(x)}{g(x)} &= \sqrt{2/\pi}\, \exp\left\{\frac{-(x^2 - 2x)}{2}\right\} \\ &= \sqrt{2/\pi}\, \exp\left\{\frac{-(x^2 - 2x + 1)}{2} + \frac{1}{2}\right\} \\ &= \sqrt{2e/\pi}\, \exp\left\{\frac{-(x - 1)^2}{2}\right\} \\ &\leq \sqrt{2e/\pi}\end{aligned} \tag{2.4}$$

Hence, we can take $c = \sqrt{2e/\pi}$ and so, from Equation (2.4)

$$\frac{f(x)}{cg(x)} = \exp\left\{\frac{-(x - 1)^2}{2}\right\}$$

Therefore, using the rejection method we can simulate the absolute value of a unit normal random variable as follows:

(a) Generate independent random variables $Y$ and $U$, $Y$ being exponential with rate 1 and $U$ being uniform on (0, 1).
(b) If $U \leq \exp\{-(Y - 1)^2/2\}$ set $X = Y$. Otherwise return to (a).

Once we have simulated a random variable $X$ having density function as in Equation 2.3 we can then generate a unit normal random variable $Z$ by letting $Z$ be equally likely to be either $X$ or $-X$.

In Step (b), the value $Y$ is accepted if $U \leq \exp\{-(Y - 1)^2/2\}$, which is equivalent to $-\log U \geq (Y - 1)^2/2$. However, in Example 2a it was shown that $-\log U$ is exponential with rate 1, and so Steps (a) and (b) are equivalent to

(a′) Generate independent exponentials with rate 1, $Y_1$ and $Y_2$.
(b′) If $Y_2 \geq (Y_1 - 1)^2/2$, set $X = Y_1$. Otherwise return to (a).

Suppose now that the foregoing results in $Y_1$ being accepted—and so we know that $Y_2$ is larger than $(Y_1 - 1)^2/2$. By how much does the one exceed the other? To answer this recall that $Y_2$ is exponential with rate 1, and so, given that it exceeds some value, the amount by which $Y_2$ exceeds $(Y_1 - 1)^2/2$ [that is, its "additional life" beyond the time $(Y_1 - 1)^2/2$] is (by the memoryless property) also exponentially distributed with rate 1. That is, when we accept Step (b′), we obtain not only $X$ (the absolute value of a unit normal) but by computing $Y_2 - (Y_1 - 1)^2/2$ we can also generate an exponential random variable (independent of $X$) having rate 1.

Hence, summing up, we have the following algorithm that generates an exponential with rate 1 and an independent unit normal random variable.

*Step 1:* Generate $Y_1$, an exponential random variable with rate 1.
*Step 2:* Generate $Y_2$, an exponential random variable with rate 1.
*Step 3:* If $Y_2 - (Y_1 - 1)^2/2 > 0$ set $Y = Y_2 - (Y_1 - 1)^2/2$ and go to Step 4. Otherwise go to Step 1.
*Step 4:* Generate a random number $U$ and set

$$Z = \begin{cases} Y_1 & \text{if } U \leq 1/2 \\ -Y_1 & \text{if } U > 1/2 \end{cases}$$

The random variables $Z$ and $Y$ generated by the foregoing are independent with $Z$ being normal with mean 0 and variance 1 and $Y$ being exponential with rate 1. (If we want the normal random variable to have mean $\mu$ and variance $\sigma^2$, just take $\mu + \sigma Z$).

REMARKS: (a) Since $c = \sqrt{2e/\pi} \approx 1.32$, the foregoing requires a geometric distributed number of iterations of Step 2 with mean 1.32.
(b) If we want to generate a sequence of unit normal random variables, then we can use the exponential random variable $Y$ obtained in Step 3 as the initial exponential needed in Step 1 for the next normal to be generated. Hence, on the average, we can simulate a unit normal by generating 1.64 ($= 2 \times 1.32 - 1$) exponentials and computing 1.32 squares. ∎

**Example 2d. Simulating Normal Random Variables—The Polar Method.** It was shown in Example 7b of Chapter 6 that if $X$ and $Y$ are independent unit normal random variables than their polar coordinates $R = \sqrt{X^2 + Y^2}$, $\oplus = \tan^{-1}(Y/X)$ are independent, with $R^2$ being exponentially distributed with mean 2 and $\oplus$ being uniformly distributed on $(0, 2\pi)$. Hence, if $U_1$ and $U_2$ are random numbers then (using the result of Example 2a) we can set

$$R = (-2 \log U_1)^{1/2}$$
$$\oplus = 2\pi U_2$$

which yields that

$$\begin{aligned} X &= R \cos\oplus = (-2 \log U_1)^{1/2} \cos(2\pi U_2) \\ Y &= R \sin\oplus = (-2 \log U_1)^{1/2} \sin(2\pi U_2) \end{aligned} \tag{2.5}$$

are independent unit normals.

The above approach to generating unit normal random variables is called the Box-Muller approach. Its efficiency suffers somewhat from its need to compute the above sine and cosine values. There is, however, a way to get around this potentially time-consuming difficulty. To begin, note that if $U$ is uniform on $(0, 1)$ then $2U$ is uniform on $(0, 2)$ and so $2U - 1$ is uniform on $(-1, 1)$. Thus, if we generate random numbers

$U_1$ and $U_2$ and set

$$V_1 = 2U_1 - 1$$
$$V_2 = 2U_2 - 1$$

then $(V_1, V_2)$ is uniformly distributed in the square of area 4 centered at (0, 0)—see Figure 10.2.

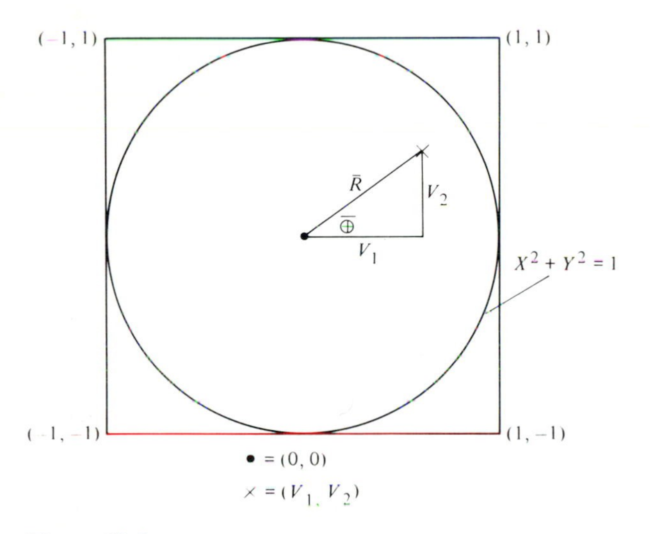


**Figure 10.2**

Suppose now that we continually generate such pairs $(V_1, V_2)$ until we obtain one that is contained in the circle of radius 1 centered at (0, 0)—that is, until $(V_1, V_2)$ is such that $V_1^2 + V_2^2 \leq 1$. It now follows that such a pair $(V_1, V_2)$ is uniformly distributed in the circle. If we let $\bar{R}$, $\bar{\oplus}$ denote the polar coordinates of this pair, then it is easy to verify that $\bar{R}$ and $\bar{\oplus}$ are independent, with $\bar{R}^2$ being uniformly distributed on (0, 1), and $\bar{\oplus}$ uniformly distributed on $(0, 2\pi)$—see Problem 13.

Since,

$$\sin\oplus = V_2/\bar{R} = -\frac{V_2}{\sqrt{V_1^2 + V_2^2}}$$

$$\cos\oplus = V_1/\bar{R} = \frac{V_1}{\sqrt{V_1^2 + V_2^2}}$$

it follows from Equation (2.5) that we can generate independent unit normals $X$ and $Y$ by generating another random number $U$ and setting

$$X = (-2 \log U)^{1/2} V_1/\bar{R}$$
$$Y = (-2 \log U)^{1/2} V_2/\bar{R}$$

In fact since (conditional on $V_1^2 + V_2^2 \leq 1$) $\overline{R}^2$ is uniform on (0, 1) and is independent of $\oplus$ we can use it instead of generating a new random number $U$; thus showing that

$$X = (-2 \log \overline{R}^2)^{1/2}\, V_1/\overline{R} = \sqrt{\frac{-2 \log S}{S}}\, V_1$$

$$Y = (-2 \log \overline{R}^2)^{1/2}\, V_2/\overline{R} = \sqrt{\frac{-2 \log S}{S}}\, V_2$$

are independent unit normals, where

$$S = \overline{R}^2 = V_1^2 + V_2^2$$

Summing up we thus have the following approach to generating a pair of independent unit normals:

*Step 1:* Generate random numbers $U_1$ and $U_2$
*Step 2:* Set $V_1 = 2U_1 - 1$, $V_2 = 2U_2 - 1$, $S = V_1^2 + V_2^2$
*Step 3:* If $S > 1$ return to step 1
*Step 4:* Return the independent unit normals

$$X = \sqrt{\frac{-2 \log S}{S}}\, V_1, \quad Y = \sqrt{\frac{-2 \log S}{S}}\, V_2$$

The above is called the polar method. Since the probability that a random point in the square will fall within the circle is equal to $\pi/4$ (the area of the circle divided by the area of the square), it follows that, on average, the polar method will require $4/\pi = 1.273$ iterations of step 1. Hence, it will, on average, require 2.546 random numbers, 1 logarithm, 1 square root, 1 division, and 4.546 multiplications to generate 2 independent unit normals. ∎

**Example 2e. Simulating a Chi-Square Random Variable.** The chi-square distribution with $n$ degrees of freedom is the distribution of $\chi_n^2 = Z_1^2 + \cdots + Z_n^2$ where $Z_i$, $i = 1, \ldots, n$ are independent unit normals. Now it was shown in Section 3 of Chapter 6 that $Z_1^2 + Z_2^2$ has an exponential distribution with rate $\frac{1}{2}$. Hence, when $n$ is even, say $n = 2k$, $\chi_{2k}^2$ has a gamma distribution with parameters $(K, \frac{1}{2})$. Hence, $-2 \log (\prod_{i=1}^k U_i)$ has a chi-square distribution with $2k$ degrees of freedom. We can simulate a chi-square random variable with $2k + 1$ degrees of freedom by first simulating a unit normal random variable $Z$ and then adding $Z^2$ to the foregoing. That is,

$$\chi_{2k+1}^2 = Z^2 - 2 \log \left( \prod_{i=1}^{k} U_i \right)$$

where $Z, U_1, \ldots, U_n$ are independent with $Z$ being a unit normal and the others being uniform (0, 1) random variables. ∎

## 3 Simulating From Discrete Distributions

All of the general methods for simulating random variables from continuous distributions have analogs in the discrete case. For instance, if we want to simulate a random variable $Z$ having probability mass function

$$P\{X = x_j\} = P_j, \qquad j = 0, 1, \ldots, \qquad \sum_j P_j = 1$$

We can use the following discrete time analog of the inverse transform technique.

To simulate $X$ for which $P\{X = x_j\} = P_j$ let $U$ be uniformly distributed over (0, 1), and set

$$X = \begin{cases} x_1 & \text{if} \quad U < P_1 \\ x_2 & \text{if} \quad P_1 < U < P_1 + P_2 \\ \vdots & \\ x_j & \text{if} \quad \sum_1^{j-1} P_i < U < \sum_i^{j} P_i \\ \vdots & \end{cases}$$

Since

$$P\{X = x_j\} = P\left\{\sum_1^{j-1} P_i < U < \sum_1^{j} P_i\right\} = P_j$$

we see that $X$ has the desired distribution.

**Example 3a. The Geometric Distribution.** Suppose that independent trials each of which results in a "success" with probability $p$, $0 < p < 1$, are continually performed until a success occurs. Letting $X$ denote the number of necessary trials, then

$$P\{X = i\} = (1 - p)^{i-1}p \qquad i \geq 1$$

which is seen by noting that $X = i$ if the first $i - 1$ trials are all failures and the $i$th is a success. The random variable $X$ is said to be a geometric random variable with parameter $p$. Since

$$\begin{aligned} \sum_{i=1}^{j-1} P\{X = i\} &= 1 - P\{X > j - 1\} \\ &= 1 - P\{\text{first } j - 1 \text{ are all failures}\} \\ &= 1 - (1 - p)^{j-1} \qquad j \geq 1 \end{aligned}$$

we can simulate such a random variable by generating a random number $U$ and then setting $X$ equal to that value $j$ for which

$$1 - (1 - p)^{j-1} < U < 1 - (1 - p)^j$$

or, equivalently, for which

$$(1 - p)^j < 1 - U < (1 - p)^{j-1}$$

Since $1 - U$ has the same distribution as $U$, we can thus define $X$ by

$$\begin{aligned} X &= \min \{j : (1 - p)^j < U\} \\ &= \min \{j : j \log(1 - p) < \log U\} \\ &= \min \left\{j : j > \frac{\log U}{\log(1 - p)}\right\} \end{aligned}$$

where the inequality changed sign since $\log(1 - p)$ is negative [since $\log(1 - p) < \log 1 = 0$]. Using the notation $[x]$ for the integer part of $x$ (that is, $[x]$ is the largest integer less than or equal to $x$), we can write

$$X = 1 + \left[\frac{\log U}{\log(1 - p)}\right]$$ ∎

As in the continuous case, special simulating techniques have been developed for the more common discrete distributions. We now present certain of these.

**Example 3b. Simulating a Binomial Random Variable.** A binomial $(n, p)$ random variable can be most easily simulated by recalling that it can be expressed as the sum of $n$ independent Bernoulli random variables. That is, if $U_1, \ldots, U_n$ are independent uniform (0, 1) variables, then letting

$$X_i = \begin{cases} 1 & \text{if } U_i < p \\ 0 & \text{otherwise} \end{cases}$$

it follows that $X \equiv \sum_{i=1}^n X_i$ is a binomial random variable with parameters $n$ and $p$. ∎

**Example 3c. Simulating a Poisson Random Variable.** To simulate a Poisson random variable with mean $\lambda$, generate independent uniform (0, 1) random variables $U_1, U_2, \ldots$ stopping at

$$N = \min \left\{n : \prod_{i=1}^{n} U_i < e^{-\lambda}\right\}$$

The random variable $X \equiv N - 1$ has the desired distribution. That is, if we continue generating random numbers until their product falls below $e^{-\lambda}$, then the number required, minus 1, is Poisson with mean $\lambda$.

That $X \equiv N - 1$ is indeed a Poisson random variable having mean $\lambda$ can perhaps be most easily seen by noting that

$$X + 1 = \min \left\{ n : \prod_{i=1}^{n} U_i < e^{-\lambda} \right\}$$

is equivalent to

$$X = \max \left\{ n : \prod_{i=1}^{n} U_i \geq e^{-\lambda} \right\} \qquad \text{where } \prod_{i=1}^{0} U_i \equiv 1$$

or, taking logarithms, to

$$X = \max \left\{ n : \sum_{i=1}^{n} \log U_i \geq -\lambda \right\}$$

or

$$X = \max \left\{ n : \sum_{i=1}^{n} -\log U_i \leq \lambda \right\}$$

However, $-\log U_i$ is exponential with rate 1 and so $X$ can be thought of as being the maximum number of exponentials having rate 1 that can be summed and still be less than $\lambda$. But by recalling that the times between successive events of a Poisson process having rate 1 are independent exponentials with rate 1, it follows that $X$ is equal to the number of events by time $\lambda$ of a Poisson process having rate 1; and thus $X$ has a Poisson distribution with mean $\lambda$. ∎

## 4 Variance Reduction Techniques

Let $X_1, \ldots, X_n$ have a given joint distribution and suppose we are interested in computing

$$\theta \equiv E[g(X_1, \ldots, X_n)]$$

where $g$ is some specified function. It sometimes turns out that it is extremely difficult to analytically compute the foregoing, and when such is the case we can attempt to use simulation to estimate $\theta$. This is done as follows: generate $X_1^{(1)}, \ldots, X_n^{(1)}$ having the same joint distribution as $X_1, \ldots, X_n$ and set

$$Y_1 = g(X_1^{(1)}, \ldots, X_n^{(1)})$$

Now simulate a second set of random variables (independent of the first set) $X_1^{(2)}, \ldots, X_n^{(2)}$ having the distribution of $X_1, \ldots, X_n$ and set

$$Y_2 = g(X_1^{(2)}, \ldots, X_n^{(2)})$$

Continue this until you have generated $k$ (some predetermined number) sets and so have also computed $Y_1, Y_2, \ldots, Y_k$. Now, $Y_1, \ldots, Y_k$ are independent and identically distributed random variables each having the same distribution of $g(X_1, \ldots, X_n)$. Thus, if we let $\bar{Y}$ denote the average of these $k$ random variables—that is

$$\bar{Y} = \sum_{i=1}^{k} \frac{Y_i}{k}$$

then

$$E[\bar{Y}] = \theta$$
$$E[(\bar{Y} - \theta)^2] = \text{Var}(\bar{Y})$$

Hence, we can use $\bar{Y}$ as an estimate of $\theta$. Since the expected square of the difference between $\bar{Y}$ and $\theta$ is equal to the variance of $\bar{Y}$ we would like this quantity to be as small as possible. [In the preceding situation, Var $(\bar{Y})$ = Var $(Y_i)/k$, which is usually not known in advance but must be estimated from the generated values $Y_1, \ldots, Y_n$]. We now present 3 general techniques for reducing the variance of our estimator.

### 4.1 *Use of Antithetic Variables*

In the foregoing situation, suppose that we have generated $Y_1$ and $Y_2$, which are identically distributed random variables having mean $\theta$. Now

$$\text{Var}\left(\frac{Y_1 + Y_2}{2}\right) = \frac{1}{4}[\text{Var}(Y_1) + \text{Var}(Y_2) + 2\,\text{Cov}(Y_1, Y_2)]$$
$$= \frac{\text{Var}(Y_1)}{2} + \frac{\text{Cov}(Y_1, Y_2)}{2}$$

Hence it would be advantageous (in the sense that the variance would be reduced) if $Y_1$ and $Y_2$ rather than being independent were negatively correlated. To see how we could arrange this, let us suppose that the random variables $X_1, \ldots, X_n$ are independent and, in addition, that each is simulated via the inverse transform technique. That is, $X_i$ is simulated from $F_i^{-1}(U_i)$ where $U_i$ is a random number and $F_i$ is the distribution of $X_i$. Hence, $Y_1$ can be expressed as

$$Y_1 = g(F_1^{-1}(U_1), \ldots, F_n^{-1}(U_n))$$

Now, since $1 - U$ is also uniform over (0, 1) whenever $U$ is a random number (and is negatively correlated with $U$), it follows that $Y_2$ defined by

$$Y_2 = g(F_1^{-1}(1 - U_1), \ldots, F_n^{-1}(1 - U_n))$$

will have the same distribution as $Y_1$. Hence, if $Y_1$ and $Y_2$ were negatively correlated, then generating $Y_2$ by this means would lead to a smaller variance than if it were generated by a new set of random numbers. (In addition, there is a computational savings since rather than having to generate $n$ additional random numbers, we need only subtract each of the previous $n$ from 1). Although we cannot, in general, be certain that $Y_1$ and $Y_2$ will be negatively correlated, this often turns out to be the case and indeed it can be proven that it will be so whenever $g$ is a monotonic function.

### 4.2 *Variance Reduction by Conditioning*

Let us start by recalling the conditional variance formula (see Section 6.4 of Chapter 7)

$$\text{Var}(Y) = E[\text{Var}(Y|Z)] + \text{Var}(E[Y|Z])$$

Now suppose we are interested in estimating $E[g(X_1, \ldots, X_n)]$ by simulating $\mathbf{X} = (X_1, \ldots, X_n)$ and then computing $Y = g(\mathbf{X})$. Now, if for some random variable $Z$ we can compute $E[Y|Z]$ then, as $\text{Var}(Y|Z) \geq 0$, it follows from the conditional variance formula above that

$$\text{Var}(E[Y|Z]) \leq \text{Var}(Y)$$

implying, since $E[E[Y|Z]] = E[Y]$, that $E[Y|Z]$ is a better estimator of $E[Y]$ than is $Y$.

**Example 4a. Estimation of $\pi$.** Let $U_1$ and $U_2$ be random numbers and set $V_i = 2U_i - 1$, $i = 1, 2$. As noted in Example 2d, $(V_1, V_2)$ will be uniformly distributed in the square of area 4 centered at (0, 0). The probability that this point will fall within the inscribed circle of radius 1 centered at (0, 0)—see Figure 10.2—is equal to $\pi/4$ (the ratio of the area of the circle to that of the square). Hence, upon simulating a large number $n$ of such pairs and setting

$$I_j = \begin{cases} 1 & \text{if the } j^{\text{th}} \text{ pair falls within the circle} \\ 0 & \text{otherwise} \end{cases}$$

it follows that $I_j$, $j = 1, \ldots, n$ will be independent and identically distributed random variables having $E[I_j] = \pi/4$. Thus, by the strong law of large numbers

$$\frac{I_1 + \cdots + I_n}{n} \longrightarrow \pi/4 \qquad \text{as} \quad n \to \infty$$

Therefore, it follows that by simulating a large number of pairs $(V_1, V_2)$ and multiplying the proportion of them that fall within the circle by 4, we can accurately approximate $\pi$.

The above estimator can, however, be improved upon by using conditional expectation. If we let $I$ be the above indicator variable for the pair $(V_1, V_2)$ then, rather than using the observed value of $I$, it is better to condition on $V_1$ and so utilize

$$\begin{aligned} E[I|V_1] &= P\{V_1^2 + V_2^2 \le 1|V_1\} \\ &= P\{V_2^2 \le 1 - V_1^2|V_1\} \end{aligned}$$

Now

$$\begin{aligned} P\{V_2^2 \le 1 - V_1^2|V_1 = v\} &= P\{V_2^2 \le 1 - v^2\} \\ &= P\{-\sqrt{1 - v^2} \le V_2 \le \sqrt{1 - v^2}\} \\ &= \sqrt{1 - v^2} \end{aligned}$$

and so

$$E[I|V_1] = E[\sqrt{1 - V_1^2}]$$

Thus, an improvement on using the average value of $I$ to estimate $\pi/4$ is to use the average value of $\sqrt{1 - V_1^2}$. Indeed, since

$$E[\sqrt{1 - V_1^2}] = \int_{-1}^{1} \tfrac{1}{2}\sqrt{1 - v^2}\, dv = \int_0^1 \sqrt{1 - u^2}\, du = E[\sqrt{1 - U^2}]$$

where $U$ is uniform over $(0, 1)$, we can generate $n$ random numbers $U$ and use the average value of $\sqrt{1 - U^2}$ as our estimate of $\pi/4$. (Problem 14 shows that this estimator has the same variance as the average of the $n$ values $\sqrt{1 - V^2}$).

The above estimator of $\pi$ can be improved even further by noting that the function $g(u) = \sqrt{1 - u^2}$, $0 \le u \le 1$ is a monotone decreasing function of $u$ and so the method of antithetic variables will reduce the variance of the estimator of $E[\sqrt{1 - U^2}]$. That is, rather than generating $n$ random numbers and using the average value of $\sqrt{1 - U^2}$ as an estimator of $\pi/4$, an improved estimator would be obtained by generating only $n/2$ random numbers $U$ and then using one-half the average of $\sqrt{1 - U^2} + \sqrt{1 - (1 - U)^2}$ as the estimator of $\pi/4$.

The following table gives the estimates of $\pi$ resulting from simulations, using $n = 10{,}000$, based on the above 3 estimators.

| *Method* | *Estimate of* $\pi$ |
|---|---|
| Use the proportion of the random points that fall in the circle | 3.1612 |
| Use the average value of $\sqrt{1 - U^2}$ | 3.128448 |
| Use the average value of $\sqrt{1 - U^2} + \sqrt{1 - (1 - U)^2}$ | 3.139578 |

A further simulation using the final approach and $n = 64{,}000$ yielded the estimate 3.143288. ∎

### 4.3 *Control Variates*

Again suppose we want to use simulation to estimate $E[g(\mathbf{X})]$ where $\mathbf{X} = (X_1, \ldots, X_n)$. But now suppose that for some function $f$ the expected value of $f(\mathbf{X})$ is known—say $E[f(\mathbf{X})] = \mu$. Then for any constant $a$ we can also use

$$W = g(\mathbf{X}) + a[f(\mathbf{X}) - \mu]$$

as an estimator of $E[g(\mathbf{X})]$. Now

$$\text{Var}(W) = \text{Var}[g(\mathbf{X})] + a^2 \text{Var}[f(\mathbf{X})] + 2a \text{Cov}[g(\mathbf{X}), f(\mathbf{X})] \tag{4.1}$$

Simple calculus shows that the foregoing is minimized when

$$a = \frac{-\text{Cov}[f(\mathbf{X}), g(\mathbf{X})]}{\text{Var}[f(\mathbf{X})]} \tag{4.2}$$

and for this value of $a$

$$\text{Var}(W) = \text{Var}[g(\mathbf{X})] - \frac{[\text{Cov}[f(\mathbf{X}), g(\mathbf{X})]]^2}{\text{Var}[f(\mathbf{X})]} \tag{4.3}$$

Unfortunately, neither $\text{Var}[f(\mathbf{X})]$ nor $\text{Cov}[f(\mathbf{X}), g(\mathbf{X})]$ is usually known so we cannot usually obtain the foregoing reduction in variance. One approach in practice is to guess at these values and hope the resulting $W$ does indeed have smaller variance than does $g(\mathbf{X})$, whereas a second possibility is to use the simulated data to estimate these quantities.

## *Problems*

**1.** The following algorithm will generate a random permutation of the elements $1, 2, \ldots, n$. It is somewhat faster than the one presented in Example 1a but is such that no position is fixed until the algorithm ends.
In this algorithm, $P(i)$ can be interpreted as the element in position $i$.
*Step 1:* Set $k = 1$
*Step 2:* Set $P(1) = 1$
*Step 3:* If $k = n$, stop. Otherwise, let $k = k + 1$.
*Step 4:* Generate a random number $U$ and let

$$P(k) = P([kU] + 1)$$
$$P([kU] + 1) = k$$

Go to Step 3.

(a) Explain in words what the algorithm is doing.
(b) Show that at iteration $k$—that is, when the value of $P(k)$ is initially set—that $P(1), P(2), \ldots, P(k)$ is a random permutation of $1, 2, \ldots, k$.

*Hint:* Use induction and argue that

$$\begin{aligned} P_k\{i_1, i_2, \ldots, i_{j-1}, k, i_j, \ldots, i_{k-2}, i\} &= P_{k-1}\{i_1, i_2, \ldots, i_{j-1}, i, i_j, \ldots, i_{k-2}\}\frac{1}{k} \\ &= \frac{1}{k!} \text{ by the induction hypothesis.} \end{aligned}$$

**2.** Develop a technique for simulating a random variable having density function

$$f(x) = \begin{cases} e^{2x} & -\infty < x < 0 \\ e^{-2x} & 0 < x < \infty \end{cases}$$

**3.** Give a technique for simulating a random variable having the probability density function

$$f(x) = \begin{cases} \frac{1}{2}(x - 2) & 2 \le x \le 3 \\ \frac{1}{2}\left(2 - \frac{x}{3}\right) & 3 < x \le 6 \\ 0 & \text{otherwise} \end{cases}$$

**4.** Present a method to simulate a random variable having distribution function

$$F(x) = \begin{cases} 0 & x \le -3 \\ \frac{1}{2} + \frac{x}{6} & -3 < x < 0 \\ \frac{1}{2} + \frac{x^2}{32} & 0 < x \le 4 \\ 1 & x > 4 \end{cases}$$

**5.** Use the inverse transformation method to present an approach for generating a random variable from the Weibull distribution

$$F(t) = 1 - e^{-at^{\beta}} \qquad t \ge 0$$

**6.** Give a method for simulating a random variable having failure rate function

(a) $\lambda(t) = c$
(b) $\lambda(t) = ct$
(c) $\lambda(t) = ct^2$
(d) $\lambda(t) = ct^3$

**7.** In the following, $F$ is the distribution function

$$F(x) = x^n \qquad 0 < x < \infty$$

(a) Give a method for simulating a random variable having distribution $F$ that uses only a single random number.
(b) Let $U_1, \ldots, U_n$ be independent random numbers. Show that

$$P\{\max (U_1, \ldots, U_n) \leq x\} = x^n$$

(c) Use part (b) to give a second method of simulating a random variable having distribution $F$.

**8.** Suppose it is relatively easy to simulate from $F_i$ for each $i = 1, \ldots, n$. How can we simulate from

(a)
$$F(x) = \prod_{i=1}^{n} F_i(x)$$

(b)
$$F(x) = 1 - \prod_{i=1}^{n} [1 - F_i(x)]$$

**9.** Suppose we have a method to simulate random variables from the distributions $F_1$ and $F_2$. Explain how to simulate from the distribution

$$F(x) = pF_1(x) + (1 - p)F_2(x) \qquad 0 < p < 1$$

Give a method for simulating from

$$F(x) = \begin{cases} \frac{1}{3}(1 - e^{-3x}) + \frac{2}{3}x & 0 < x \leq 1 \\ \frac{1}{3}(1 - e^{-3x}) + \frac{2}{3} & x > 1 \end{cases}$$

**10.** In Example 2c we simulated the absolute value of a unit normal by using the rejection procedure on exponential random variables with rate 1. This raises the question of whether we could obtain a more efficient algorithm by using a different exponential density—that is, we could use the density $g(x) = \lambda e^{-\lambda x}$. Show that the mean number of iterations needed in the rejection scheme is minimized when $\lambda = 1$.

**11.** Use the rejection method with $g(x) = 1, 0 < x < 1$, to determine an algorithm for simulating a random variable having density function

$$f(x) = \begin{cases} 60x^3(1 - x)^2 & 0 < x < 1 \\ 0 & \text{otherwise} \end{cases}$$

**12.** Explain how you could use random numbers to approximate $\int_0^1 k(x)\,dx$ where $k(x)$ is an arbitrary function.
*Hint:* If $U$ is uniform on (0, 1), what is $E[k(U)]$?

**13.** Let $(X,Y)$ be uniformly distributed in the circle of radius 1 centered at the origin. Its joint density is thus

$$f(x,y) = 1/\pi, \qquad 0 \leq x^2 + y^2 \leq 1$$

Let $R = (X^2 + Y^2)^{1/2}$ and $\oplus = \tan^{-1}(Y/X)$ denote its polar coordinates. Show that $R$ and $\oplus$ are independent with $R^2$ being uniform on (0, 1) and $\oplus$ being uniform on $(0, 2\pi)$.

**14.** In Example 4a we have shown that

$$E[(1 - V^2)^{1/2}] = E[(1 - U^2)^{1/2}] = \pi/4$$

when $V$ is uniform $(-1, 1)$ and $U$ is uniform (0,1). Show that

$$\text{Var}\ [(1 - V^2)^{1/2}] = \text{Var}\ [(1 - U^2)^{1/2}]$$

and find their common value.

**15.** (a) Verify that the minimum of (4.1) occurs when $a$ is as given by (4.2).
(b) Verify that the minimum of (4.1) is given by (4.3).

**16.** Let $X$ be a random variable on (0, 1) whose density is $f(x)$. Show that we can estimate $\int_0^1 g(x)\,dx$ by simulating $X$ and then taking $g(X)/f(X)$ as our estimate. This method, called importance sampling, tries to choose $f$ similar in shape to $g$ so that $g(X)/f(X)$ has a small variance.

# Answers to Selected Problems

## Chapter 1

### *Theoretical Exercises*

**2.** $\sum_{1}^{m} n_i$ **3.** $\frac{n!}{(n-r)!}$ **17.** $\binom{n+r-1}{r}$

### *Problems*

**1.** (a) 67,600,000 (b) 19,656,000 **2.** 24,4
**3.** 5184 **4.** (a) 720 (b) 72 (c) 144 (d) 72
**5.** (a) 120 (b) 1260 (c) 34,650 **6.** 27,720
**7.** (a) 40,320 (b) 10080 (c) 1152 (d) 2880 (e) 384
**8.** (a) 720 (b) 72 (c) 144
**9.** (a) 720 (b) 672 (c) 384 (d) 216 (e) 576
**10.** (a) 24,300,000 (b) 17,100,720
**11.** 2,598,960 **12.** 600 **13.** 120,110 **14.** 36, 26
**15.** 48 **17.** $\frac{(52)!}{[(13)!]^4}$ **19.** 27,720 **20.** 210
**21.** (a) 165 (b) 35 **22.** Assume teachers are distinct: (a) 65,536 (b) 2520
**23.** 1287, 14,112 **24.** 1,852,200
**25.** (a) 12,600 (b) 945
**26.** 564,480 **27.** (a) 220 (b) 552

## Chapter 2

### *Theoretical Exercises, Sections 1 and 2*

**6.** (a) $EF^cG^c$ (b) $EGF^c$ (c) $E \cup F \cup G$ (d) $EF \cup EG \cup FG$
(e) $EFG$ (f) $E^cF^cG^c$ (g) $EF^cG^c \cup E^cFG^c \cup E^cF^cG \cup E^cF^cG^c$
(h) $(EFG)^c$ (i) $EFG^c \cup EF^cG \cup E^cFG$ (j) $S$

**7.** (a) $E$ (b) $EF$ (c) $EG \cup F$

### *Problems, Sections 1–2*

**1.** $S = \{RR, RG, RB, GR, GG, GB, BR, BG, BB\}$
$S = \{RG, RB, GR, GB, BR, BG\}$

**4.** $\{A \text{ wins}\} = \{1, 0001, 0000001, \ldots, \underbrace{00\ldots0}_{3n}1, \ldots\}$

$(A \cup B)^c = \{000\ldots, 001, 000001, \ldots \underbrace{00\ldots0}_{3n+2}1, \ldots\}$

### *Problems, Sections 3–6*

**1.** 20,000, 12,000, 11,000, 10,000
**2.** 1.057
**3.** .0769, .03116
**4.** (a) .0020 (b) .4226 (c) .0475 (d) .0211 (e) .00024

**6.** $9.10946 \times 10^{-6}$ **7.** $\dfrac{32}{663}$ **9.** $\frac{2}{5}$ **10.** .492929

**11.** .58333 **12.** $P\{\text{different}\}$ = .2477 without replacement, = .2099 if with replacement

**13.** $\frac{1}{2}$ **14.** $\frac{2}{9}, \frac{1}{9}$ **15.** $\dfrac{g}{g+b}$ **16.** $\dfrac{70}{323}$ **17.** $\dfrac{12}{25}$

**18.** $\frac{1}{64}, \frac{21}{64}, \frac{36}{64}, \frac{6}{64}$ **19.** .5177 **20.** $1 - \left(\dfrac{35}{36}\right)^n, n \geq 25$

**21.** $\dfrac{2}{n}, \dfrac{2}{n-1}$ **23.** $\dfrac{1}{n}, \dfrac{(n-1)^{k-1}}{n^k}$

**24.** $1.0604 \times 10^{-3}$ **25.** .4329 **26.** $2.6084 \times 10^{-6}$

**27.** $\dfrac{\binom{n}{m}(N-1)^{n-m}}{N^n}$ **28.** (a) .09145 (b) .4268

**29.** $\frac{36}{63} = \frac{4}{7}$ **30.** $\frac{12}{35}$ **31.** .0511

**32.** (a) $\dfrac{4\binom{50}{11} - 6\binom{48}{9} + 4\binom{46}{7} - \binom{44}{5}}{\binom{52}{13}} = .2198$

(b) $\dfrac{13\binom{38}{9} - \binom{13}{2}\binom{44}{5} + \binom{13}{3}\binom{40}{1}}{\binom{52}{13}} = .0342$

## Chapter 3

### *Theoretical Exercises*

**8.** (b) $\dfrac{bg}{(r+b)(r+b+g)} + \dfrac{bg}{(r+g)(r+b+g)}$

**17.** $P_n = \alpha_n p + (1 - \alpha_n)p'$

**18.** (b) $P_{n,m} = \dfrac{n-m}{n+m}$

(c) $P_{n,m} = \dfrac{n}{n+m} P_{n-1,m} + \dfrac{m}{n+m} P_{n,m-1}$, *last* vote

### *Problems*

**1.** $\frac{1}{3}$

**2.** $P\{6 \mid \text{sum of } 6+i\} = \dfrac{1}{7-i}, i = 1, \ldots, 6$

**3.** .339
**4.** $P\{\text{at least one} \mid 12\} = 1$. Otherwise, twice the probabilities in 2
**5.** $\frac{6}{91}$ **6.** $\frac{1}{2}$ **7.** $\frac{2}{3}$ **8.** $\frac{1}{2}$ **9.** $\frac{7}{11}$
**10.** (a) .1818, .2845 (b) .4073, .2532
**11.** $\frac{11}{50}$ **12.** $\frac{35}{768}$ **13.** $\frac{4}{9}, \frac{1}{2}$ **15.** $\frac{1}{3}, \frac{1}{2}$ **17.** $\frac{20}{21}, \frac{40}{41}$
**18.** $\frac{7}{12}, \frac{3}{5}$ **19.** $\frac{5}{11}$ **20.** $\frac{4}{5}$ **21.** $\frac{54}{62}$ **22.** $\frac{3}{4}$
**23.** $\frac{1}{2}$ **24.** $\frac{1}{3}, \frac{1}{5}, 1$ **25.** $\frac{12}{37}$ **26.** $\frac{46}{185}$ **27.** $\frac{3}{13}, \frac{5}{13}, \frac{5}{52}, \frac{15}{52}$

**28.** $\dfrac{b}{b+r+c}$ **29.** $\frac{43}{459}$ **30.** $\frac{1000}{29}$ percent **31.** $\frac{4}{9}$

**33.** $\frac{1}{11}$ **35.** $\frac{2}{3}$ **36.** $\frac{19}{268}$ **37.** 17.5 percent, $\frac{38}{165}, \frac{17}{33}$

**39.** 9 **42.** (c) $\sum_{i=k}^{n} \binom{n}{i} p^i (1-p)^{n-i}$

**43.** $\frac{9}{128}, \frac{9}{128}, \frac{18}{128}, \frac{110}{128}; \frac{1}{32}, \frac{1}{32}, \frac{1}{16}, \frac{15}{16}$

**44.** $\frac{1}{9}, \frac{1}{18}$ **45.** $\frac{38}{64}, \frac{13}{64}, \frac{13}{64}$ **47.** $\frac{1}{16}, \frac{1}{32}, \frac{10}{32}, \frac{1}{4}, \frac{31}{32}$

**48.** $\frac{1}{2-p}, \frac{p(1-p)^{i-1}}{1-(1-p)^k}$ **49.** $\frac{P_1}{P_1 + P_2 - P_1P_2}$

**50.** $\frac{P_1(1-P_2)(1-P_3) + (1-P_1)P_2P_3}{P_1 + P_2 + P_3 - P_1P_2 - P_1P_3 - P_2P_3}$

**52.** $\frac{3}{10}, \frac{1 - (\frac{2}{3})^3}{1 - (\frac{2}{3})^{10}}$ **53.** .5550

**54.** (a) $\begin{cases} (\frac{1}{2})^i & i < n \\ (\frac{1}{2})^{n-1} & i = n \end{cases}$ (b) $(\frac{1}{2})^{n-1}$ **55.** .9530

**56.** (a) $\frac{P_1^2}{P_1^2 + P_2^2 - P_1^2P_2^2}$ (c) same as (a) except $P_1^3$ replaces $P_i^2$ throughout

**57.** $\frac{1}{2}, \frac{3}{5}, \frac{4}{5}$ **58.** (a) $\frac{9}{19}, \frac{6}{19}, \frac{4}{19}$ (b) $\frac{77}{165}, \frac{53}{165}, \frac{35}{165}$

**60.** $\frac{i^n}{\sum_{j=1}^{k} j^n}$ **62.** $\frac{97}{142}, \frac{15}{26}, \frac{33}{102}$

## Chapter 4

### *Theoretical Exercises*

**3.** $1 - \lim_{h \to 0} F(a-h)$ **4.** Not true.

**5.** $F\left(\frac{x-\beta}{\alpha}\right)$ when $\alpha > 0$. **8.** $k/n$ **13.** $k$

### *Problems*

**1.** $P(4) = \frac{6}{91}, P(2) = \frac{8}{91}, P(1) = \frac{32}{91}, P(0) = \frac{1}{91}, P(-1) = \frac{16}{91}, P(-2) = \frac{28}{91}$

**3.** $P(3) = P(18) = \frac{1}{216}$ $P(7) = P(14) = \frac{15}{216}$
$P(4) = P(17) = \frac{3}{216}$ $P(8) = P(13) = \frac{21}{216}$
$P(5) = P(16) = \frac{6}{216}$ $P(9) = P(12) = \frac{25}{316}$
$P(6) = P(15) = \frac{10}{216}$ $P(10) = P(11) = \frac{27}{316}$

**4.** $\frac{1}{2}, \frac{5}{18}, \frac{5}{36}, \frac{5}{84}, \frac{5}{252}, \frac{1}{252}, 0, 0, 0, 0$
**5.** $n - 2i,\ i = 0, 1, \ldots, n$
**6.** $P(3) = P(-3) = \frac{1}{8},\ P(1) = P(-1) = \frac{3}{8}$
**8.** (a) $P(6) = \frac{11}{36},\ P(5) = \frac{9}{36},\ P(4) = \frac{7}{36},\ P(3) = \frac{5}{36},\ P(2) = \frac{3}{36},$
$P(1) = \frac{1}{36}$
(d) $P(5) = \frac{1}{36},\ P(4) = \frac{2}{36},\ P(3) = \frac{3}{36},\ P(2) = \frac{4}{36},\ P(1) = \frac{5}{36},$
$P(0) = \frac{6}{36},\ P(-j) = P(j)$
**12.** (a) $P(4) = \frac{1}{16},\ P(3) = \frac{1}{8},\ P(2) = \frac{1}{16},\ P(0) = \frac{1}{2},\ P(-i) = P(i)$
(b) $P(0) = 1$
**13.** $\frac{1}{4}, \frac{1}{6}, \frac{1}{12}, \frac{1}{2}$ **15.** $\frac{1}{2}, \frac{1}{10}, \frac{1}{5}, \frac{1}{10}, \frac{1}{10}$
**16.** $\frac{3}{8}$ **17.** $\frac{11}{243}$ **18.** $\frac{11}{64}$ **19.** $P \geq \frac{1}{2}$
**22.** 3 **27.** (a) .5768, (b) .6070
**29.** .3935, .3033, .0902
**30.** .8886
**31.** .4082
**33.** (a) .0821 (b) .2424
**35.** (a) $(.9)^n$ (b) $P(1) = 1 - P(n + 1) = (.9)^n$
**36.** (a) .3935 (b) .2293 (c) .3935 (d) $1 - \exp\{(i - 500)/1000\}$
**37.** (a) .1500 (b) .1012
**38.** $\binom{i-1}{3}(.6)^4(.4)^{i-4}$
**39.** (a) $\frac{32}{243}$ (b) $\frac{4864}{6561}$ (c) $\frac{160}{729}$ (d) $\frac{160}{729}$
**40.** $P(i) = \binom{9+i}{9}\left(\frac{1}{2}\right)^{10+i},\ i \geq 0$
**41.** $\binom{N_1 + N_2 - k}{N_1}\left(\frac{1}{2}\right)^{N_1+N_2-k+1}$
$+\binom{N_1 + N_2 - k}{N_2}\left(\frac{1}{2}\right)^{N_1+N_2-k+1}$
**43.** $\dfrac{18(17)^{n-1}}{(35)^n}$
**45.** $P\{\text{rej}|1\} = \frac{3}{10},\ P\{\text{rej}|4\} = \frac{5}{6},\ P\{4|\text{rej}\} = \frac{75}{138}$
**46.** .3439

## Chapter 5

### *Problems*

**1.** $\frac{3}{4},\ F(x) = \dfrac{3x}{4} - \dfrac{x^3}{4} + \dfrac{1}{2},\ -1 < x < 1$ **2.** $\frac{7}{2}e^{-5/2}$

**3.** no; no **4.** (a) $\frac{1}{2}$, (b) $\dfrac{y-10}{y}$, $y > 10$

**5.** $1-(.01)^{1/5}$ **6.** $\frac{2}{3}, \frac{2}{3}$ **7.** $\frac{2}{5}$

**9.** $\frac{2}{3}, \frac{1}{3}$ **11.** $(.9938)^{10}$ **13.** 9.5 percent, .0019

**14.** .9258, .1762 **17.** .0606, .0525

**20.** $e^{-1}, e^{-1/2}$ **21.** $e^{-1}$ **22.** $e^{-1}, \frac{1}{3}$

**26.** $\frac{3}{5}$ **27.** $f(y) = \exp\{y - e^y\}$ **28.** $f(y) = \dfrac{1}{y}$, $1 < y < e$

**29.** $F(x) = \dfrac{1}{\pi}(\text{arc sin}\,(x/A) + \pi/2)$, $-A \le x \le A$

## Chapter 6

### Problems

**1.** (a) $P\{X = i, Y = j\} = \begin{cases} \frac{1}{36} & \text{if } 1 \le i \le 6, j = 2i \\ \frac{1}{18} & \text{if } 2 \le i \le 6, i + 1 \le j < 2i \\ 0 & \text{otherwise} \end{cases}$

(b) $P\{X = i, Y = j\} = \begin{cases} i/36 & \text{if } 1 \le i \le 6, j = i \\ \frac{1}{36} & \text{if } 1 \le i < j \le 6 \\ 0 & \text{otherwise} \end{cases}$

(c) $P\{X = i, Y = j\} = \begin{cases} \frac{1}{36} & \text{if } 1 \le i = j \le 6 \\ \frac{1}{18} & \text{if } 1 \le i < j \le 6 \end{cases}$

**2.** $p(i, j) = \frac{1}{10}$, $i = 1, 2, 3, 4$, $j = 1, 2, \ldots, 5 - i$

**3.** $p(i, j) = (1-p)^{i+j}p^2$, $i, j = 0, 1, \ldots$

**4.** (a) $c = \frac{1}{8}$ (b) $f_Y(y) = \frac{1}{6}y^3e^{-y}$, $f_X(x) = \frac{1}{4}e^{-|x|}(1 + |x|)$

**5.** (b) $f(x) = \frac{6}{7}(2x^2 + x)$ (c) $\frac{15}{56}$ (d) .8625 **6.** (a) $\frac{1}{2}$ (b) $1 - e^{-a}$

**7.** .1458 **8.** $(39.3)e^{-5}$ **9.** $\frac{1}{6}, \frac{1}{2}$

**11.** (a) $f_X(x) = f_Y(x) = 1, 0 < x < 1$ (b) Yes (c) $1 - \pi/16$

**12.** $\frac{1}{3}$ **13.** $\frac{7}{9}$ **14.** $\frac{1}{2}$ **16.** $e^{-1}/i!$

**18.** $F_{X+Y}(a) = \begin{cases} a - 1 + e^{-a}, a < 1 \\ 1 - e^{-a}(e-1), a > 1 \end{cases}$

$F_{X/Y}(a) = a(1 - e^{-1/a})$

**19.** $F_Z(a) = \dfrac{\lambda_1 a}{\lambda_1 a + \lambda_2}, \dfrac{\lambda_1}{\lambda_1 + \lambda_2}$

**21.** $P\{X = j \mid Y = i\}$

| $i \backslash j$ | 1 | 2 | 3 | 4 | 5 |
|---|---|---|---|---|---|
| 1 | .438 | .219 | .146 | .1095 | .0876 |
| 2 | 0 | .3896 | .2597 | .1948 | .1558 |
| 3 | 0 | 0 | .4255 | .3191 | .2553 |
| 4 | 0 | 0 | 0 | .5556 | .4444 |
| 5 | 0 | 0 | 0 | 0 | 1 |

**22.** $P\{Y = j \mid X = i\} = \begin{cases} \dfrac{1}{2i-1} & \text{if } 6 \geq i = j \geq 1 \\ \dfrac{2}{2i-1} & \text{if } 6 \geq i > j \geq 1 \end{cases}$

**24.** $f_{X|Y}(x \mid y) = (y+1)^2 x e^{-x(y+1)}$ $\quad f_{Y|X}(y \mid x) = x e^{-xy}$ $\quad f_Z(x) = e^{-x}$

**25.** $F_{Y|X}(y \mid x) = \frac{1}{2} + \dfrac{3y}{4x} - \frac{1}{4}\dfrac{y^3}{x^3}$ $\quad$ **28.** $\left(\dfrac{L-2d}{L}\right)^3$ $\quad$ **29.** .79297

**30.** (a) $1 - e^{-5\lambda a}$ $\quad$ (b) $(1 - e^{-\lambda a})^5$

**32.** $f(r, \theta) = \dfrac{r}{\pi}, 0 < r < 1, 0 < \theta < 2\pi$

**33.** $f(r, \theta) = r, 0 < r \sin\theta < 1, 0 < r\cos\theta < 1, 0 < \theta < \dfrac{\pi}{2}, 0 < r < \sqrt{2}$

**35.** (b) $f(u) = \dfrac{1}{u^2}\log u, u \geq 1, f(v) = \dfrac{1}{2v^2}, v > 1, f(v) = \frac{1}{2}, 0 < v < 1$

**36.** $f(u, v) = \dfrac{u}{(v+1)^2}, 0 < uv < 1 + v, 0 < u < 1 + v$

**38.** $f(y_1, y_2) = \dfrac{\lambda^2 e^{-\lambda y_1}}{y_2}, 1 \leq y_2, y_1 \geq \log y_2$

## Chapter 7

*Theoretical Exercises*

**9.** $2p^k(1-p) + (n-k-1)p^k(1-p)^2$ $\quad$ **10.** $\dfrac{k}{n}$

**11.** $\sum_{j=1}^{r} (1 - p_j)^n$ **12.** $\lambda$ **13.** $\frac{(1 - p)}{p^2}$

**14.** $\frac{1}{\lambda^2}$ **15.** $\frac{(b - a)^2}{12}$

**19.** mean $= \sum_1^n P_j$, variance $= \sum_1^n P_j(1 - P_j)$; independence only needed for variance

**23.** maximum when $P_i = \frac{\mu}{n}$; minimum when $P_i = 1, i = 1, \ldots [\mu]$, $P_{[u]+1} = u - [u]$

**26.** $\frac{r(1 - p)}{p^2}$ **27.** $\frac{\sigma_2^2}{(\sigma_1^2 + \sigma_2^2)}$

**40.** $\frac{x}{n}$ **45.** $M_{3,5} = 1.75$ **55.** $\phi_Y(t) = e^{tb}\phi_X(at)$

*Problems*

**1.** (a) $\frac{49}{32}$, (b) $10 - 5e^{-1/8} - 2e^{-9/8} - 3e^{-25/8}$
**2.** $-\frac{17}{216}$ **3.** 3.5 **4.** $p = \frac{11}{18}$, maximin $= \frac{23}{72} =$ minimax

**6.** (a) $\frac{11}{2}$, (b) $\frac{17}{5}$ **7.** $A(p + \frac{1}{10})$ **8.** $\frac{3}{5}$ **9.** $C_1 \leq \frac{p}{1 - p} C_2$

**12.** (a) 4, (b) 0, (c) $\infty$ **13.** $a = \frac{3}{5}, b = \frac{6}{5}$ **14.** $\frac{2}{\alpha}$

**15.** $\frac{n}{n+1}, \frac{1}{n+1}$ **16.** $\frac{1}{n+1}$ **17.** $p^*$ **18.** $\frac{A}{2}, \frac{\log 2}{\lambda}$, median

**19.** 96 **22.** (c) $\frac{2M - 1}{8}$ **23.** $\frac{1}{4}\log(\frac{3}{2})$ **24.** $\sqrt{2 \log\left(\frac{6}{\sqrt{2\pi}}\right)}$

**25.** $110 - 100(.9)^{10}$ **26.** $\frac{109}{60}$ **29.** 4

**30.** (b) $\sum_{j \neq 1} P_j/(P_j + P_1)$ **31.** $\sum_{j \neq 1} \frac{W(j)}{W(1) + W(j)}$

**32.** (a) .9301 (b)87.5757
**33.** 14.7 **34.** $\frac{147}{110}$ **36.** $\frac{20}{19}, \frac{360}{361}$

**37.** 21.2, 18.929, 49.214 **38.** $\frac{-n}{36}$ **39.** 0 **40.** $\frac{1}{8}$

**43.** 6, $\frac{336}{99}$ **44.** $\frac{100}{19}, \frac{(900)(18)}{(19)^2(17)}, \frac{10}{19}, \frac{(180)}{(19)^2}\frac{(18)}{(17)}$ **47.** $\frac{1}{2}$, 0

**49.** 6, 7, 5.81920 **50.** $\frac{9}{5}, \frac{6}{5}, \frac{3}{5}$, 0 **51.** $2y^2$

**52.** $\frac{y^3}{4}$ **53.** 12 **55.** $N[1 - e^{-10/N}]$ **56.** 12.5

**57.** 218 **59** $x[1 + (2p - 1)^2]^n$ **61.** $\frac{1}{2}, \frac{1}{16}, \frac{2}{81}$

**62.** $\frac{1}{2}, \frac{1}{3}$ **63.** $u$; $1 + \sigma^2$; yes; $\sigma^2$

## Chapter 8

### Problems

**1.** $p \geq \frac{19}{20}$ **2.** (a) $\frac{15}{17}$ (b) $p \geq \frac{3}{4}$ (c) $n \geq 10$ **3.** $n \geq 3$

**4.** (a) $p \leq \frac{20}{15}$ (a useless bound) (b) $p \approx .8686$ (the exact value is .8435 and the continuity correction gives .8438)

**5.** .1416 **6.** .9431 **7.** .3085 **8.** .6932

**9.** $.01\sqrt{n} = 2.58$

**10.** 117

**11.** $p \geq .6$

**14.** $p \leq .2$

**16.** (a) $E[X^3] \geq 15{,}625$ (b) $E[\sqrt{X}] \leq 5$ (c) $E[\log X] \leq \log 25$ (d) $E[e - X] \geq e^{-25}$

## Chapter 9

### Problems

**1.** $\frac{1}{9}, \frac{5}{9}$

**3.** .0265, .0902, .2642, .4422

**10.** (b) $\frac{1}{6}$

**14.** 2.585, .5417, 3.1267

**15.** 5.5098

# Index